ELEMENTS OF
POLYMER
DEGRADATION

ELEMENTS OF POLYMER DEGRADATION

LEO REICH, Ph.D.
Polymer Research Branch, Picatinny Arsenal
Dover, New Jersey

SALVATORE S. STIVALA, Ph.D.
Department of Chemistry and Chemical Engineering
Stevens Institute of Technology, Hoboken, New Jersey

McGRAW-HILL BOOK COMPANY
New York   St. Louis   San Francisco   Düsseldorf   Johannesburg
Kuala Lumpur   London   Mexico   Montreal   New Delhi
Panama   Rio de Janeiro   Singapore   Sydney   Toronto

ELEMENTS OF POLYMER DEGRADATION

Copyright © 1971 by McGraw-Hill, Inc. All Rights Reserved.
Printed in the United States of America. No part of this
publication may be reproduced, stored in a retrieval system,
or transmitted, in any form or by any means, electronic,
mechanical, photocopying, recording, or otherwise, without
the prior written permission of the publisher.
Library of Congress Catalog Card Number 75-148995

07-051760-6

1234567890 MAMM 754321

To my wife Doris, daughter Margery, and mother Rose
Leo Reich

To my wife Virginia, my children Victoria and Richard,
and mother Betty
Salvatore S. Stivala

CONTENTS

PREFACE

In connection with the graduate course, Special Topics in Polymer Science, offered at Stevens Institute of Technology, the authors introduced Polymer Degradation as the topic for 1969. No satisfactory textbook was available for introducing the various ramifications embodying general principles of polymer degradation. In this context may be mentioned that although several books were available, these were about 15 years old. Since then much work has been reported on polymer degradation, making the need for an up-to-date book on the subject highly desirable. Furthermore, these early books were either too quantitative or too qualitative, and those of more recent vintage not only fall in one of these categories but also are not general in scope. Therefore, the purpose of the present book is not only to discuss various recent polymer degradation processes in a general sense but also to bridge the hiatus between the highly quantitative and qualitative aspects of the subject. Accordingly, this book is an outgrowth from the development of lecture material presented by the authors.

The book is intended primarily for those who wish to enter the field of polymer degradation but also may be of value as a reference for those working in the area. It may also be useful as a supplement or a textbook in a course in polymer chemistry. The book treats mechanisms and kinetics of various types of polymer degradation, such as thermal, oxidative, chemical, radiative, mechanical, and biological degradation. Generally, final equations are developed through a series of steps rather than introduced in final form. In this manner the reader can understand how the final relationships were obtained. Illustrative examples are given throughout the text. Further, a chapter is devoted largely to qualitative aspects of polymer stability. The book also describes relatively recent techniques used in polymer degradation, e.g., infrared spectroscopy (IR), dynamic thermogravimetric analyses (TGA), differential thermal analyses (DTA), and gel permeation chromatography (GPC), including theories and applications of these methods.

The treatment of the various areas of degradation is not confined to any particular class of polymers, e.g., polyolefins. Various types of polymers have been selected mainly because they are treated relatively extensively in the literature in regards to kinetics and mechanisms and thereby illustrate various principles which are described in this book.

Following this preface is a brief introduction which discusses informally various basic aspects of polymer degradation in order to set the pace for the reader's venture into the book. Chapter 1 covers in order: thermal, oxidative, radiative, mechanical, chemical, and biological degradation. Since both thermal and oxidative degradations have been treated extensively in the literature, the authors feel that these topics are worthy of separate chapters, i.e., Chaps. 4 and 5, respectively. Nevertheless, they are included in Chap. 1 mainly as introductory concepts to subsequent topics in the chapter. Radiative degradation includes photochemical and ionizing radiation while mechanical includes degradation due to machining and ultrasonics. The topic on chemical degradation emphasizes the effects of air pollutants, e.g., nitrogen dioxide, sulfur dioxide, and ozone, as well as oxidizing and hydrolytic agents, e.g., concentrated nitric acid, weak acids, and alkali. The above topics obviously do not include all chemical agents which attack polymers but are primarily mentioned because of the authors' preference.

Chapters 2 and 3 present the theories and practices of various instrumental techniques employed in polymer degradation. Thus, TGA and DTA are treated in Chap. 2 while ancillary methods such as IR, GPC and viscosity are treated in Chap. 3. Chapter 6 is devoted to a qualitative and semiquantitative discussion of various factors affecting polymer stability.

The chapters have been so written as to be essentially independent of one another (except for occasional cross-referencing). Thus, Chap. 6 may be read initially for a qualitative survey of factors affecting polymer degradation.

Inasmuch as there have appeared in the literature numerous symbols designating a given parameter and, conversely, a given symbol to designate several parameters, the authors have attempted to compromise this condition. Therefore, when a given symbol has been used for a number of different parameters, or the converse, these symbols are redefined where mentioned throughout the text.

The authors would like to express their gratitude to L. Z. Pollara and Miss A. Randy (Stevens Institute of Technology), and to D. W. Levi and J. D. Matlack (Picatinny Arsenal) for their assistance. The authors are also grateful to Eutectic Welding Co. for making available to one of us (S.S.S.) funds which were used, in part, for the typing of the manuscript.

Leo Reich
Salvatore S. Stivala

INTRODUCTORY CONCEPTS TO DEGRADATION

Degradation is generally looked upon as a deleterious process. Thus, to degrade a substance is to impair it in respect to some physical property or to reduce its complexity, as in a chemical compound. However, degradation is not always viewed as harmful, as may be seen in the mastication of rubber where mechanical forces, e.g., shearing, are employed to lower its molecular weight to such levels as to render the rubber processible and commercially useful (cf. Mechanical Degradation in Chap. 1). The causes of degradation of polymers may be (1) environmental, e.g., heat, light, man-made atmospheric pollutants, (2) induced, as in rubber mastication, or (3) an inadvertert consequence of processing due to heat and/or air.

The exposure of polymers to the influence of environmental factors over a period of time generally leads to deterioration in physical properties. The degree, or discernibility, of deterioration of physical properties depends on the extent of the degradation as well as on the nature of the chemical processes involved. Often the course of degradation reactions under various environmental conditions is generally complex. Accordingly, there are instances where the complete elucidation of the mechanism has yet to be achieved (cf. Oxidative, Radiative, and Chemical Degradations in Chap. 1; and Chap. 5).

In regard to environmental factors we should like to mention in particular the germane topic of air pollution. Among the principal man-made pollutants prevalent in air are: (1) carbon monoxide, which evolves chiefly from domestic fires and exhaust fumes from various transportation vehicles; (2) sulfur dioxide (cf. Chemical Degradation in Chap. 1), emitted from the combustion of coals and other sulfur-containing fuels (through oxidation in the presence of moist air, it may generate sulfuric acid micromist which may comprise the so-called "acid smog" experienced by large cities such as London and Pittsburgh); (3) nitrogen oxides which evolve from the exhaust fumes of transportation vehicles [the nitrogen dioxide (cf. Chemical Degradation in Chap. 1) is responsible for the formation of the "oxidizing smog" rich in ozone, and organic peroxides prevalent in sunny regions such as Los Angeles]. These pollutants not only affect the general health of man but can also attack synthetic and natural polymeric materials as discussed under Chemical Degradation. For example, fabrics made from nonresin treated cotton, viscose rayon, and high-wet-modulus rayon exhibit lower breaking strength when exposed to air containing 0.1 ppm of sulfur dioxide compared to fabrics exposed to air alone. Besides air pollution, there is today wide concern over water pollution caused, in some cases, by the resistance of the polyelectrolytes, that comprise detergents, to degrade biologically. The susceptibility of the synthetic and natural polymers to biological degradation is discussed in the last section of Chap. 1.

The contribution that induced degradation can make in solving some problems of pollution by waste materials is of great interest. Since a modest percentage of waste consists of plastics and other polymeric substances, the possibility of degrading them to harmless and biologically useful chemical compounds is a challenging thought. The contribution of plastics, and similar materials, to the ever-increasing problem of pollution comes about largely because they are not readily degradable by the type of microorganisms which normally attack most other forms of organic matter, thus returning them to the biological life cycle. Processes recently have been developed for attaching several sensitizer groups, capable of absorbing ultraviolet light, along the backbone of the polymer chain. These sensitizer groups, which may be introduced during synthesis of the polymer, are in low concentrations so as not to appreciably affect the general physical properties of the polymer. When these groups absorb ultraviolet light, chain scission occurs, causing the material to decrease markedly in physical strength. The resulting brittleness in the polymer will enable it to be broken into smaller fragments by natural erosion, wind, rain, etc. The small pieces then become part of the soil and are in a form presumably more suitable for attack by microorganisms. The object is to select those sensitizer groups that do not absorb visible light but only ultraviolet radiation, for obvious practical reasons.

Broadly, degradation of polymers may be considered any type of modification of a polymer chain involving the main-chain backbone or side groups or both. These modifications are often of a chemical nature, i.e., requiring the breaking of primary valence bonds. This can lead to lower molecular weights, crosslinking, and cyclization (cf. Chap. 6). Main-chain scission may occur in one of two ways, or both, corresponding approximately to the reverse of the two types of polymerization processes. Thus, scission may occur randomly (random degradation) where chain breaking occurs at random points along the chain, or at the terminal of a polymer radical, or polymer, where monomer units are successively released (unzipping). In the former case, the molecular weight decreases continuously with the extent of reaction (as in hydrolysis of a condensation polymer), and the fragments formed which are large relative to monomer unit, are mixtures having molecular weights up to several hundred. In the latter case of depolymerization, the production of monomer is appreciable. Random degradation may be regarded roughly as the reverse of condensation polymerization, whereas chain depolymerization is the reverse of the propagation step in vinyl polymerization. (A distinguishing characteristic between the two types of degradation is that generally in the random process the molecular weight falls rapidly but considerably more slowly than in depolymerization.)

On the other hand, the breaking of secondary valence bonds within a chain, e.g., hydrogen bonds in proteins, leads to conformational changes which constitute a degradative process called *denaturation*, influenced by heat, pH changes, chemical agents, etc. In this connection it is interesting to mention that in the treatment of sickle cell anemia it was important to denature hydrophobic bonds of hemoglobin S, the red blood cell molecule involved in sickle cell anemia. This was achieved by intravenous injection of urea, a denaturant, in an osmotically balanced solution of invert sugar. This resulted in alleviation of a period of excrutiating pain brought on by periodic blockage of small vessels by the crescent-shaped sickled cells. This demonstrates again a case of a beneficial effect of a degradative process.

This book is mainly concerned with the degradative process which affects the breaking of primary valence bonds, essentially in the chain backbone, with a concomitant decrease in the polymer molecular weight. This process may be achieved by the following types of degradation: (a) thermal (cf. Chaps. 1, 4 and 6); (b) oxidative (cf. Chaps. 1, 5 and 6); (c) radiative (cf. Chaps. 1 and 6); (d) mechanical (cf. Chap. 1); (e) chemical (cf. Chap. 1) and (f) biological (cf. Chap. 1). Since polymers are used successfully in countless number of applications, e.g., packaging, electronics, building, boats, housewares, appliances, coatings, adhesives, fibers, elastomers, plastics, sealants, etc., their response to external factors affecting their properties are of paramount importance. Degradative symptoms may be described as hardening, embrittlement, softening, cracking, crazing, discoloration or alteration of certain properties, e.g., dielectric constant.

Numerous polymer physical properties such as mechanical strength (e.g., tensile, impact), elasticity, solution viscosity, softening point, correlate with not only chemical composition but with molecular weight (MW), molecular weight distribution (MWD), branching, crystallinity, crosslinking, conformation, chain flexibility, etc. (cf. Chap. 6). Since these parameters can be affected by degradation it would then follow that various properties of the polymer will be altered. For example, reduction of the MW of the polymer, which correlates with interchain forces, may lead to increased softness and tackiness, lower mechanical strength and softening points. Crosslinking introduces strong primary valence restraints, therefore, separation of the main chains from one another is not possible. Consequently, such properties as solubility, fusibility, flow, moldability are now eliminated. The introduction of these crosslinks invariably increase stiffness, creep resistance, brittleness. Further, a high degree of crosslinking generally leads to low ultimate strength because of interference with chain alignment and uniform load distribution. Similar properties, as in crosslinking, may result from cyclization reactions.

Different polymers can show wide variation in their response to degradative agents. Further, these differences can often be correlated with differences in their structure, e.g., crystallinity, long side chain. Chapter 6 discusses qualitatively various factors affecting polymer stability.

It was stated earlier that degradation may be environmental, a result of processing, or induced. The first two are almost always not wanted since they lead to undesirable breakdown of properties. Therefore, in commercial products stabilizers are incorporated into the polymer in small quantities that either prevent, minimize, or postpone the onset of degradation. The efficacy of these stabilizers depend on concentration, combination with other compounds, and type. Effective stabilizers against thermal, oxidative and ultraviolet attack have been widely used. The stabilization of polymers, though mentioned here, is not a main topic in this book.

1

VARIOUS TYPES OF
POLYMER DEGRADATION

Thermal and oxidative degradation processes are discussed in greater detail in Chaps. 4 and 5 of this book than in this chapter. Yet it appeared advisable to introduce these types of degradation at this point since they are mentioned from time to time in discussions of radiative and mechanochemical degradations which follow.

THERMAL DEGRADATION

This book deals primarily with degradation in the bulk phase. Furthermore, it is assumed that any monomer formed during thermal degradation volatilizes rapidly from the reaction medium. Hence, transfer reactions due to monomer and/or solvent will not be considered. Also, reactions that lead to crosslinking will not be treated.

The following equations represent a general scheme for thermal degradation which is employed later in the book. The first step will invariably involve an initiation process. This process may take place by two different modes, random and chain-end scission, namely,

$$
\begin{array}{c}
\quad\;\; \text{H}\;\;\; \text{H}\;\;\; \text{H}\;\;\; \text{H} \\
\sim\!\!\sim \text{C}\!-\!\text{C}\!-\!\text{C}\!-\!\text{C}\!\sim\!\!\sim \\
\quad\;\; \text{H}\;\;\; \text{X}\;\;\; \text{H}\;\;\; \text{X}
\end{array}
\longrightarrow
\begin{array}{c}
\quad\;\; \text{H}\;\;\; \text{H} \\
\sim\!\!\sim \text{C}\!-\!\text{C}\cdot \\
\quad\;\; \text{H}\;\;\; \text{X}
\end{array}
+
\begin{array}{c}
\quad\;\; \text{H}\;\;\; \text{H} \\
\sim\!\!\sim \text{C}\!-\!\text{C}\cdot \\
\quad\;\; \text{X}\;\;\; \text{H}
\end{array}
$$

$$(1)$$

and

$$\underset{\substack{H \;\; X \;\; H \;\; X}}{\overset{\substack{H \;\; H \;\; H \;\; H}}{\sim\!\!\sim\!\!\sim C-C-C-CH}} \longrightarrow \underset{\substack{H \;\; X}}{\overset{\substack{H \;\; H}}{\sim\!\!\sim\!\!\sim C-C\cdot}} + \underset{\substack{X \;\; H}}{\overset{\substack{H \;\; H}}{HC-C\cdot}} \qquad (2)$$

The first equation represents the random type of scission while Eq. (2) denotes the chain-end type of scission. It should be noted here that in quantitative treatments the monomeric radical in Eq. (2) is often assumed to volatilize readily.

TABLE 1   Monomer Yields during Thermal Degradation [1]

Polymer	Monomer yield, %	Minimum zip length
Polyethylene	0.03	0.01
Polystyrene.	40	3
Poly(α-methylstyrene)	> 95	> 200

The initiation step is followed by a depropagation step in which monomer fragments split off stepwise along the chain, namely,

$$\underset{\substack{H \;\; X}}{\overset{\substack{H \;\; H}}{\sim\!\!\sim\!\!\sim C-C\cdot}} \longrightarrow \underset{\substack{H \;\; X}}{\overset{\substack{H \;\; H}}{\sim\!\!\sim\!\!\sim C-C\cdot}} + \underset{\substack{H \;\; X}}{\overset{\substack{H \;\; H}}{C=C}} \qquad (3)$$

This step is the reverse of the propagation process encountered during addition polymerization and is often referred to as the *unzipping reaction*. When the polymer zip length (the number of zips that occur per polymer molecule) is relatively high, it would be anticipated that monomer yields should also be high (cf. Table 1 [1]), during thermal degradation.

It may be mentioned here that when the zip length (or the kinetic chain length) is much greater than the initial degree of polymerization of a particular polymer, then it would be anticipated that the initiation process would be followed by a rapid depropagation and that the number-average molecular weight of the polymer will not change much with conversion. However, when the zip length of the polymer is much less than its degree of polymerization, then the number-average degree of polymerization of the polymer will decrease with increasing conversion. This behavior represents a method for estimating zip lengths (cf. Fig. 1.1). In this figure, curve 1 denotes a case where zip length is much greater than degree of polymerization D_P whereas curve 2 denotes the opposite. In Fig. 1.1, $D_{P,0}$ and D_P denote degree of polymerization at times zero and t, respectively.

It should be mentioned that an elaborate mathematical treatment will also take into account the volatilization of fragments larger than monomer.

After the depropagation step, we may consider transfer processes. These may involve intra- and intermolecular reactions, namely,

$$\underset{\substack{| \\ X}}{\overset{\substack{H \\ |}}{\sim\!\!\sim C}}-\underset{\substack{| \\ H}}{\overset{\substack{H \\ |}}{C}}-\underset{\substack{| \\ X}}{\overset{\substack{H \\ |}}{C}}-\underset{\substack{| \\ H}}{\overset{\substack{H \\ |}}{C}}-\underset{\substack{| \\ X}}{\overset{\substack{H \\ |}}{C}}\cdot \longrightarrow \underset{\substack{| \\ X}}{\overset{\substack{H \\ |}}{\sim\!\!\sim C}}\cdot + \underset{\substack{| \\ H}}{\overset{\substack{H \\ |}}{C}}=\underset{\substack{| \\ X}}{\overset{\substack{H \\ |}}{C}}-\underset{\substack{| \\ H}}{\overset{\substack{H \\ |}}{C}}-\underset{\substack{| \\ H}}{\overset{\substack{H \\ |}}{CX}}$$

(4)

and

$$\underset{\substack{| \\ X}}{\overset{\substack{H \\ |}}{\sim\!\!\sim C}}\cdot + \underset{\substack{| \\ H}}{\overset{\substack{H \\ |}}{\sim\!\!\sim C}}-\underset{\substack{| \\ X}}{\overset{\substack{H \\ |}}{C}}-\underset{\substack{| \\ H}}{\overset{\substack{H \\ |}}{C}}-\underset{\substack{| \\ X}}{\overset{\substack{H \\ |}}{C}}\sim\!\!\sim \longrightarrow$$

$$\underset{\substack{| \\ H}}{\overset{\substack{H \\ |}}{\sim\!\!\sim CX}} + \underset{\substack{| \\ H}}{\overset{\substack{H \\ |}}{\sim\!\!\sim C}}-\underset{\substack{| \\ X}}{\overset{\substack{H \\ |}}{C}}=\underset{\substack{| \\ H}}{\overset{\substack{H \\ |}}{C}} + \underset{\substack{| \\ X}}{\overset{\substack{H \\ |}}{\sim\!\!\sim C}}\cdot$$

(4a)

The intramolecular transfer reaction (4), if restricted to the terminal positions of the chain, would tend to make the apparent zip length longer than the actual length (without transfer). However, if intramolecular transfer occurred randomly along the chain, then the kinetics of the degradation would tend to be random. Random degradation would also be favored if intermolecular transfer [step (4a)] were operative. When depropagation is predominant during degradation, the molecular weight tends to be maintained, high monomer yields result, and rate (of conversion)

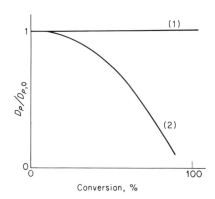

FIG. 1.1 Zip length versus percent conversion during thermal degradation.

decreases throughout the reaction. On the other hand, when transfer effects are dominant, the molecular weight decreases rapidly, volatile products contain larger chain fragments, and conversion rate exhibits a maximum during the reaction.

The termination of chain reaction may occur by various mechanisms such as unimolecular and bimolecular. The unimolecular is more difficult to depict but is invoked from time to time to explain kinetic data. This first-order termination may be portrayed as

$$\sim\sim \overset{\overset{\displaystyle H}{|}}{\underset{\underset{\displaystyle H}{|}}{C}}-\overset{\overset{\displaystyle H}{|}}{\underset{\underset{\displaystyle X}{|}}{C}}-\overset{\overset{\displaystyle H}{|}}{\underset{\underset{\displaystyle H}{|}}{C}}\cdot \; + \; Z \longrightarrow \sim\sim \overset{\overset{\displaystyle H}{|}}{\underset{\underset{\displaystyle H}{|}}{C}}-\overset{\overset{\displaystyle H}{|}}{\underset{\underset{\displaystyle X}{|}}{C}}-\overset{\overset{\displaystyle H}{|}}{\underset{\underset{\displaystyle H}{|}}{C}}Z \tag{5}$$

where Z may denote any molecule present during degradation whose concentration remains essentially constant. This type of termination may also be written as

$$\sim\sim \overset{\overset{\displaystyle H}{|}}{\underset{\underset{\displaystyle H}{|}}{C}}-\overset{\overset{\displaystyle H}{|}}{\underset{\underset{\displaystyle X}{|}}{C}}-\overset{\overset{\displaystyle H}{|}}{\underset{\underset{\displaystyle H}{|}}{C}}\cdot \longrightarrow PRD\,(g)\,(+\,Products) \tag{5a}$$

where PRD(g) denotes products which may form *after* a relatively low-molecular-weight radical diffuses out of the reaction zone into the gaseous phase. This type of termination has been postulated during photochemical degradations wherein reactions near the surface layer are occurring primarily. Presumably, the relatively low-molecular-weight radical possesses sufficient mobility so that it can diffuse into the gas phase and thereby escape from the reaction medium [2].

Besides unimolecular termination, there are two main types of bimolecular termination, i.e., termination by combination and by disproportionation. These are depicted respectively as follows:

$$\sim\sim C-C-C\cdot \; + \; \cdot C-C-C \sim\sim \longrightarrow$$

$$\sim\sim C-C-C-C-C-C \sim\sim \tag{6}$$

and

$$\sim\sim C-C-C\cdot \; + \; \cdot C-C-C \sim\sim \longrightarrow$$

$$\sim\sim C-C-CX \; + \; C=C-C \sim\sim \tag{6a}$$

OXIDATIVE DEGRADATION [3]

To better understand the oxidative degradation processes involved for polymers (especially polyolefins), it is necessary to treat initially the autoxidation of simple saturated hydrocarbons. In the following will be presented schemes involving the autoxidation of hydrocarbons in absence of and in presence of additives (metal salts or inhibitors) [3]. (See Derivation of Rate Expressions, page 9, for mathematical treatments.)

Hydrocarbon Oxidation

In absence of additives. The autoxidation process, like the thermal degradation process, generally involves an initiation step. Thus, we may write for this step

$$X \longrightarrow \text{Production of radicals} \qquad (7)$$

where X may denote a free-radical initiator such as α,α-azobisisobutyronitrile, or hydroperoxide molecules (RO_2H) which may undergo unimolecular or bimolecular decomposition into radicals, depending upon their concentration. Thus, at low values of $[RO_2H]$,

$$RO_2H \longrightarrow \text{Radicals} \qquad (7a)$$

whereas at high values of $[RO_2H]$,

$$2RO_2H \longrightarrow \text{Radicals} \qquad (7b)$$

Hydroperoxide molecules may be present because of their addition to the reaction medium or as a result of their formation during the oxidative process. It may also be noted here that various oxidation products, e.g., ketones and alcohols, may arise from the decomposition of hydroperoxides during autoxidation of hydrocarbons. It may be further mentioned that X may also denote the reaction of substrate with oxygen to form radicals (in the absence of hydroperoxides or free-radical initiators). Thus, this type of initiation process has been depicted as occurring by two different modes [3], namely,

$$RH + O_2 \longrightarrow \text{Free radicals} \qquad (7c)$$

and

$$2RH + O_2 \longrightarrow \text{Radicals + hydroperoxide} \qquad (7d)$$

After the initiation step, two propagation steps may occur:

$$R\cdot + O_2 \longrightarrow RO_2\cdot \qquad (8a)$$

and

$$RO_2\cdot + RH \longrightarrow RO_2H + R\cdot \qquad (8b)$$

Reaction (8a) generally is very rapid, having an activation energy E of about zero value. Reaction (8b) is less rapid and possesses a value of E in the vicinity of 7 kcal mole^{-1}. It may be mentioned here that when RH denotes an olefinic substrate, another reaction is possible:

$$RO_2 \cdot \; + \; >C{=}C< \;\longleftrightarrow\; RO_2\overset{|}{\underset{|}{C}}{-}\overset{|}{\underset{|}{C}}\cdot \tag{8c}$$

Also, it has been indicated, based upon polarity factors, that reaction (8b) may involve an intermediate complex,

$$RO_2\cdot \; + \; R'H \longrightarrow \left[RO_2:\overset{\ominus}{H}\cdot\overset{\oplus}{R'} \longleftrightarrow RO_2\cdot H:R' \right]$$
$$\text{(I)} \qquad\qquad \text{(II)}$$
$$\longrightarrow RO_2H + R'\cdot \tag{8d}$$

Structure (I) will be favored by electron-donating substituents and destabilized by electron-attracting groups in R'. Structure (II) accounts for the effects of the stability of the incipient radical upon carbon-hydrogen bond reactivity. Thus, when both polar and radical stability effects complement each other, the reactivity of a carbon-hydrogen bond toward a peroxy radical can be very high.

After the propagation steps, three types of termination steps can ensue. These are

$$2R\cdot \longrightarrow \text{Products} \tag{9a}$$

$$RO_2\cdot \; + \; R\cdot \longrightarrow \text{Products} \tag{9b}$$

$$2RO_2\cdot \longrightarrow \text{Products} + O_2 \tag{9c}$$

These termination reactions are generally very rapid and possess values of E of about 3 kcal mole^{-1}. However, in the presence of oxygen (>100 mm), steps (9a) and (9b) may be neglected. Thus, termination step (9c) is the most important termination process during oxidation. However, it should be noted that this termination process involves only primary or secondary peroxy radicals. For instance, during the autoxidation of ethylbenzene, the termination step is considered to be

$$2 \; \overset{CH_3}{\underset{\phi}{\diagdown}}CHO_2\cdot \longrightarrow \left[\overset{CH_3}{\underset{\phi}{\diagdown}}\overset{O-O}{\underset{H}{\diagup}}\overset{}{\underset{\phi CHCH_3}{\diagdown O}} \right] \longrightarrow \begin{array}{l} \phi COCH_3 \; + \\ CH_3 \\ \overset{}{\underset{\phi}{\diagdown}}CHOH + O_2 \end{array} \tag{10}$$

However, when tertiary peroxy radicals are involved in termination, more complex reactions may occur, such as cage formation, namely [4],

$$2RO_2 \cdot \xrightarrow{K} RO_4R \xrightarrow{k_1} [2RO \cdot + O_2]_{cage} \qquad (11)$$

$$[2RO \cdot + O_2]_{cage} \begin{cases} \xrightarrow[k_6]{collapse} RO_2R + O_2 & (12a) \\ \\ \xrightarrow[k_6']{diffusion} 2RO \cdot + O_2 & (12b) \end{cases}$$

$$RO \cdot (e.\,g., \; cumyl) \xrightarrow{O_2} \phi COCH_3 + CH_3O_2 \cdot \qquad (13)$$

$$RO_2 \cdot + CH_3O_2 \cdot \longrightarrow Products \qquad (14)$$

$$RO \cdot + RH \xrightarrow{O_2} ROH + RO_2 \cdot \qquad (15)$$

$$CH_3O_2 \cdot + RH \xrightarrow{O_2} CH_3O_2H + RO_2 \cdot \qquad (16)$$

$$RO \cdot + RO_2H \longrightarrow ROH + RO_2 \cdot \qquad (17)$$

$$CH_3O_2 \cdot + RO_2H \longrightarrow CH_3O_2H + RO_2 \cdot \qquad (18)$$

It may be mentioned that in the presence of excess hydroperoxide, steps (13) to (16) are suppressed, and the apparent termination process involves cage collapse, reaction (12a). Under these conditions, the apparent termination constants k_t may be expressed as a combination of various terms (cf. Derivation of Rate Expressions, page 9), i.e.,

$$k_t = \frac{k_1}{K} f \qquad (19)$$

where $f = k_6/(k_6 + k_6')$. It was recently reported that changes in the value of k_t as a result of changes in the type of R moiety were due primarily to changes in the value of k_1 [5].

It should also be mentioned here that oxidation products may derive from hydroperoxides under conditions wherein they decompose (more severe experimental conditions in which relative short kinetic chain lengths result). Thus, the following scheme has been postulated [3] for *n*-decane oxidation at 140°C:

$$Hydroperoxides \begin{cases} \longrightarrow Alcohols \\ \quad\quad\quad\;\downarrow \\ \longrightarrow Ketones \longrightarrow Acids \end{cases} \qquad (19a)$$

To account for the above sequence, it has been proposed that hydroperoxides can decompose to form alkoxy radicals which then abstract hydrogen to form alcohols [cf. Eq. (15)]. Hydroperoxides may also react as follows:

$$R\cdot + R_1 R_2 CHOOH \longrightarrow R_1 R_2 \dot{C}OOH + RH$$
$$\longrightarrow R_1 R_2 C = O + HO\cdot \quad (19b)$$

The alkoxy radical may also react as follows:

$$R(CH_3)CHCH_2 O\cdot \longrightarrow R\dot{C}HCH_3 + CH_2 O \quad (19c)$$

and

$$R(CH_3)CHO\cdot \longrightarrow R\cdot + CH_3 CHO \quad (19d)$$

In presence of free-radical inhibitors (AH) [3]. During inhibited autoxidation, there will be reactions in addition to those mentioned in absence of additives. Thus, we may include

$$AH + O_2 \longrightarrow Radicals \quad (20)$$

$$RO_2\cdot + AH \underset{k_{-21}}{\overset{k_{21}}{\rightleftharpoons}} RO_2 H + A\cdot \quad (21)$$

$$RO_2\cdot + AH \longleftrightarrow [AH \to RO_2] \overset{RO_2\cdot}{\longrightarrow} Products \quad (22)$$

$$RO_2\cdot + A\cdot \longrightarrow Products \quad (23)$$

$$2A\cdot \longrightarrow Products \quad (24)$$

$$A\cdot + RH \longrightarrow AH + R\cdot \quad (25)$$

The various steps (20) to (25) have been invoked in order to explain various kinetic dependencies during inhibited autoxidation. Thus, various workers employed step (22) in explaining the dependency of oxidation rate upon the terms $[I]^{1/2} [AH]^{-1/2}$ where, $[I]$ denotes concentration of the initiator used. The type of kinetic expression obtained may often depend on the type of inhibitor used. Thus, for strongly hindered phenols (steric factor), e.g., 2,6-di-*tert*-butylphenol, steps (−21) and (25) may be neglected. The predominant termination step will be either (23) or (24). In either case, the oxidation rate is a function of $[I] [AH]^{-1}$, as observed experimentally. [Step (−21) may also be excluded from the scheme since the addition of high concentrations of $RO_2 H$ had little effect upon the inhibited rates of oxygen absorption.] In nonhindered phenols (e.g., phenol itself), the addition of hydroperoxide to the reaction medium can affect the inhibited oxidation rate, implying that step (–21) can no longer be neglected. Also, step (25) must be included. The inclusion of these steps leads to complex oxygen-absorption rate expressions.

Finally, it may be remarked that the unimolecular termination step,

$$RO_2 \cdot \longrightarrow Products \tag{26}$$

has been invoked in order to explain certain kinetic dependencies obtained (see Derivation of Rate Expressions, page 9).

In presence of metal catalysts [3]. Metal catalysts can react with hydroperoxides during autoxidation,

$$RO_2H + Me^{n+} \longrightarrow RO\cdot + Me^{(n+1)+} + OH^- \tag{27}$$

and

$$RO_2H + Me^{(n+1)+} \longrightarrow RO_2\cdot + Me^{n+} + H^+ \tag{28}$$

Whether reaction (27) or (28) or both will preferentially occur will depend mainly on the type of metal catalyst. Other reactions of metal catalysts during autoxidation include those with oxygen and substrate (e.g., olefin) and with oxy radicals. A more important reaction is that with peroxy radicals. Thus, in this case, manganese salts (MnA_2) may react with peroxy radicals as follows:

$$MnA_2 + RO_2\cdot \longrightarrow RO_2MnA_2 \tag{29}$$

Such reactions help explain various rate data. Thus, at relatively high metal-catalyst concentrations, it was observed that after a critical catalyst concentration, the oxidation rate dropped catastrophically for a certain oxidation system (cf. Derivation of Rate Expressions).

Derivation of Rate Expressions for Autoxidation

In absence of additives [3]. Assume that the basic autoxidation scheme (BAS) applies, i.e.,

$$X \xrightarrow{k_i} Radicals \tag{R-1}$$

$$R\cdot + O_2 \xrightarrow{k_2} RO_2\cdot \tag{R-2}$$

$$RO_2\cdot + RH \xrightarrow{k_3} RO_2H + R\cdot \tag{R-3}$$

$$2R \cdot \xrightarrow{k_4} \left. \begin{array}{c} \\ \\ \end{array} \right\} \text{Products} \tag{R-4}$$

$$RO_2 \cdot + R \cdot \xrightarrow{k_5} \tag{R-5}$$

$$2RO_2 \cdot \xrightarrow{k_6} \text{Products} + O_2 \tag{R-6}$$

where X denotes a source of free radicals.

At relatively high kinetic chain lengths (CL),

$$k_2 [R \cdot][O_2] = k_3 [RO_2 \cdot][RH] \tag{R-7}$$

and rate of oxidation ρ_{ox} becomes

$$\rho_{ox} = k_3 [RO_2 \cdot][RH] \tag{R-8}$$

At relatively high oxygen pressures (usually above 100 torrs), steps (R-4) and (R-5) may be neglected since $[RO_2 \cdot] \gg [R \cdot]$, and from steady-state conditions,

$$k_i [X] \equiv R_i = k_6 [RO_2 \cdot]^2 \tag{R-9}$$

Upon combining Eqs. (R-8) and (R-9),

$$\rho_{ox} = \frac{k_3 [RH] R_i^{1/2}}{k_6^{1/2}} \tag{R-10}$$

When relatively low oxygen pressures are employed (<100 torrs), steps (R-5) and (R-6) may be neglected since $[R \cdot] \gg [RO_2 \cdot]$, and

$$R_i = k_4 [R \cdot]^2 \tag{R-11}$$

Upon combining Eqs. (R-7), (R-8), and (R-11),

$$\rho_{ox} = \frac{k_2 [O_2] R_i^{1/2}}{k_4^{1/2}} \tag{R-12}$$

It should be noted that the BAS applies to primary and secondary peroxy radicals. When tertiary peroxy radicals are involved, termination can no longer be simply represented by Eq. (R-6) [cf. Eqs. (11) to (18)]. However, the mathematical treatment may be considerably simplified when an excess of hydroperoxide is added to the system, thereby suppressing steps (13) to (16), and the apparent termination rate constant k_t involves essentially cage collapse [cf. Eq. (12a)]. Under these conditions and assuming steady state for the cage radicals,

$$\frac{k_1}{K} [RO_2 \cdot]^2 = (k_6 + k_6')[\ ]_{cage} \tag{R-13}$$

and

$$k_6'[\ ]_{cage} = \frac{k_6}{(k_6 + k_6')} \frac{k_1}{K} [RO_2 \cdot]^2$$

$$= k_t [RO_2 \cdot]^2 \tag{R-14}$$

From Eq. (R-14), it can be seen that

$$k_t = \frac{k_1}{K} f \tag{R-15}$$

where $f = k_6/(k_6 + k_6')$ [cf. Eq. (19)].

The apparent rate constant k_t may be obtained from expressions such as (R-10) ($k_6 \equiv k_t$ in this equation).

In presence of antioxidants. In addition to the steps in the BAS [Eqs. (R-1) to (R-6)], the following steps are included:

$$RO_2 \cdot + AH \underset{k_{-16}}{\overset{k_{16}}{\rightleftharpoons}} RO_2H + A \cdot \tag{R-16}$$

$$RO_2 \cdot + AH \xleftarrow{\hspace{1cm}} [AH \rightarrow RO_2 \cdot] \xrightarrow{RO_2 \cdot} Products \tag{R-17}$$

$$RO_2 \cdot + A \cdot \xrightarrow{k_{18}} Products \tag{R-18}$$

$$2 A \cdot \xrightarrow{k_{19}} Products \tag{R-19}$$

$$A \cdot + RH \xrightarrow{k_{20}} AH + R \cdot \tag{R-20}$$

A rate expression which involves the ratio $[I]^{1/2} [AH]^{-1/2}$ may be obtained by assuming that the termination step (R-17) applies. Then we may write

$$R_i = k_2 [R \cdot][O_2] - k_3 [RO_2 \cdot][RH] \tag{R-21}$$

$$R_i = 2ek_i[I] = k[AH][RO_2 \cdot]^2 \tag{R-22}$$

where e = efficiency of initiator I and k = constant. From Eqs. (R-21) and (R-22),

$$R \equiv \rho_{ox} - 2ek_i[I] = k_3 [RH] \left(\frac{2ek_i}{k}\right)^{1/2} \left(\frac{[I]}{[AH]}\right)^{1/2} \tag{R-23}$$

For strongly hindered antioxidants, e.g., 2,6-di-*tert*-butylphenol, steps (–16) and (R-20) may be neglected. Then, we may write

$$R_i = 2k_{16} [AH][RO_2 \cdot] \tag{R-24}$$

assuming that the predominant termination step is (R-18) and that it rapidly follows step (R-16). When the main termination step is (R-19),

$$R_i = k_{16} [AH][RO_2 \cdot] \tag{R-25}$$

By using Eqs. (R-21) and either (R-24) or (R-25), the rate expression for the oxidation of hydrocarbons in the presence of strongly hindered phenols is

$$R = \frac{k_3 [RH] R_i}{n k_{16} [AH]} \tag{R-26}$$

where n = 1 or 2, and $R_i = f[I]^1$.

When nonhindered phenols are considered, the rate expression becomes much more complex since steps (–16) and (R-20) must now be considered (see Ref. 3, pp. 226ff.).

In presence of metal catalysts. In addition to the steps in the BAS [Eqs. (R-1) to (R-6)] , consider

$$RO_2 H + Me^{n+} \longrightarrow RO \cdot + Me^{(n+1)+} + OH^- \tag{R-27}$$

$$RO_2 H + Me^{(n+1)+} \longrightarrow RO_2 \cdot + Me^{n+} + H^+ \tag{R-28}$$

and

$$RO \cdot + RH \longrightarrow ROH + R \cdot \tag{R-29}$$

Generally, instead of using both steps (R-27) and (R-28), the reaction between $RO_2 H$ and metal catalyst is represented by a single step, e.g., in the case of manganese, the rate-controlling step is

$$RO_2H + Mn \xrightarrow{k_{30}} Radical \xrightarrow{RH} R \cdot \qquad (R\text{-}30)$$

where [Mn] or [Me] = total concentration of metal catalyst, and, at steady-state conditions, $k_{27}[Me]^{n+} = k_{28}[Me]^{(n+1)+} = (k_{30}/2)[Me]$. The catalyst may also participate in termination steps, e.g.,

$$RO_2 \cdot + MnA_2 \xrightarrow{k_{31}} RO_2MnA_2 \qquad (R\text{-}31)$$

$$RO_2 \cdot + MnA_2 \xleftarrow{K_{32}} RO_2MnA_2 \qquad (R\text{-}32)$$

$$RO_2MnA_2 + RO_2 \cdot \xrightarrow{k_{33}} Products \qquad (R\text{-}33)$$

For the rate of oxidation,

$$\rho_{ox} = k_2[R \cdot][O_2] - \frac{k_6}{2}[RO_2 \cdot]^2 \qquad (R\text{-}34)$$

Also, from steady-state conditions for [R·],

$$k_2[R \cdot][O_2] = k_3[RH][RO_2 \cdot] + k_{29}[RO \cdot][RH] \qquad (R\text{-}35)$$

and

$$\rho_{ox} = k_3[RH][RO_2 \cdot] + k_{29}[RO \cdot][RH] - \frac{k_6}{2}[RO_2 \cdot]^2 \qquad (R\text{-}36)$$

When steady-state conditions apply to $[RO_2H]$, $\rho_{ox} \to \rho_{ox,\infty}$ (maximum oxidation rate) and

$$\left(k_{27}[Me]^{n+} + k_{28}[Me]^{(n+1)+}\right)[RO_2H] = k_{30}[Me][RO_2H]$$
$$= k_3[RO_2 \cdot][RH] = k_6[RO_2 \cdot]^2 \qquad (R\text{-}37)$$

Since

$$k_{27}[Me]^{n+}[RO_2H] = \frac{k_{30}}{2}[Me][RO_2H] = k_{29}[RO \cdot][RH]$$
$$(R\text{-}38)$$

we obtain upon employing (R-36) to (R-38)

$$P_{ox,\infty} = \frac{(k_3[RH])^2}{k_6} \tag{R-39}$$

and

$$[RO_2H]_\infty = \frac{(k_3[RH])^2}{k_6 k_{30}[Me]} \tag{R-40}$$

When the termination step (R-31) is included ($k_{27}[Me]^{n+} \neq k_{28}[Me]^{(n+1)+}$),

$$\left(k_{27}[Me]^{n+} + k_{28}[Me]^{(n+1)+}\right)[RO_2H] = k_3[RO_2\cdot][RH]$$
$$= k_6[RO_2]^2 + k_{31}[Me][RO_2\cdot] \tag{R-37'}$$

From Eq. (R-37′),

$$[RO_2\cdot]_\infty = \frac{k_3[RH] - k_{31}[Me]}{k_6} \tag{R-41}$$

Also,

$$k_{27}[RO_2H][Me]^{n+} = k_{29}[RO\cdot][RH] \tag{R-38'}$$

and

$$k_{28}[RO_2H][Me]^{(n+1)+}$$
$$= k_{27}[RO_2H][Me]^{n+} + k_{31}[RO_2\cdot][Me]^{n+} \tag{R-42}$$

Upon combining Eqs. (R-37′) and (R-42),

$$2k_{27}[RO_2H][Me]^{n+} = k_6[RO_2\cdot]^2 \tag{R-43}$$

When Eqs. (R-36), (R-38′), (R-41), and (R-43) are combined,

$$P_{ox,\infty} = k_3[RH]\left(\frac{k_3[RH] - k_{31}[Me]^{n+}}{k_6}\right) \tag{R-44}$$

When the termination steps (R-32) and (R-33) are employed with the exclusion of step (R-31),

$$\left(k_{27}[Me]^{n+} + k_{28}[Me]^{(n+1)+}\right)[RO_2H] = k_3[RO_2\cdot][RH]$$
$$= k_6[RO_2\cdot]^2 + k_{33}K_{32}[RO_2\cdot]^2[Me]^{n+} \qquad (R-37'')$$

From (R-37''),

$$[RO_2\cdot]_\infty = \frac{k_3[RH]}{k_6 + k_{33}K_{32}[Me]^{n+}} \qquad (R-45)$$

Also,

$$k_{28}[Me]^{(n+1)+}[RO_2H] = k_{27}[Me]^{n+}[RO_2H]$$
$$+ k_{33}K_{32}[RO_2\cdot]^2[Me]^{n+} \qquad (R-42')$$

Upon combining Eqs. (R-36), (R-37''), (R-38'), (R-42'), and (R-45),

$$P_{ox,\infty} = k_3[RH]\frac{k_3[RH]}{k_6 + k_{33}K_{32}[Me]^{n+}} \qquad (R-46)$$

When termination steps such as (R-31) to (R-33) become important (at relatively high catalyst concentrations), Eqs. (R-44) and/or (R-46) may become applicable. Thus, it has been observed [3] that when the cobalt(II) decanoate concentration was increased during the autoxidation of Tetralin in chlorbenzene solution at 65°C, a critical catalyst concentration was reached, after which the maximum oxidation rate dropped catastrophically. If Eq. (R-44) is utilized, it can be seen that the critical concentration may be expressed as

$$[Me]_c^{n+} = \frac{k_3[RH]}{k_{31}} \qquad (R-47)$$

When $[Me]_c^{n+} = f[Me]$, Eq. (R-47) was found to hold at relatively high values of $[RH]$. (This relationship between $[Me]_c^{n+}$ and $[Me]$ may not apply at relatively low values of $[RH]$.) At the lower values of $[RH]$, an additional unimolecular termination step (R-48) was postulated to account for deviations observed for Eq. (R-47):

$$RO_2\cdot \longrightarrow Products \qquad (R-48)$$

RADIATIVE DEGRADATION

Radiative degradation in polymers generally results in two types of reactions, namely, chain scission and crosslinking. The former gives rise to the formation of low-molecular-weight moieties whereas the latter results in network structures that are insoluble and infusible. Generally, competition ensues between the two types of reaction mechanisms.

Degradation of polymers by radiant energy is conveniently subdivided into two types, depending on the mode of action, which in turn is dictated by the amount of radiant energy involved. Accordingly, we divide radiative degradation into:

1. *Photolysis (photochemical radiation)*, e.g., ultraviolet light having wavelength λ in the range of 10^4 to 10^2 Å, and imparting energy in the range of 10^2 to 10^3 kcal mole^{-1}.

2. *Radiolysis (ionizing radiation)*, a form of high radiant energy, e.g., X rays, electron beams, γ rays, having λ in the range of 10^2 to 10^{-3} Å, and imparting energy in the range of 10^5 to 10^{10} kcal mole^{-1}, depending on type. Thus, for X rays, electron beams, and γ rays, energy is of the order 10^5 to 10^7, 10^7 to 10^9, and 10^7 to 10^{10} kcal mole^{-1}, respectively.

Fundamentally, the two types of radiant energies differ in mode of attack on a chain. Degradation of a polymer chain by ultraviolet light (UV) is due to the absorption of energy in discrete units by specific functional groups (chromophores) that may be present in the chain. On the other hand, these specific groups are not needed for the absorption of high radiant energy since the energy is transferred directly to the electrons that are in the path of the high-energy photons. Furthermore, UV radiation tends to only excite an electron within a specific functional group to a higher energy state, whereas high radiant energy tends, more often than not, to completely eject an electron from a molecule. Because of the latter result, high radiation energy is often referred to as "ionizing radiation."

The photochemical aspect is emphasized in the following.

Photochemical Radiation

In absence of oxygen. Photochemical radiation is capable of cleaving C—C bonds. For example, light at 2800 Å, having an energy of 100 kcal mole^{-1}, may in principle cleave the C—C bond, whose bond energy is about 80 kcal mole^{-1}. During photodegradation, generally chain scission, crosslinking, and monomer production, including other small molecular-weight fractions, can occur. Chain scission is generally of the random type which may be mathematically expressed as follows (cf. Chap. 4):

$$S = \frac{D_{P,0}}{D_P} - 1 = D_{P,0}\alpha \tag{30}$$

where S = number of scissions per original chain; $D_{P,0}$ and D_P denote number-average degree of polymerization initially and at time t, respectively;

and α denotes fraction of bonds broken per original chain. Further, we may write [cf. Chap. 4, Eq. (18')]

$$\alpha = 1 - e^{-kt} \tag{31}$$

From Eqs. (30) and (31) we obtain

$$\frac{1}{D_P} - \frac{1}{D_{P,0}} = 1 - e^{-kt} \tag{32}$$

At low conversions, Eq. (32) becomes

$$\frac{1}{D_P} - \frac{1}{D_{P,0}} \simeq kt \tag{32a}$$

When photodegradations occur in solution, we may write [6]

$$-\frac{d[D_P]}{dt} = \phi I_a \tag{33}$$

where $[D_P]$ denotes concentration of chain links, ϕ = quantum yield for chain scission, I_a = average number of quanta absorbed per unit volume per unit time. From Eq. (33) we can write

$$[D_{P,0}] - [D_P] = \phi I_a t \tag{33a}$$

Also at low conversions

$$[D_{P,0}] - [D_P] = \frac{[D_{P,0}]}{D_P} \cdot S = \phi I_a t \tag{34}$$

From Eqs. (30) and (34) we may write

$$\alpha = \frac{\phi I_a t}{[D_{P,0}]} \tag{35}$$

and, therefore,

$$\frac{1}{D_P} - \frac{1}{D_{P,0}} = \frac{\phi I_a t}{[D_{P,0}]} \tag{36}$$

In addition (for low conversion), from the Beer-Lambert law,

$$I = I_0 e^{-k_2[D_{P,0}]l} \tag{37}$$

where k_2 is an optical constant, l is optical path length, and I and I_0 are intensities of light at any time t and initially, respectively. We may write from Eq. (37)

$$I_a \equiv I_0 - I = I_0 \left(1 - e^{-k_2[D_{P,0}]l}\right) \tag{38}$$

Substituting Eq. (38) into Eq. (36), we obtain

$$\frac{1}{D_P} - \frac{1}{D_{P,0}} = \frac{\phi I_0 \left(1 - e^{k_2[D_{P,0}]l}\right)t}{[D_{P,0}]} \tag{39}$$

For low degree of light absorption (low conversion) we may write [cf. Eq. (38)]

$$\frac{I_a}{I_0} = 2.3 \log \frac{I_0}{I} = 2.3E \tag{40}$$

where E denotes optical density. Substituting Eq. (40) into Eq. (36), we obtain

$$\frac{1}{D_P} - \frac{1}{D_{P,0}} \approx \frac{2.3\phi I_0 Et}{[D_{P,0}]} \equiv k_e t \tag{41}$$

It should be mentioned at this point that $E = f([D_{P,0}])$ [cf. Eq. (37)], and therefore at low conversion $(1/D_P - 1/D_{P,0})$ should be independent of $[D_{P,0}]$.

It can be seen from Eq. (41) that the degree of degradation $S/D_{P,0}$ or $1/D_P - 1/D_{P,0}$ and the experimental rate constant k_e should be proportional to the first power of I_0 (intensity exponent equal to unity). Table 2 lists cases of polymer degradation in solution in which the intensity exponent was found to be unity [6].

In the following a treatment of the photodegradation of polymeric films will be presented. This is somewhat more complex than in solution since degradation is affected by film thickness. The intensity of the light varies within the film depending on its thickness. Thus, degradation throughout the film will be nonuniform. Therefore, only average values of molecular weights and other parameters will be obtained.

For a monodisperse polymeric film, the following expression has been derived [7]:

$$\bar{\alpha} = \frac{1}{D_P} - \frac{1}{D_{P,0}} = \frac{KI_0 t \left(1 - e^{-al}\right)}{l[D_{P,0}]} \tag{42}$$

TABLE 2   Photodegradation of Various Polymers in Solution

Polymer	Experimental conditions	Value of intensity exponent
Polymethyl vinyl ketone	25–250°C, dioxane solution, 3130 Å	1 (implies random chain scission)
Polyacrylonitrile	25°C, ethylene carbonate-propylene carbonate solutions, 2537 Å	1 (implies random chain scission)

where the bar denotes average values, K is a proportionality constant, and a is an optical constant.

For small light absorption Eq. (42) becomes

$$\bar{\alpha} = \frac{K'I_0 t}{[D_{P,0}]} = kt = \frac{1}{D_P} - \frac{1}{D_{P,0}} \tag{43}$$

where $K' = aK$. From Eq. (43) k should be proportional to the first power of I_0 (intensity exponent equals unity).

TABLE 3   Photodegradation of Various Polymeric Solids

Polymer	Experimental conditions	Value of intensity exponent
Poly(α-methylstyrene)	27 and 115°C; vacuum; films, 1.5–14 μ thick; 2537–2804 Å	1 (chain scission and monomer formation)
Cellulose acetate	25°C; air; powder; 2540–4340 Å	1 (chain scission)

Table 3 lists cases of photodegraded polymeric films (or powder) where the intensity exponent was found to be unity. Although these expressions [cf. Eqs. (41) and (43)] were derived for cases involving essentially pure chain scission, there are various other cases in which similar expressions can be derived (cf. Chap. 4). For example, for a photodecomposition involving random initiation, depropagation, and termination by disproportionation (cf. Chap. 4, case 7) an equation such as

$$\frac{1}{D_P} - \frac{1}{D_{P,0}} \simeq kt \tag{41a}$$

may be obtained under the assumption of $1 + 2L \ll D_P$, where L is the largest length of the volatile fragment. In this case, the formation of volatile products is considered, e.g., monomer.

In the following will be given the details of the kinetics and mechanisms of the photodegradation of polyacrylonitrile in solution [8] and of poly(α-methylstyrene) in the bulk phase.

Polyacrylonitrile (in solution). The degradation of polyacrylonitrile followed Eq. (41) (intensity exponent equal to unity; cf. Table 2). The following mechanism was suggested:

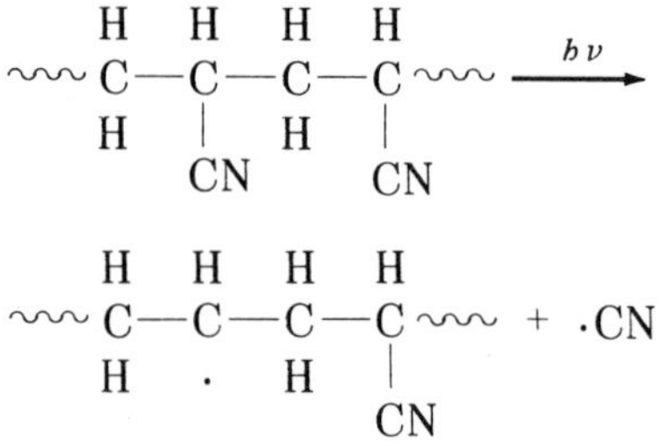

$$\tag{44}$$

(III)

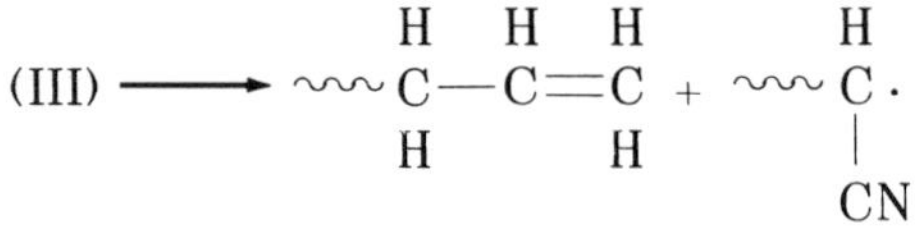

$$\tag{44a}$$

Equations (44) and (44a) represent initial chain-scission reactions which are followed by reactions of the liberated ·CN radical, namely,

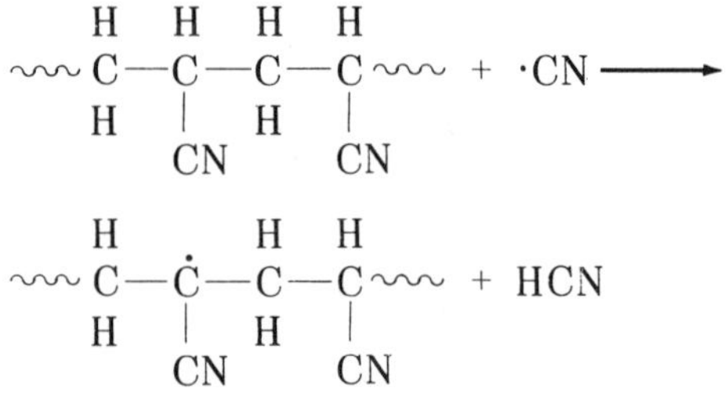

$$\tag{45}$$

(IV)

$$(IV) \xrightarrow[\text{scission}]{\text{chain}} \quad \sim\!\!\sim\!\!\sim \underset{\underset{H}{|}}{\overset{\overset{H}{|}}{C}} - \underset{\underset{CN}{|}}{C} = \underset{\underset{H}{|}}{\overset{\overset{H}{|}}{C}} \quad + \quad \sim\!\!\sim\!\!\sim \underset{\underset{CN}{|}}{\overset{\overset{H}{|}}{C}} \cdot \tag{45a}$$

$$(V)$$

The terminal double bond, containing the CN moiety, would be expected to increase the absorption at 2160 Å, as was found experimentally. Finally, in order to account for the production of other small molecules, e.g., H_2, the following was postulated:

$$(V) \left\{ \begin{array}{l} \xrightarrow{h\nu} \quad \sim\!\!\sim\!\!\sim \underset{\underset{CN}{|}}{C} = \underset{\underset{H}{|}}{C} - \underset{\underset{CN}{|}}{C} = \overset{\overset{H}{|}}{C} \quad + \; H_2 \tag{46} \\[2em] \xrightarrow{h\nu} \quad \sim\!\!\sim\!\!\sim \underset{\underset{H}{|}}{C} = \underset{\underset{H}{|}}{C} - \underset{\underset{CN}{|}}{C} = \overset{\overset{H}{|}}{C} \quad + \; HCN \tag{46a} \end{array} \right.$$

The conjugated structures in Eqs. (46) and (46a) should give rise to optical absorptions at 2950 Å, as observed.

Poly(α-methylstyrene) (in bulk). As may be noted from Table 3, the photodegradation of poly(α-methylstyrene), (PMS), presumably involves an initiation step of random chain scission [6, 9]. Based on monomer formation this initiation step could be followed by a depropagation step. Furthermore, in order to explain the kinetic behavior of the degradation, a unimolecular termination step was included [cf. Chap. 4, Eq. (4)]. Besides the formation of monomer during the PMS photodegradation, small amounts of various volatile products were also formed, e.g., H_2, CO_2, CO.

The following mechanism was postulated:

Initiation

$$P_0 \xrightarrow{h\nu} R_{\dot{m}} + R_{\dot{n}-m} \tag{47}$$

Depropagation

$$R_{\dot{m}} \xrightarrow{k_2} R_{\dot{m}-1} + M \tag{48}$$

Termination

$$R_{\dot{m}} \xrightarrow{k_3} P_m \tag{49}$$

where P_0 is an original polymer chain, R· denotes radicals, and P_m represents terminated polymer. From the preceding mechanism we may write for the breaking of links, [cf. Eq. (33)]

$$-\frac{d[D_P]}{dt} = k_1 I_a \tag{50}$$

where k_1 is the specific rate constant for the initiation step, Eq. (47). Using steady-state conditions for [R·] (it is assumed that all radical species have similar reactivities), we may write

$$[R·] = \frac{2k_1 I_a}{k_3} \tag{51}$$

where [R·] denotes radical concentration. Furthermore, the rate of formation of volatile products, as monomer, may be written as

$$\frac{d(M)}{V\,dt} \equiv \frac{d[M]}{dt} = k_2[R·] = \frac{2k_2 k_1 I_a}{k_3} \tag{52}$$

where (M) denotes amount of monomer, $[M]$ is monomer concentration, and V equals volume of reaction medium.

Upon integration of Eq. (52), the amount of monomer liberated is

$$[M] = \frac{2k_2 k_1 I_a t}{k_3} \tag{52a}$$

Letting the kinetic chain length be denoted as (CL) [cf. Chap. 4, Eq. (62)]

$$(CL) = \frac{k_2[R·]}{k_3[R·]} = \frac{k_2}{k_3} \tag{53}$$

Upon combining Eqs. (52a) and (53), there is obtained

$$[M] = 2k_1 (CL) I_a t \tag{54}$$

Now, if we let ϕ_M denote monomer quantum yield, we may write

$$\phi_M = \frac{k_2[R·]}{I_a} = \frac{2k_2 k_1}{k_3} \tag{55}$$

From Eqs. (53) to (55) we obtain

$$[M] = \phi_M I_a t \qquad (56)$$

In terms of amount of monomer formed (M) per given weight of polymer W, Eq. (56) becomes

$$\frac{(M)}{W} = \frac{\phi_M I_a t}{\rho} \qquad (56a)$$

where ρ is density of the polymer (which should be approximately constant over a large range of conversion). An expression similar to Eq. (56a) was found to agree well with experimental data.

Using the mechanisms represented by Eqs. (47) to (49), we may derive an expression relating number-average molecular weight M_n as a function of time [cf. Chap. 4, Eqs. (8) to (10)]. From the following definition, $M_n = W/(n)$ where (n) is number of polymer molecules, we can obtain

$$\frac{dM_n}{dt} = \frac{M_n}{W}\frac{dW}{dt} - \frac{M_n^2}{W}\frac{d(n)}{dt} \qquad (57)$$

We can also write, using Eq. (52) and the expression

$$\frac{d(M)}{dt} = \frac{1}{M_m}\left(-\frac{dW}{dt}\right) \qquad (58)$$

where M_m is the monomeric molecular weight,

$$-\frac{dW}{dt} = \frac{2k_1 k_2 M_m I_a V}{k_3} \qquad (59)$$

Furthermore,

$$\frac{d(n)}{V\,dt} = -k_1 I_a + k_3 [R\cdot] = k_1 I_a \qquad (60)$$

Combining Eqs. (57) to (59) and (60), we obtain

$$\frac{dM_n}{dt} = -\frac{2k_1 k_2 M_n I_a M_m}{k_3 \rho} - \frac{k_1 I_a M_n^2}{\rho} \qquad (61)$$

When $2k_2/k_3 (= \phi_M/k_1) \ll D_P$, (e.g., ϕ_M/k_1 was calculated to have values of 7 and 25, at 27 and 115°C, respectively), Eq. (61) becomes

$$\frac{dM_n}{dt} = - \frac{M_n^2 k_1 I_a}{\rho} \tag{62}$$

Upon integration, Eq. (62) yields

$$\frac{1}{M_n} - \frac{1}{M_{n,0}} = \frac{k_1 I_a t}{\rho} \tag{63}$$

At low conversions, Eq. (63) reduces to

$$\frac{1}{D_P} - \frac{1}{D_{P,0}} = \frac{k_1 I_a t}{[D_{P,0}]} \tag{63a}$$

Equation (63a) was found to be in good agreement with experimental data.

Based on the mechanism [Eqs. (47) to (49)], the following reaction scheme was proposed for PMS:

Initiation

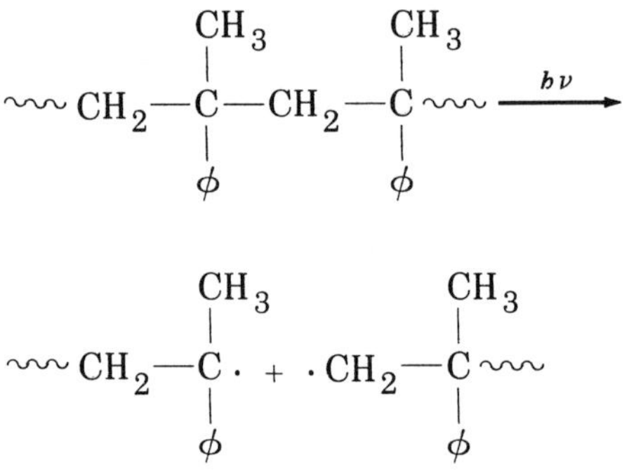

$$\tag{64}$$

Depropagation

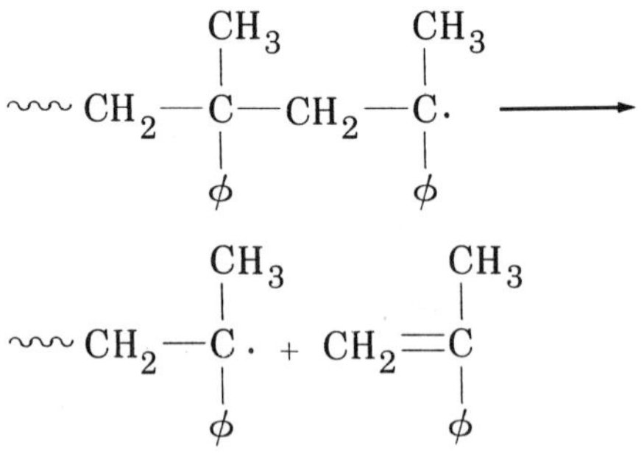

$$\tag{65}$$

Termination

$$\underset{\underset{\phi}{|}}{\overset{CH_3}{\overset{|}{\sim\!\sim CH_2\!-\!C\cdot}}} + \underset{\underset{\phi}{|}}{\overset{CH_3}{\overset{|}{CH_2\!=\!C}}}$$

$$\longrightarrow \sim\!\sim CH=\underset{\underset{\phi}{|}}{\overset{CH_3}{\overset{|}{C}}} + CH_3\!-\!\underset{\underset{\phi}{|}}{\overset{\cdot}{C}}\!-\!CH_3 \qquad (66)$$

$$\longrightarrow \sim\!\sim CH_2\!-\!\underset{\underset{\phi}{|}}{CH}\!-\!CH_3 + CH_2\!=\!\underset{\underset{\phi}{|}}{\overset{\cdot}{C}}\!-\!CH_2 \qquad (66a)$$

As indicated previously, the termination step was assumed to be unimolecular in order to account for the observed kinetic dependencies. Thus, in the termination steps (66) and (66a), it must be assumed that monomer concentration in the film remains virtually constant during the reaction. Accordingly, the reaction is depicted as pseudo-unimolecular termination.

In order to account for various volatiles detected, the following additional steps were postulated:

$$\underset{\underset{\phi}{|}}{\overset{CH_3}{\overset{|}{\sim\!\sim CH_2\!-\!C}}} \sim\!\sim \xrightarrow{h\nu}$$

$$\longrightarrow \sim\!\sim \overset{\cdot}{CH}\!-\!\underset{\underset{\phi}{|}}{\overset{CH_3}{\overset{|}{C}}}\sim\!\sim + \text{H}\cdot \qquad (67)$$

$$\longrightarrow \sim\!\sim CH_2\!-\!\underset{\underset{\phi}{|}}{\overset{\overset{\cdot}{CH_2}}{\overset{|}{C}}}\sim\!\sim + \text{H}\cdot \qquad (67a)$$

$$\longrightarrow \sim\!\sim CH_2\!-\!\underset{\underset{\phi}{|}}{\overset{\cdot}{C}}\sim\!\sim + \cdot CH_3 \qquad (67b)$$

$$\cdot CH_3 + \text{H}\cdot \longrightarrow CH_4 \qquad (68)$$

$$2\cdot CH_3 \longrightarrow C_2H_6 \qquad (69)$$

In presence of oxygen. In the following will be treated the photochemical oxidation of polymers in the bulk phase. Although saturated hydrocarbons are transparent to the usual light radiation used, i.e., 2000 Å, photochemical initiation may occur owing to the initial presence of light-absorbing groups, such as double bonds and carbonyl (which may be present as impurities). In this connection, it may be mentioned that the attachment of a few sensitizer groups along the backbone of a polymer chain can cause rapid breakage of links. These sensitizer groups have the property of strongly absorbing ultraviolet light. A possible general scheme to account for the photochemical oxidation process has been advanced by Ershov, Kuzina, and Neiman [10]. They postulated the following:

$$S + h\nu \xrightarrow{k_{70}} S^* \tag{70}$$

$$S^* \xrightarrow{k_{71}} S + h\nu' \tag{71}$$

$$S^* \xrightarrow{k_{72}} T \tag{72}$$

$$S^* + A \xrightarrow{k_{73}} S + A^* \tag{73}$$

$$T \xrightarrow{k_{74}} \text{a transformation} \tag{74}$$

$$S^* \xrightarrow{k_{75}} 2R \cdot \tag{75}$$

$$(M - M) + R \cdot \xrightarrow{k_{76}} R \cdot + P_f \tag{76}$$

$$R \cdot (\text{condensed phase}) \xrightarrow{k_{77}} R \cdot (\text{gas}) \tag{77}$$

$$2R \cdot \xrightarrow{k_{77a}} C \tag{77a}$$

$$R \cdot + O_2 \xrightarrow{k_{78}} P_0 + R \cdot \tag{78}$$

In the above mechanism S and S^* denote polymer with absorbing groups in the ground singlet state and in the excited singlet state, respectively; $h\nu'$ represents fluorescence; T denotes excited triplet state; A is an acceptor molecule, whereas A^* is the acceptor molecule at the excited state; $M - M$ is a portion of the polymer molecule which does not contain a light-absorbing group; C denotes crosslinked product; and P_0 represents products obtained from peroxy radicals. It would appear from step (78) that only $R \cdot + O_2$ occurs but that the reaction

$$RO_2 \cdot + RH \longrightarrow RO_2H + R \cdot \tag{79}$$

does not. This presumably is due to the high immobility of the peroxy radicals in the bulk phase (cf. Chap. 5). The formation of products P_0 from peroxy radicals is not without precedent. Thus, Kamiya and Ingold [11] postulated formation of inactive products from peroxy radicals during the autoxidation of Tetralin.

Polystyrene. The above mechanism has been used in explaining the data obtained by Grassie and Weir [12] for the photochemical oxidation of polystyrene (PS). They obtained the following for PS films at 28°C and $\lambda = 2537$ Å: H_2O and CO_2 were the main volatile products; rate of oxidation ρ_{ox} is a function of I_0 (initial light intensity) to the first power at relatively high oxygen pressures of 20 to 600 mm Hg, but below 20 mm Hg the ρ_{ox} was found to be independent of I_0; ρ_{ox} is independent of the polymer molecular weight but is a function of the oxygen concentration to the first power; and the free-radical inhibitor, 2,6-di-*tert*-butyl-4-methylphenol, was found to have no influence on ρ_{ox}. Further, it was found that the following expression was valid:

$$\rho_{ox} = f\left(1 - e^{-\alpha l}\right) \tag{80}$$

where α is a constant, and l is the film thickness. Equation (80) may be derived as shown in Fig. 1.2.

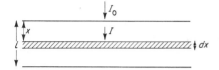

FIG. 1.2 Schematic representation of polymeric film exposed to radiation.

Figure 1.2 shows a film of thickness l, upon which radiation of initial intensity I_0 is striking the surface. We select an element of thickness dx having a radiation intensity I. According to the Beer-Lambert law, we may write

$$I_0 - I = I_0\left(1 - e^{-ax}\right) \tag{81}$$

where a is an optical constant. From Eq. (81), the number of quanta absorbed per unit area at a layer of thickness dx may be written as

$$\frac{d(I_0 - I)}{dx} = -\frac{dI}{dx} = I_0 a e^{-ax} \tag{82}$$

Over the entire film thickness l, the radiation absorbed is thus given by

$$\int_0^l I_0 a e^{-ax}\, dx = I_0\left(1 - e^{-al}\right) \tag{83}$$

If $\rho_{ox} = f(I_{abs})$, where the subscript abs denotes absorbed, then

$$\rho_{ox} = K I_0\left(1 - e^{-al}\right) \tag{84}$$

Equation (84) was observed to be valid.

The ineffectiveness of free-radical inhibitors was attributed to the lack of mobility of radicals in the bulk phase. This supports step (78) of the general scheme as a main oxidation step. Step (77) may also be assumed to apply since the photochemical oxidation was found to occur mainly near the surface of the film; and mobile radicals, e.g., $H\cdot$ and $\cdot CH_3$, could presumably diffuse out of the film.

From the scheme [Eqs. (70) to (78)] we may write

$$-\frac{d[O_2]}{dt} \equiv \rho_{ox} = k_{71}[O_2][R\cdot] \tag{85}$$

Also, assuming steady-state conditions,

$$\frac{d[R\cdot]}{dt} = k_{75}[S^*] - k_{77}[R\cdot] = 0 \tag{86}$$

where S^* could denote benzene-ring-absorbing centers, e.g.,

$$\sim\!\!\sim \underset{\underset{\phi}{|}}{\overset{\overset{\textstyle H}{|}}{C}}\!-\!CH_2 \sim\!\!\sim + h\nu \xrightarrow{\ k_{70}\ } \sim\!\!\sim \underset{\underset{\phi^*}{|}}{\overset{\overset{\textstyle H}{|}}{C}}\!-\!CH_2 \sim\!\!\sim (S^*) \tag{87}$$

Apparently, the activation represented by Eq. (87) occurs more favorably at $\lambda = 2537$ Å then at $\lambda = 3560$ Å. Then,

$$(S^*) \xrightarrow{\ k_{75}\ } H\cdot + \sim\!\!\sim \underset{\underset{\phi}{|}}{\overset{\textstyle \cdot}{C}}\!-\!CH_2 \sim\!\!\sim \tag{88}$$

Since H_2 was not detected as one of the oxidation products, presumably the reaction $H\cdot + O_2 \longrightarrow HO_2\cdot$ occurred.

Assuming steady-state conditions for $[S^*]$, we may write, neglecting step (73),

$$\frac{d[S^*]}{dt} = k_{70}[S]I_0 - (k_{71} + k_{72} + k_{75})[S^*] = 0 \tag{89}$$

It follows from Eq. (89) that

$$[S^*] = \frac{k_{70}[S]I_0}{k_q} \tag{90}$$

where $k_q \equiv k_{71} + k_{72} + k_{75}$.

Using Eqs. (86) and (90), we obtain

$$[R \cdot] = \frac{k_{75} [S^*]}{k_{77}} = \frac{k_{75} k_{70} [S] I_0}{k_{77} k_q} \tag{91}$$

Substituting Eq. (91) into Eq. (85), there is obtained

$$P_{ox} = \frac{k_{78} [O_2] k_{75} k_{70} [S] I_0}{k_{77} k_q} \tag{92}$$

Upon integrating, the amount of oxygen consumed, $-\Delta[O_2]$, may be obtained as

$$-\Delta[O_2] \equiv \int P_{ox} dt = \frac{k_{78} k_{75} k_{70} [S] [O_2] I_0 t}{k_{77} k_q} \tag{93}$$

In the above $[O_2]$ is assumed to be virtually constant. The above expressions agree well with experimental data obtained by Grassie and Weir. At low oxygen pressures (low oxygen solubility in the PS film), rate of oxygen absorption would be determined primarily by rate of diffusion, and is consequently independent of I_0, which was observed.

The formation of water (one of the main products) was postulated as forming from

$$HO_2 \cdot + RH \longrightarrow R \cdot + H_2 O_2$$
$$\searrow H_2 O \tag{94}$$
$$2HO_2 \cdot \longrightarrow O_2 + H_2 O_2 \nearrow$$

No clear explanation was provided for the formation of CO_2.

Ershov and coworkers [10] have also used the general scheme, Eqs. (70) to (78) to explain mathematically the photolytic oxidation of poly(methyl methacrylate) at 2540 Å at 30°C. In this case the initiation step involves the absorption of a photon followed by the transition of a p electron of the carbonyl group from the nonbonding orbital of the oxygen atom to the antibonding orbital ($n \to \pi^*$ transition). Thus,

$$\sim\sim CH_2 \overset{\displaystyle CH_3}{\underset{\displaystyle \underset{OCH_3}{\overset{|}{C}}=O}{\overset{|}{\underset{|}{C}}}}\sim\sim + h\nu \longrightarrow \sim\sim CH_2 \overset{\displaystyle CH_3}{\underset{\displaystyle \underset{OCH_3}{\overset{|}{C}}=O^*}{\overset{|}{\underset{|}{C}}}} \tag{95}$$

The excitation energy may then be transmitted by an intramolecular transition without emission to an ester C—O bond or a C—C bond, leading to the photodissociation of the bond [step (75) in the general scheme Eqs. (70) to (78)]:

$$\sim\!\!\sim CH_2-\underset{\underset{\underset{CH_3}{|}}{\overset{|}{O}}}{\overset{\overset{CH_3}{|}}{\underset{|}{C}}}\sim\!\!\sim \longrightarrow CH_3O\cdot + \sim\!\!\sim CH_2-\underset{\underset{\cdot C=O}{|}}{\overset{\overset{CH_3}{|}}{C}}\sim\!\!\sim \qquad (96)$$

or

$$\sim\!\!\sim CH_2-\underset{\underset{\underset{CH_3}{|}}{\overset{|}{O}}}{\overset{\overset{CH_3}{|}}{\underset{|}{C}}}\sim\!\!\sim \longrightarrow \sim\!\!\sim CH_2-\underset{\cdot}{\overset{\overset{CH_3}{|}}{C}}\sim\!\!\sim + CH_3O-\dot{C}=O$$

$$(97)$$

Reaction (97) represents a Norrish type I process, which will be discussed later in this section. Unlike vinyl polymers, e.g., polystyrene and poly-(methyl methacrylate), polyolefins do not contain inherent light-absorbing centers. Therefore, the kinetics and mechanisms of the saturated polyolefins would be expected to be more complex. Ershov and coworkers also mathematically treated low-pressure polyethylene. However, owing to spurious and skittery factors which can affect the photolytic oxidation of saturated polyolefins, this mathematical treatment will not be given. These factors are considered qualitatively in the discussion to follow.

Saturated polyolefins [16]. Polyolefins degrade photochemically by virtue of the presence of impurities, e.g., during processing (formation of carbonyl, hydroperoxide, and similar functional groups from oxidation.) Once initiated, photoxidation may proceed through a free-radical chain mechanism analogous to that usually used in explaining thermal oxidation (cf. Chap. 5).

Two types of initiation processes of photoxidation may be considered, i.e., primary and secondary initiations.

Primary initiation. This can occur by the absorption of ultraviolet light by the oxidizing substance (oxygen) or by possible oxygen-substrate complexes. A possible initiation involves the excitation of oxygen by light from the $^3\Sigma_g^-$ ground state to the $^3\Sigma_u^+$ state by means of a Herzberg forbidden transition. Another possible method of initiation is the oxygen-perturbed transition from the singlet (S) to the triplet (T) state. This applies generally for olefinic and aromatic hydrocarbons, but little is known about the possible initiation of aliphatic hydrocarbons. Free radicals may also be obtained by photolysis of the charge-transfer complexes between oxygen and various sites along the polymer chain, particularly where there are functional groups or double bonds. The latter are probably present in polyolefins synthesized by Ziegler-Natta catalysts. Thus, charge-transfer complexes initially formed may lead to active intermediates which may yield oxidation products. Furthermore, the triplet state of oxygen may be quenched by the presence of excited carbonyl groups which may lead to the formation of singlet oxygen. These singlet oxygen molecules are then capable of reacting with double bonds, leading to the formation of oxidation products, e.g., hydroperoxides. These products can react further by means of photolyses, thereby continuing the oxidation process.

Secondary initiation. This can occur when ultraviolet light is absorbed by oxidation products of the substrate, e.g., peroxides, carbonyls, or other forms of impurities that may be present. This type of initiation appears to be more important in connection with the photosensitivity of polyolefins.

Metallic impurities present in polyolefins synthesized by Ziegler-Natta catalysts are capable of acting as sensitizers which promote the photoxidation of polyolefins. Thus, it was found that the degradation of polypropylene induced by ultraviolet light depends on oxygen concentration and on the residues of the polymerization catalysts [13]. In general, the transition-metal residues absorb light, producing free radicals through a photoexcited electron transfer from the anion to the cation. These radicals then commence the oxidation of the substrate. It should be mentioned that the nature of the anion was found to be important in the resistance of a polymer to photoxidation [14]. Thus, with increasing electron affinity of the anion, the absorption of the metal compound tends to shift toward shorter wavelengths than those normally prevalent in sunlight at the level of the earth. Accordingly, the photosensitizing action of metal residues should decrease.

Oxidation products (hydroperoxides and carbonyl) arising from processing may also initiate photoxidation of polyolefins. Thus, hydroperoxides may absorb light, leading to the breakage of the O—O bonds creating the two free radicals, RO$\cdot$ and $\cdot$OH. These radical pairs can then initiate the free-radical chain. In this connection may be mentioned the work of Ershov and coworkers [15] on the photolysis of atactic polypropylene hydroperoxide. They postulated, in order to account for kinetic data obtained, the formation of RO$\cdot$ and $\cdot$OH radicals. This type of decomposition will be discussed later in connection with polypropylene photoxidation.

Relative to the presence of carbonyl, two important processes have been advanced to account for their photochemical degradation. These are:

Norrish type I process

$$-\overset{|}{\underset{|}{C_1}}-\overset{O}{\overset{||}{C}}-\overset{|}{\underset{|}{C_2}}-\ \xrightarrow{\ h\nu\ }\ -\overset{|}{\underset{|}{C_1}}-\overset{O}{\overset{||}{C}}\cdot\ +\ \cdot\overset{|}{\underset{|}{C_2}}- \qquad (98)$$

or

$$-\overset{|}{\underset{|}{C_1}}\cdot\ +\ CO\ +\ \cdot\overset{|}{\underset{|}{C_2}}- \qquad (98a)$$

The type I reaction leads to the formation of free radicals which should reach thermal equilibrium rapidly. It has been shown [16] that the precursors of this reaction are both the excited singlet and excited triplet states of ketones.

Norrish type II process

$$-\overset{H}{\underset{|}{\underset{|}{C_3}}}-\overset{|}{\underset{|}{C_2}}-\overset{|}{\underset{|}{C_1}}-\overset{O}{\overset{||}{C}}-\ \xrightarrow{\ h\nu\ }\quad \qquad (99)$$

$$-\overset{|}{\underset{|}{C_3}}{=}\overset{|}{\underset{|}{C_2}}\ +\ \overset{|}{\underset{|}{C_1}}{=}\overset{OH}{\overset{|}{\underset{|}{C}}}- \qquad (99a)$$

$$\downarrow$$

$$H-\overset{|}{\underset{|}{C_1}}-\overset{O}{\overset{||}{C}}- \qquad (99b)$$

The type II reaction may occur, assuming that the ketone has at least one hydrogen atom on the carbon atom in the γ-position relative to the carbonyl group. The type II reaction proceeds by intramolecular hydrogen transfer, yielding one olefin and one enol, ultimately rearranging to ketone of smaller size. In contrast to the type I reaction, it is noted that the type II reaction does not directly produce free radicals. Norrish types I and II reactions also apply to aldehyde-containing compounds analogous to ketonic compounds, namely,

$$-\overset{|}{\underset{|}{C}}-CHO\ \xrightarrow{\ h\nu\ }\ -\overset{|}{\underset{|}{C}}\cdot\ +\ \cdot CHO \qquad (100)$$

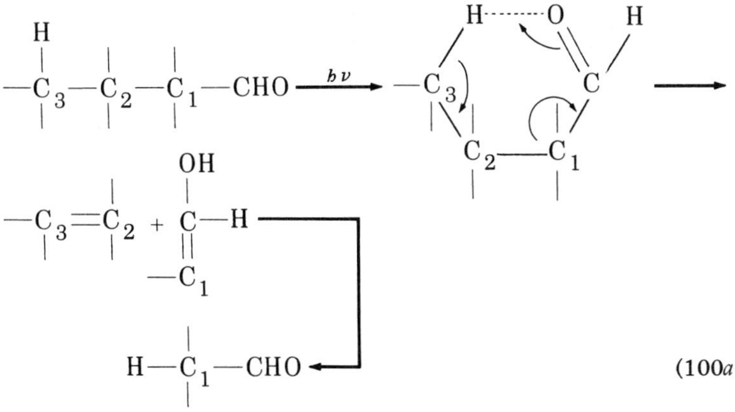

(100*a*)

The Norrish type II is not as important as type I since it involves cleavage to form nonradical fragments, whereas the type I cleavage results in the formation of radicals which can induce further oxidative degradation.

Another process which may be mentioned, along with the Norrish type, is

$$\rangle C{=}O \xrightarrow{h\nu} \rangle\dot{C}{-}O\cdot \xrightarrow{RH} \rangle\dot{C}{-}OH + R\cdot \qquad (101)$$

This reaction (intermolecular hydrogen transfer) does not occur to the same extent as the Norrish type II (intramolecular hydrogen transfer) for comparable ketones.

The various reactions described above will now be applied to the photoxidation of saturated polyolefins such as polyethylene and polypropylene.

Polyethylene. A mechanism was proposed by Trozzolo and Winslow [17] for the oxidative photodegradation of polyethylene. This mechanism stresses the Norrish type II reaction. The experimental evidence of their mechanism was based on established photochemistry of model compounds and on the known products of the photoxidation of polyethylene. The mechanism involves the following: (1) excitation of carbonyl groups by light absorption; (2) Norrish type II cleavage involving (n, π^*) excited states of carbonyl groups, as illustrated below:

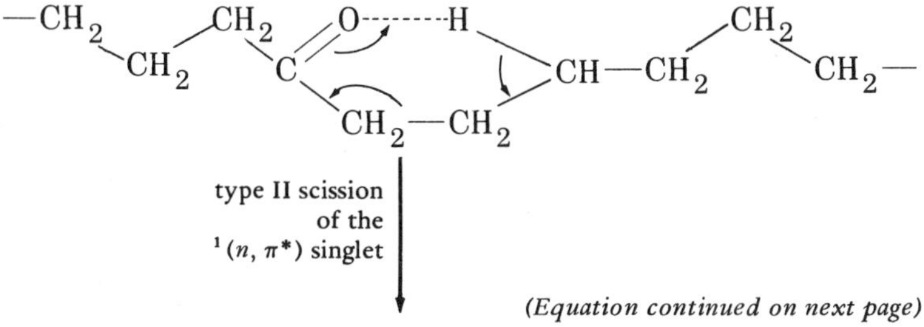
type II scission
of the
$^1(n, \pi^*)$ singlet

(Equation continued on next page)

$$\text{Ketone} \qquad + \qquad \text{Olefin} \qquad (102)$$

(3) formation of singlet oxygen by quenching of the $^3(n, \pi^*)$ triplet state of carbonyl groups, namely,

$$\text{Ketone } ^3(n, \pi^*) + {}^3O_2 \longrightarrow \text{Ketone}(S_0) + {}^1O_2 \qquad (103)$$

and, finally, (4) reaction of singlet oxygen molecules with vinyl groups formed in the Norrish type II cleavage, namely:

$$(104)$$

Further reactions of hydroperoxides formed in Eq. (104), induced by light, could lead to additional carbonyl groups which may undergo further scission.

Polypropylene. Similar mechanisms previously discussed for polyethylene should also apply to polypropylene, as illustrated in the steps below:

First phase (initiation)

type II cleavage
of $^1(n, \pi^*)$ singlet

(Equation continued on next page)

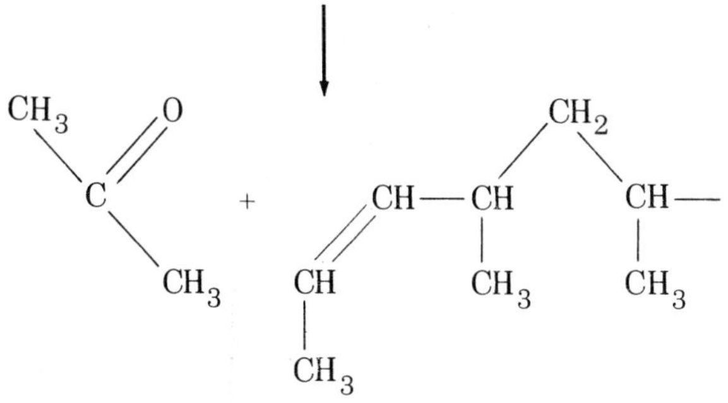

$$\text{Acetone} \qquad\qquad \text{Olefin} \tag{105}$$

$$\text{Ketone}^3\,(n, \pi^*) + {}^3O_2 \longrightarrow \text{Ketone}\,(S_0) + {}^1O_2 \tag{103}$$

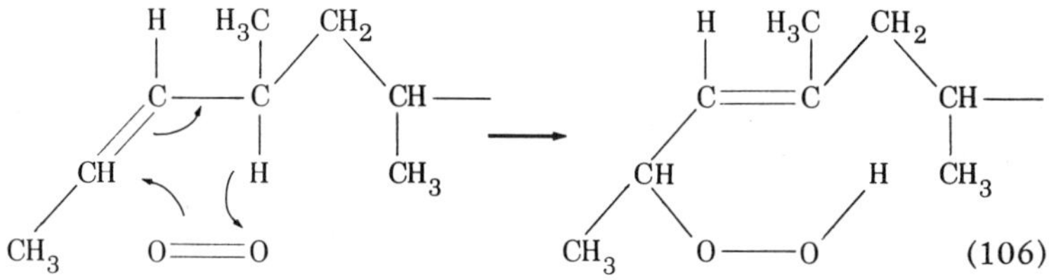

$$\tag{106}$$

Second phase (degradation)

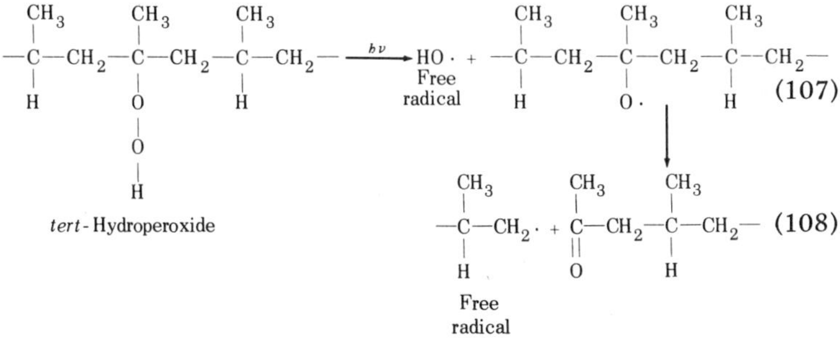

$$\xrightarrow{h\nu} \text{HO} \cdot + \quad \text{(Free radical)} \tag{107}$$

tert-Hydroperoxide

$$\tag{108}$$

Free radical

$$\text{Free radicals} \xrightarrow{\;O_2,\ RH\;} \textit{tert}\text{-Hydroperoxides} \tag{109}$$

Referring to step (105), we may note that in a recent report by Carlsson and Wiles [18] they exposed powdered, isotactic, unstabilized polypropylene to air at 225°C for 1 to 5 min, corresponding roughly to commercial processing. These workers found that the principal products formed were two ketones having the following structures (observed at two infrared-absorption peaks indicated below):

$$\underset{\text{H}}{\overset{\text{CH}_3}{\underset{|}{\overset{|}{\text{C}}}}}-\text{CH}_2-\overset{\text{O}}{\overset{||}{\text{C}}}-\text{CH}_2-\underset{\text{H}}{\overset{\text{CH}_3}{\underset{|}{\overset{|}{\text{C}}}}}$$

(A) 1718 cm^{-1} absorption maximum

and

$$\underset{\text{H}}{\overset{\text{CH}_3}{\underset{|}{\overset{|}{\text{C}}}}}-\text{CH}_2-\overset{\text{O}}{\overset{||}{\text{C}}}-\text{CH}_3$$

(B) 1726 cm^{-1} absorption maximum

Presumably both structures (A) and (B) can be formed by cleaving tertiary oxy radicals, such as

$$\underset{\text{H}}{\overset{\text{CH}_3}{\underset{|}{\overset{|}{\text{C}}}}}-\text{CH}_2-\underset{\text{O} \cdot}{\overset{\text{CH}_3}{\underset{|}{\overset{|}{\text{C}}}}}-\text{CH}_2-\underset{\text{H}}{\overset{\text{CH}_3}{\underset{|}{\overset{|}{\text{C}}}}}$$

(C)

Upon photolysis of the oxidized sample in vacuo at 30°C the main products detected were acetone and CO. Only photolysis of structure (B) will yield acetone and an unsaturated structure [cf. step (105), which is a Norrish type II cleavage]. If structure (A) should undergo a Norrish type II cleavage, then ketones similar to structure (B) should form. However, because of the immediate drop of the 1726 cm^{-1} absorption maximum on irradiation, this precludes the significance of such a step. Accordingly, it may be assumed that structure (A) photolyzes predominantly by a Norrish type I process to give two macroradicals and CO.

The presence of acetaldehyde and CH_4 in the photolysis products indicates that some Norrish type I photolysis of ketone of structure (B) does occur. The experimental data were insufficiently precise to exclude the Norrish type II scission in the photolysis of ketone (A). Equations (103) and (106) to (109) should still be valid in spite of the preceding. In a subsequent paper [19], these workers found that polypropylene hydroperoxides were readily photodegraded to give various products [cf. step (107)].

Thus, one of the major products formed was water. Presumably, the hydroxyl radical from the hydroperoxide photolysis abstracted hydrogen to form water. Further, the tertiary alkoxy radicals, formed concurrently, underwent scission to produce ketonic products (A) and (B). Structure (A) is not considered in Eqs. (107) and (108) of the scheme. However, structure (B) is included in step (108).

A study was conducted by Hatton and coworkers [20] on the crosslinking of samples of atactic and isotactic polypropylene, swollen in a solution of benzophenone in allyl acrylate, by irradiation with ultraviolet light. The mechanism proposed for the crosslinking process included a step similar to step (101), namely,

$$\phi_2 C{=}O + h\nu \longrightarrow \phi_2 C{=}O^* \tag{110}$$
$$(D)$$

$$(D) + P \longrightarrow R \cdot + \underset{\underset{OH}{|}}{\phi_2 C} \cdot \tag{111}$$

where P represents polymer.

Light stabilization of polymers may be achieved by the following methods:

1. Ultraviolet absorbers. One of the additives used for this purpose is carbon black [21]. It not only acts as an ultraviolet light filter but also presumably quenches excited states. Other pigments in addition to carbon black have been found effective.

2. Quenchers. These are compounds that tend to transfer electronic energies from the excited state of a sensitizer molecule to that of an acceptor molecule, the quencher. Ni(II) chelates appear able to quench excited molecules both in the singlet and triplet states. Chelates of other transition metals have also been used for polyolefin stabilization [22].

3. Antioxidants.

4. Metal deactivators. These deactivators, e.g., organic chelates, tend to suppress the interaction between hydroperoxides and the metal catalyst present in the polyolefin.

Ionizing Radiation (High-energy Electrons, X Rays, γ Rays, etc.) [23]

In presence and absence of oxygen. Atomic radiation is not as selectively absorbed as ultraviolet radiation. The absorption of energy from atomic radiation (γ rays) is a function of electron density in the path of the radiation and is not appreciably affected by the manner in which atoms are linked. However, once a random ray strikes an atom and energy transfer occurs, radicals may form and the oxidation chain reaction may proceed as via ultraviolet radiation (or via thermal methods). Additionally, ions (including electrons), excited molecules, etc., may form.

Once initiation has proceeded, unsaturation in chemical bonding, chain scission, and crosslinking can occur, and a relatively large number of volatile fragments form, e.g.,

$$
\begin{array}{ccc}
\text{\textasciitilde CH \textasciitilde} & & \\
| & & \\
H & & \text{\textasciitilde CH \textasciitilde} \\
+ & \longrightarrow & \quad | \qquad + H_2 \\
H & & \text{\textasciitilde CH \textasciitilde} \\
| & & \\
\text{\textasciitilde CH \textasciitilde} & &
\end{array}
\tag{112}
$$

Generally, all polymers when in the path of very high doses of ionizing radiation will degrade to form low-molecular-weight fragments. However, some polymers will crosslink at a more rapid rate to form very high molecular-weight substances when subjected to relatively low radiation dose, e.g., $<10^8$ to 10^9 rads (1 rad = 100 ergs of ionizing radiation absorbed per gram of polymer) [24].

Generally, vinyl polymers that have long zip lengths (cf. Chap. 4) and high monomer yields in thermal degradation tend to degrade upon radiolysis whereas those with low monomer yields will crosslink. Furthermore, polymers that exhibit high heats of polymerization tend to crosslink while those with low heats of polymerization tend to degrade (a correlation often also observed in photolysis) (see Table 4). It is noted from Table 4

TABLE 4   Correlation of Radiation Crosslinking
and Degradation with Heats of Polymerization
and Monomer Yield on Pyrolysis [25]

Predominance of radiation effects	Polymer	H_p, kcal mole^{-1}	Monomer yield, wt. %
Crosslink	Tetrafluoroethylene	46	100
Crosslink	Ethylene	22.3	0.025
Crosslink	Propylene	16.5	2
Crosslink	Methacrylate	19	2
Crosslink	Acrylic acid	18.5	
Crosslink	Acrylonitrile	17.3	2
Crosslink	Styrene	17	40
Degrade	Methylacrylic acid	15.8	
Degrade	Isobutylene	12.6	20
Degrade	Methyl acrylonitrile	15.3	85
Degrade	Methyl methacrylate	12.9	100
Degrade	α-Methylstyrene	8	100

that as the heat of polymerization H_p decreases, the tendency for degradation (chain scission) increases while that for crosslinking decreases (with the exception of tetrafluoroethylene). Polymers formed from cyclic monomers, including step growth and natural polymers, fail to correlate into some simple classification scheme. It may be mentioned that, in general, cellulosics are primarily degrading polymers whereas polyesters, polyamides, and polysiloxanes tend to crosslink [26]. Additional factors that can affect type of products formed are polymer crystallinity, polymer orientation, and heat treatment before irradiation [26a].

Ionizing radiation effects on specific polymers, e.g., polyethylene, polystyrene, poly(α-methylstyrene), and poly(methyl methacrylate), will be considered.

Polyethylene. Charlesby [27] in 1952 observed that polyethylene could be crosslinked, by irradiation, in a controlled manner. This perhaps served as an inducement to other investigators to study the effects of ionizing radiation on polymers. A considerable amount of gas is generally evolved (80 to 95 percent being H_2) in the irradiation of simple aliphatic hydrocarbons, which is closely paralleled by linear and branched polyethylene. Furthermore, in the case of polymers, crosslinking sites occur, which parallels the formation of new C—C bonds in the simple aliphatic hydrocarbons [28]. Harlen [29] observed that the amount of volatile hydrocarbons produced on irradiation increases while the H_2 yield decreases, as the degree of chain branches increases in polyethylene. The reason for this is perhaps that in low-density polyethylene the short branches are cleaved intact, appearing as saturated hydrocarbons in the condensable gases. This may be shown in the following sequence of reactions:

$$\sim\!\!\sim CH_2\,CHCH_2\sim\!\!\sim \quad\xrightarrow{\quad\sim\!\!\sim\quad}\quad \sim\!\!\sim\overset{\cdot}{C}HCHCH_2\sim\!\!\sim \; + \; H\cdot$$
$$|\qquad\qquad\qquad\qquad\qquad\qquad\qquad | $$
$$(CH_2)_n\,CH_3 \qquad\qquad\qquad\qquad (CH_2)_n\,CH_3 \qquad (113)$$

$$\xrightarrow{\qquad}\quad \sim\!\!\sim CH\!\!=\!\!CHCH_2\sim\!\!\sim \; + \; CH_3(CH_2)_{n-1}\overset{\cdot}{C}H_2 \qquad (113a)$$

$$\qquad\qquad\qquad\qquad\qquad\qquad H$$
$$CH_3(CH_2)_{n-1}\overset{\cdot}{C}H_2 \; + \; \sim\!\!\sim\; C \sim\!\!\sim \xrightarrow{\qquad}$$
$$\qquad\qquad\qquad\qquad\qquad\qquad H$$

$$CH_3(CH_2)_{n-1}CH_3 \; + \; \sim\!\!\sim\overset{\cdot}{C}\sim\!\!\sim \qquad\qquad (114)$$
$$\qquad\qquad\qquad\qquad\qquad\qquad H$$

The formation of hydrogen and volatile hydrocarbons is accompanied by both an increase of unsaturation in the polymer chain [Eq. (113a)] (essentially transvinylene) and an increase in crosslinking density by recombination of polymeric chains with increasing number of radical sites, namely,

$$2\sim\!\!\sim\overset{\cdot}{C}HCHR\sim\!\!\sim \xrightarrow{\qquad} \sim\!\!\sim CHCHR\sim\!\!\sim \qquad\qquad (115)$$
$$\qquad\qquad\qquad\qquad\qquad\qquad |$$
$$\qquad\qquad\qquad\qquad\qquad \sim\!\!\sim CHCHR\sim\!\!\sim$$

It was shown [30, 31] that chain scission also occurs with crosslinking, though to a lesser extent; for example, the ratio of chain-scission to crosslinking reactions is approximately 0.3 for polyethylene. In this connection may be mentioned the work of Charlesby and Pinner [31] who have derived the following expression for a polymer with an initial random molecular-weight distribution:

$$S + \sqrt{S} = \frac{B}{X} + \frac{1}{XM_n \gamma} \tag{116}$$

where S = fraction of polymer still soluble after a radiation dose γ, M_n = initial number-average molecular weight, and B and X denote the relative probabilities of chain scission and crosslinking, respectively. Values of B and X can be obtained from a plot of Eq. (116). Such a plot was constructed for relatively thick polymer sheets irradiated in the presence of air. In this manner, the following values of B/X were obtained for the following polymers: 0.10, polyvinyl acetate; 0.35, polyvinyl chloride; 0.32, polyethylene; and 1.0, polypropylene. From the preceding, polyvinyl acetate shows a relative high propensity toward crosslinking. This may be partially due to crosslinking through the acetate side chains. It should be mentioned that the mode of the crosslinking step [Eq. (115)] in the polymer solid is still in doubt because of the high immobility of chains in the bulk phase. A more reasonable mechanism for the crosslinking step would appear to exist where radical sites are initially generated in close proximity to one another, namely,

$$\begin{array}{ccc}
\sim\!\!\sim CH_2\,CH_2\!\sim\!\!\sim & \quad & \sim\!\!\sim \overset{\displaystyle .}{C}HCH_2\!\sim\!\!\sim \\[4pt]
\xrightarrow{\hspace{2cm}} & & + \; 2H\cdot \\[4pt]
\sim\!\!\sim CH_2\,CH_2\!\sim\!\!\sim & & \sim\!\!\sim \overset{\displaystyle .}{C}HCH_2\!\sim\!\!\sim
\end{array} \tag{117}$$

$$\xrightarrow{\hspace{1.5cm}} \begin{array}{c}
\sim\!\!\sim CHCH_2\!\sim\!\!\sim \\
| \\
\sim\!\!\sim CHCH_2\!\sim\!\!\sim
\end{array} \; + \; H_2$$

as opposed to the concept of the polymeric radicals undergoing a sequence of intermolecular or intramolecular chain-transfer reactions in the bulk phase until such time as the radicals are in close proximity for recombination [32], namely,

$$\begin{array}{ccc}
\sim\!\!\sim \overset{\displaystyle .}{C}HCH_2\!\sim\!\!\sim & \quad & \sim\!\!\sim CH_2\,CH_2\!\sim\!\!\sim \\[4pt]
+ & \xrightarrow{\hspace{2cm}} & + \\[4pt]
\sim\!\!\sim CH_2\,CH_2\!\sim\!\!\sim & & \sim\!\!\sim \overset{\displaystyle .}{C}HCH_2\!\sim\!\!\sim
\end{array} \tag{118}$$

The reaction described in Eq. (118) would require a high activation energy for each transfer step, ≈ 10 kcal mole^{-1} [26] , making it less plausible relative to the reaction of Eq. (117).

Average G values (number of chemical events occurring per 100 eV of absorbed energy) may be calculated for the formation of the various degradation products. These values tend to depend on such factors as type of radiation, temperature during irradiation, and structural nature of polymer, e.g., linearity. Table 5 lists various G values for polyethylene. Values

TABLE 5 G Values for Formation of Degradation Products of Polyethylene

G value	Type of product	Conditions	References
4.1 ± 0.8	H_2	Room temperature	26
2.0	Cl	Room temperature	26
$1.8 \pm .05$	—CH=CH—	Room temperature	26
3.2 ± 0.2	H_2	Below glass-transition temperature	26
1.0 ± 0.2	Cl	Below glass-transition temperature	
1.8 ± 0.5	—CH=CH—	Below glass-transition temperature	
$\approx 2^*$	Crosslinking (Cl)		32–34
$\approx 2^*$	Double-bond formation (—CH=CH—)		

*Depending on temperature and linearity.

of $G(Cl)$ can increase to as high as 5 for irradiation at 150°C. If the only molecular events that occur are the formation of H_2, unsaturation, and crosslinking, the following may be written (assuming no appreciable yield of volatile hydrocarbons):

$$G(H_2) = G(Cl) + G(—CH=CH—) \qquad (119)$$

The above relationship has generally not been observed. Discrepancies of about 0.5 often exist, with values of $G(H_2)$ exceeding the sum of $G(Cl)$ and $G(—CH=CH—)$. Since G values for volatile hydrocarbon formation are usually less than 0.1, this does not contribute appreciably to the observed discrepancy [26]. It should be mentioned that chain scission is generally not considered in calculating G values even though it does occur. The reason is that chain scission is not a termination step and therefore the reaction would not be of importance to the overall balance of G values.

Polystyrene. Generally, aromatic compounds tend to be more stable to ionizing radiation than aliphatic compounds. Accordingly, it is not surprising to observe that approximately 2,000 eV is needed to produce one crosslink in polystyrene as opposed to 50 eV for polyethylene.

Crosslinking predominates over chain scission, the ratio of chain scission to crosslinking lying between 0 and 0.2 [35]. The $G(Cl)$ is ≈0.4, depending on type of ionizing radiation [26]. Hydrogen is produced almost exclusively, with $G(H_2)$ values ranging generally between 0.013 to 0.026 [26]. Studies of γ-irradiated polystyrene, using electron-spin resonance, support the premise that —$CH_2 \dot{C}\phi$— radicals are formed, and the principal crosslinking reactions occur through the combination of these radicals [24]. Cyclohexadienyl radicals, formed by the addition of a hydrogen atom to the phenyl ring, were also detected from electron-spin resonance [36, 37].

Poly(α-methylstyrene). Poly(α-methylstyrene) undergoes random chain scission followed by depropagation to monomer when subjected to γ rays at 25°C in vacuo or in oxygen. Oxygen was found to inhibit cleavage (cf. Fig. 1.3). Main volatile products are H_2 and CH_4.

Poly(methyl methacrylate). The radiolysis of poly(methyl methacrylate) results in a linear rate of decrease in molecular weight that intensifies with radiation [38]. The main volatile products formed are: H_2, CO_2, CO, CH_4, propane, and methyl methacrylate monomer, depending essentially on temperature and type of ionizing radiation [26, 35]. The main reaction is random chain scission with crosslinking occurring to a negligible or very small extent. The G values for chain scission and total gas are 1.9 ± 0.3 [26, 35] and ≈1.5, respectively.

The relative quantities of H_2, CO, CO_2, and CH_4 formed appear to be equivalent roughly to the composition of an ester side group, —$COOCH_3$. Furthermore, they are evolved to an extent that is equivalent to one ester

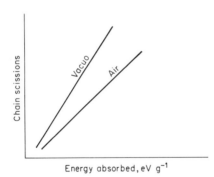

FIG. 1.3 Plot of chain scission versus energy absorbed, electron volt per gram of poly(methyl methacrylate).

side chain decomposed per chain rupture [38]. The above were observed for irradiation of poly(methacrylate) much below its ceiling temperature.

A number of mechanisms have been advanced to account for the cleavage of the main chain and of a side group [24, 39]. One such mechanism is described below.

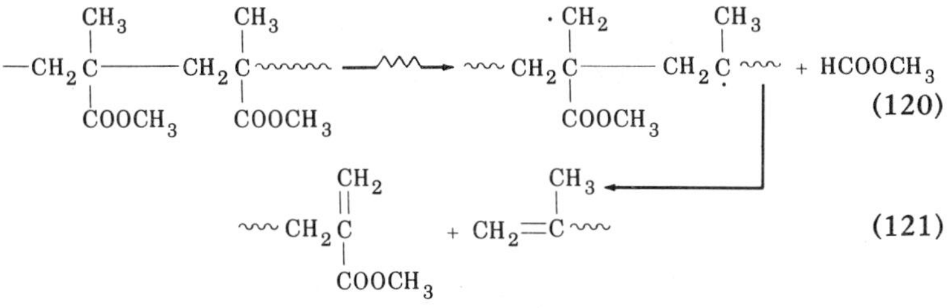

$$\underset{\substack{|\\ COOCH_3}}{-CH_2\,\overset{CH_3}{\underset{|}{C}}}\text{------}\underset{\substack{|\\ COOCH_3}}{CH_2\,\overset{CH_3}{\underset{|}{C}}}\sim\sim\;-\text{\char`\~}\hspace{-0.3em}\Lambda\hspace{-0.3em}\text{\char`\~}\!\!\rightarrow\;\sim\sim CH_2\,\underset{\substack{|\\ COOCH_3}}{\overset{\cdot CH_2}{\underset{|}{C}}}\text{------}\underset{\cdot}{CH_2\,\overset{CH_3}{\underset{|}{C}}}\sim\sim\;+\;HCOOCH_3$$

$$\tag{120}$$

$$\sim\sim CH_2\,\underset{\substack{|\\ COOCH_3}}{\overset{CH_2}{\overset{||}{C}}}\quad+\quad CH_2\!\!=\!\!\overset{CH_3}{\underset{|}{C}}\sim\sim\hspace{4em}\tag{121}$$

In the reaction sequence shown above, the formation of an unstable 1,4-diradical is postulated, with subsequent rearrangement and chain scission forming the two stable fragments having unsaturated end groups. The following mechanism involving hydrogen abstraction from a methylene group, —CH$_2$—, in the main chain was also proposed:

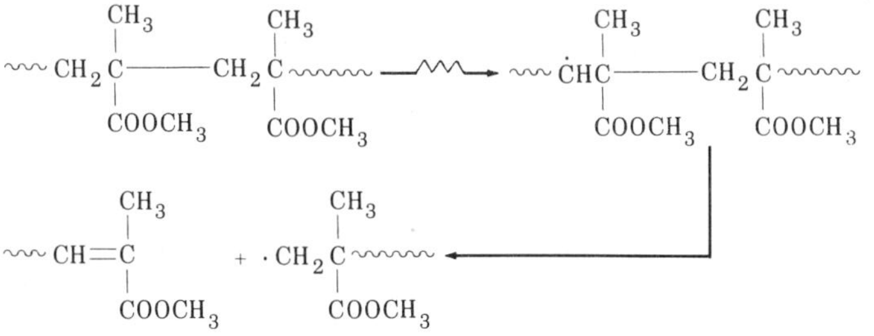

$$\sim\sim CH_2\,\underset{\substack{|\\ COOCH_3}}{\overset{CH_3}{\underset{|}{C}}}\text{------}\underset{\substack{|\\ COOCH_3}}{CH_2\,\overset{CH_3}{\underset{|}{C}}}\sim\sim\;-\text{\char`\~}\hspace{-0.3em}\Lambda\hspace{-0.3em}\text{\char`\~}\!\!\rightarrow\;\sim\sim \overset{\cdot}{C}H\underset{\substack{|\\ COOCH_3}}{\overset{CH_3}{\underset{|}{C}}}\text{------}\underset{\substack{|\\ COOCH_3}}{CH_2\,\overset{CH_3}{\underset{|}{C}}}\sim\sim\;+\;H\cdot$$

$$\tag{122}$$

$$\sim\sim CH\!\!=\!\!\underset{\substack{|\\ COOCH_3}}{\overset{CH_3}{\underset{|}{C}}}\quad+\quad \cdot CH_2\,\underset{\substack{|\\ COOCH_3}}{\overset{CH_3}{\underset{|}{C}}}\sim\sim\sim\hspace{4em}\tag{123}$$

$$\cdot CH_2\,\underset{\substack{|\\ COOCH_3}}{\overset{CH_3}{\underset{|}{C}}}\sim\sim\sim\;\longrightarrow\;CH_2\!\!=\!\!\overset{CH_3}{\underset{|}{C}}\sim\sim\;+\;\cdot COOCH_3\hspace{2em}\tag{124}$$

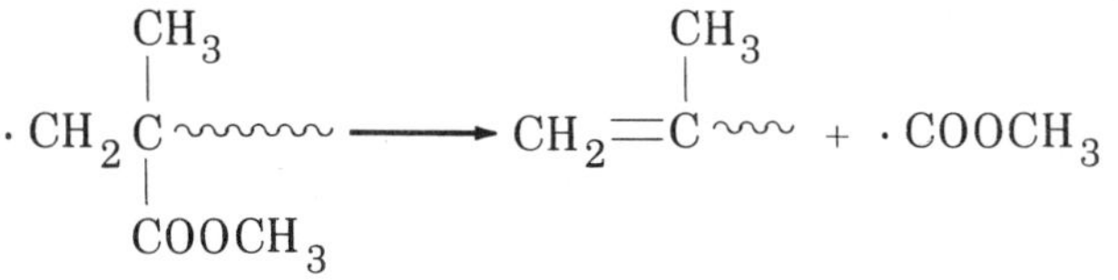

Figure 1.3 shows the degree of chain scission that accompanies ionizing radiation absorbed, in electron volts (eV) per gram of polymer. The plot is for poly(methyl methacrylate) when irradiated by γ rays in a finely dispersed state. As may be anticipated from Fig. 1.3, viscosity decreases should be greater in vacuo than in presence of air.

Increasing alkyl side chains in poly(-n-alkyl methacrylates) favors crosslinking relative to chain scission [crosslinking predominates beyond poly-(-n-hexyl methacrylate) for ionizing radiation] [40]. The following schematic sequence of reactions may be for irradiation in absence and presence of air.

$$
\begin{array}{c}
\qquad\qquad R \\
H \quad | \quad H \quad R \\
\sim\!\!\sim\!\!\sim C - C - C - C \sim\!\!\sim\!\!\sim \\
H \quad X \quad H \quad X
\end{array}
$$

vacuo   Air

$$
\begin{array}{cc}
R \qquad\qquad R & R \\
\sim\!\!\sim CH_2 - C - CH_2 - C \sim\!\!\sim + X\cdot & \sim\!\!\sim CH_2 - C - CH_2 \sim\!\!\sim + X\cdot \\
\qquad\qquad\qquad X
\end{array}
$$

R· / scission P_n O_2 O_2

Crosslinking $\sim\!\!\sim CH_2 - C = CH_2$

$X + P_n$

$$
\sim\!\!\sim CH_2 - C - CH_2 \sim\!\!\sim \qquad \text{Volatile}
$$
$$
\overset{|}{O} - O\cdot \qquad\qquad \text{products}
$$

$$
\begin{array}{c}
R \\
+ \sim\!\!\sim C\cdot \\
X
\end{array}
$$

X· (or H·)

$$
\begin{array}{c}
R \\
\sim\!\!\sim CH_2 C - CH_2 \sim\!\!\sim \\
OOH
\end{array}
$$

Products   (125)

The kinetics of ionizing radiation of polymeric systems have not been extensively studied. Accordingly, mechanisms are not fully elucidated, and perhaps to a lesser degree than for photolytic processes.

MECHANOCHEMICAL DEGRADATION

Thomas Hancock in 1820 observed [41] that natural rubber coalesced into a soft and coherent mass when passed between spiked rollers. The rubber did not shred as might have been anticipated. Instead, as was not known to Hancock at the time, the rubber degraded to a lower molecular weight. Staudinger later showed this to be the case by measuring molecular weights of raw and masticated rubber [42]. He also observed that forcing polymer solutions through fine jets resulted in mechanical degradation [43].

In the discussion to follow, mechanochemical degradation for convenience will be divided into two parts: Machining Processes will cover various aspects of polymer breakdown resulting primarily from machine-type operations, e.g., mastication, grinding, ball milling, roll milling. Ultrasonic Degradation is devoted to breakdown resulting from ultrasonic vibrations.

Machining Processes [44]

Viscoelastic behavior [45] of linear amorphous polymers may be divided into five areas (Fig. 1.4). In terms of a relaxation modulus (obtained by measuring stress in a sample maintained at a fixed stretched length) these areas are as follows:

1. *Glassy region.*   Chain segments are frozen in fixed positions and vibrate about these positions with little diffusional motion. In this state the polymers are generally brittle, and deform according to Hooke's law, i.e., $\sigma = ES$, where σ is the stress, S is the strain, and E is the proportionality constant referred to as the *modulus of elasticity.* The diffusional motion becomes more significant above a characteristic temperature Tg (glass-transition temperature).

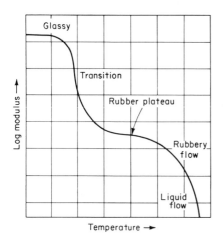

FIG. 1.4 Five regions of visco-elastic behavior for an amorphous polymer.

2. *Transition region.*   In this region polymeric chain segments are undergoing short-range diffusional motion, and the modulus is changing rapidly with temperature. As a result of applied forces, alterations occur in chain configurations, and the orientation of the chains is in the direction of the forces.

3. *Rubbery plateau.*   In this region the short-wave diffusional motions of the polymer segments are very rapid. However, the motions of the chains as a whole are retarded because of entanglements between them, which act as temporary crosslinks. The kinetic theory of rubbery elasticity may be applied to this region. The stress is time-dependent.

4. *Rubbery flow.*   The motion of the molecule as a whole is now becoming of some importance. Slippages of long-range entanglements are occurring which result in major conformation changes. Applied stresses may cause permanent deformation owing to chain slippages.

5. *Liquid flow.*   In this region the long-range conformational changes of the chains occur in a very short time interval. Deformation of such chains is irreversible and time-dependent.

Mechanochemical degradation may occur in all the regions except that of the liquid flow. Scission of polymers in the rubbery plateau region is not necessarily accompanied by their comminution, e.g., rubbers. However, in the glassy and transition regions degradation is associated with comminution.

Elementary processes. Two reactions occur during mechanochemical degradation, as follows:

Primary reaction. Formation of macroradicals occurs as a direct result of the applied mechanical forces. However, the location of sites leading to radicals depend on the manner in which the forces are applied. It should be mentioned that both the rate of degradation and resulting molecular-weight distribution of the polymer may depend on the site of scission. In general, nonrandom scission occurs in mechanochemical reactions. This is in contrast to many thermal reactions where scission occurs with equal probability along the polymer chain:

$$\sim\!\!\sim \xrightarrow[\text{forces}]{\text{mechanical}} \sim\!\!\sim \cdot \; + \; \cdot \sim\!\!\sim \tag{126}$$

Subsequent reactions. Once radicals are formed, as in the primary reaction, the following reactions may take place:

$$R\cdot \; + \; \text{Monomer } (M) \xrightarrow{\text{addition}} RM\cdot \tag{127}$$

$$R\cdot \; + \; M \xrightarrow[\text{transfer}]{\text{monomer}} \text{Polymer} \; + \; M\cdot \tag{128}$$

$$R\cdot \; + \; \text{Solvent (s)} \xrightarrow[\text{transfer}]{\text{solvent}} \text{Polymer} \; + \; S\cdot \tag{129}$$

$$R\cdot \; + \; R'XH \xrightarrow[\text{transfer}]{\text{scavenger}} \text{Polymer} \; + \; R'X\cdot \tag{129a}$$

$$R\cdot \; + \; O_2 \longrightarrow RO_2\cdot \tag{130}$$

$$RO_2\cdot \; + \; RH \longrightarrow RO_2H \; + \; R\cdot \tag{131}$$

$$RO_2H \longrightarrow RO\cdot \; + \; \cdot OH \tag{132}$$

$$RO_2\cdot \; + \sim\!\!\sim CH\!=\!CH \sim\!\!\sim \longrightarrow \sim\!\!\sim \underset{\underset{O_2R}{|}}{CH}\!-\!\overset{\cdot}{CH} \sim\!\!\sim \tag{133}$$

The radical formed in step (133) can undergo reactions which lead to branching and crosslinking. Also, when radical fragments are relatively short (mobility is relatively high), they may undergo termination by combination or disproportionation:

$$2\sim\!\!\sim\cdot \longrightarrow \sim\!\!\sim \text{ (combination)} \tag{134}$$

$$2\sim\!\!\sim\cdot \longrightarrow \sim\!\!\sim \; + \; \sim\!\!\sim \text{ (disproportionation)} \tag{135}$$

Effects of mechanical degradation on polymer properties. Degradation generally leads to chain scission and consequently to a reduction in molecular weight. Figure 1.5 depicts generally the change in molecular weight that accompanies time of comminution for various polymers: poly(methyl methacrylate), polystyrene, poly(vinyl acetate), natural rubber, chloroprene, etc. Generally, polymers under mechanical shear initially degrade at a relatively rapid rate and level off to a limiting, though still quite high, molecular weight in a relatively short time.

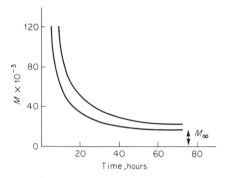

FIG. 1.5 Molecular weight (M) changes of polymers with comminution time.

The extent of mechanical scission will depend essentially on inter- and intramolecular forces for a given mechanical intensity. Consideration of these forces can lead to an expression which involves M_∞ (limiting molecular weight resulting from mechanical degradation), as in the following treatment. Letting q denote an energy function of the intermolecular force of attraction, S_2 equal intermolecular distance when rupture of intermolecular bonds is occurring, E denote energy function of principal valency bonds, M_m equal monomer molecular weight, S_1 denote limiting intramolecular distance during bond rupture, then when chain rupture is no longer occurring (intermolecular and intramolecular forces F equal to each other) we may write

$$F = \frac{M_\infty q}{S_2} = \frac{EM_m}{S_1} \tag{136}$$

or

$$M_\infty = \frac{EM_m S_2}{qS_1} \tag{136a}$$

Limiting molecular weights are generally observed, as indicated in Fig. 1.5. These molecular weights may be utilized as follows, assuming average values for all molecular weights.

If the molecular weights at time t and initially are denoted as M_t and M_0, respectively, we may write for the rate of chain scission R_s,

$$R_s = k \frac{M_t - M_\infty}{M_\infty} = - \frac{d}{dt} \frac{M_t - M_\infty}{M_\infty} \tag{137}$$

From the above equation may be obtained, upon integration,

$$\ln \frac{M_t - M_\infty}{M_0 - M_\infty} = -kt \tag{138}$$

or

$$M_t \equiv M_\infty + (M_0 - M_\infty)\, e^{-kt} \tag{138a}$$

When Eq. (138a) is applied to data, as represented by Fig. 1.5, various values of k were obtained as shown in Table 6.

In connection with molecular-weight changes during mechanical degradation, various indices have been described. Thus, for example,

$$\phi_1 = \frac{M_0 - M_t}{M_0 - M_\infty} \tag{139}$$

Combining Eq. (138) with Eq. (139) there is obtained

$$\phi_1 = 1 - e^{-kt} \tag{140}$$

where ϕ_1 ranges from 0 to 1.

TABLE 6   Values of k Obtained from Eq. (138a) for Various Polymers [44]

Polymer	M_∞	k, hr^{-1}
Poly(vinyl alcohol)	4,000	0.0237
Polystyrene	7,000	0.0945
Poly(methyl methacrylate).	9,000	0.1200
Poly(vinyl acetate)	11,000	0.0468

Recently, Baranwal [46] studied the effect of mechanochemical degradation on an ethylene-propylene terpolymer (EPDM) from intrinsic viscosity measurements. If $[\eta] \cong KM$, then Eq. (138a) becomes

$$[\eta]_t = [\eta]_\infty + b e^{-kt} \tag{141}$$

where b is a constant, and

$$\phi_1 = \frac{[\eta]_0 - [\eta]_t}{[\eta]_0 - [\eta]_\infty} \qquad (142)$$

Such expressions were found to be valid. Furthermore, it was observed that as the mastication temperature was increased in air (30 min of mastication), the value of $[\eta]$ rose to a maximum at 315°F and then decreased rapidly thereafter. This behavior was partially explained as follows. Above 315°F extensive chain scission occurred owing to extensive oxidation (as detected by infrared), whereas below this temperature oxidative processes were much less pronounced. Below 315°F the curve possessed two distinct regions. Between 68 and 150°F the change in $[\eta]$ with temperature was much more rapid than for the temperature range of 150 to 315°F (see Fig. 1.6).

From Bueche's theory [47], $F \propto \nu\gamma M_e^{1/2}/\rho$, where F = breaking tension at center of polymer chain; ν and γ denote melt viscosity and shearing rate, respectively; ρ = density; and M_e = molecular weight between entanglements. In the first region, where the temperature is sufficiently low, ν is high, and therefore the breaking force is high (assuming that γ, ρ, and M_e remain constant). However, as the temperature increases, ν decreases

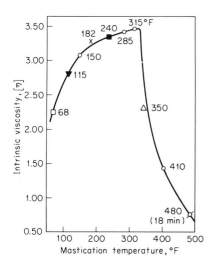

FIG. 1.6 Changes in intrinsic viscosity with mastication temperature [46].

and F becomes less, leading to a higher value of $[\eta]$. In the second region, the polymer becomes soft (there is a large decrease in ν). Because of the relatively low value of F that exists in this region, the value of $[\eta]$ does not change as much with temperature in this region.

Baranwal [46] also masticated EPDM polymer at 75°F in the presence of the free-radical scavenger, diphenylpicrylhydrazine (DPPH), in a nitrogen atmosphere by means of a Uni-Rotor. The percentage of combined DPPH was obtained by infrared spectroscopy (cf. Fig. 1.7). Thus, radicals are forming up to at least 60 min.

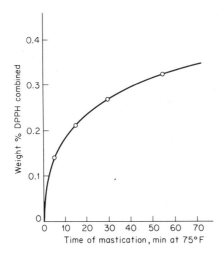

FIG. 1.7 **Weight percent DPPH combined versus time of mastication** [46].

The rate of formation of free-radical molecules for m grams of polymer can be expressed as [cf. Eq. (137)]

$$R = K \frac{M_t - M_\infty}{M_\infty} \frac{m}{M_t} \tag{143}$$

This expression may be converted to [see Eq. (138)]

$$R = \frac{K'e^{-kt}}{ae^{-kt} + M_\infty} \tag{144}$$

where K and K' are constants and $a = M_0 - M_\infty$. During the later stages of the reaction it is assumed that $M_\infty \gg ae^{-kt}$ (during the mastication of EPDM at 68°F, it was observed that after about 100 min the intrinsic viscosity decreased from 3.5 to only about 2.9), and, therefore,

$$R = K''e^{-kt} \tag{145}$$

From the equation, a plot of $\log R$ versus t should afford a linear relationship. Such a plot was obtained for values of t between 20 and 50 min and gave a value of $k = 3 \times 10^{-2}$ min^{-1} at 75°F.

In connection with molecular-weight reduction the following properties are also affected: solubility [e.g., poly(vinyl chloride) becomes more soluble in acetone]; molecular-weight distribution (generally becomes narrower); tensile strength (decreases); crosslinking (is destroyed); crystallinity (decreases); and plasticity (increases). As an example of the effect of molecular weight, the changes in plasticity as a function of time are mathematically treated below.

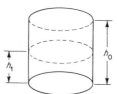

FIG. 1.8 Measurement of "Karrer" plasticity [48].

The "plasticity" of a substance is dependent on the method used to measure it. In the case of the "Karrer" [48] plasticity, a cylindrical sample is placed between two plates (1 cm^2 in area) and compressed under a definite load for a definite time. The load is then removed, and the height h_2 is measured after 30 sec. The plasticity is calculated as

$$Pl = \frac{h_0 - h_2}{h_0 + h_1} \tag{146}$$

where the symbols are as defined in Fig. 1.8 (h_1 is height immediately after compression). Values of Pl range from zero to unity.

The following expression was obtained for various rubbers at 20 to 30°C, such as natural rubber, butadiene-styrene rubber, and butadiene-acrylonitrile rubber:

$$Pl = (Pl)_0 + A(1 - e^{-kt}) \tag{147}$$

where $(Pl)_0$ is plasticity of the rubber before mastication, and A is a constant. Equation (147) may be semiempirically derived as follows. Writing Eq. (138) in the following form,

$$M_t = M_0 - a(1 - e^{-kt}) \tag{138b}$$

where $a = M_0 - M_\infty$, and assuming that plasticity is inversely proportional to molecular weight of the substance, i.e., $Pl = K'/M_t$, then Eq. (138b) becomes

$$\frac{1}{(Pl)_0} - \frac{1}{Pl} = \frac{a}{K'}(1 - e^{-kt}) \tag{148}$$

Equation (148) may be rewritten as

$$Pl = \frac{(Pl)_0}{1 - \dfrac{a(Pl)_0}{K'}(1 - e^{-kt})} \tag{148a}$$

and assuming $1 \gg [a(Pl)_0/K'](1 - e^{-kt})$, then Eq. (148a) becomes approximately

$$Pl \approx (Pl)_0 [1 + b'(1 - e^{-kt})] \tag{149}$$

where $b' = a(Pl)_0/K'$.

Equation (149) can be rewritten as Eq. (147) when $A = b'(Pl)_0$.

Some factors influencing mechanochemical degradation.   There are a number of factors that can generally affect the rate of mechanochemical degradation [49]. Thus, the following may be considered.

1. *Temperature.* Raising the temperature under a given rate of shear for a polymer generally does not produce the Arrhenius dependence but rather a much smaller change. On the contrary, a decrease in the rate of degradation may be observed with a rise in temperature since the accompanying drop in viscosity tends to reduce the work required in mechanically deforming the polymer (cf. page 44).

2. *Viscosity.* Factors that affect viscosity of a polymer, such as temperature, molecular weights, and plasticizers, can influence the rate of degradation. The importance of viscosity is manifested in the work required in deforming the polymer, as with temperature discussed above. Higher rates of degradation are observed in polymers of higher molecular weights, whereas lower degradation rates occur in polymers containing plasticizers, because of high and low viscosities, respectively. In this connection may be mentioned the mechanochemical degradation of two types of gelatin, I and II (I had a much higher initial molecular weight than II). Gelatin I degraded much more rapidly initially than II, and the limiting molecular weights were of about the same magnitude [44].

3. *Chain configuration.* Linear chains are more readily ruptured than chains of coiled polymers. The more globular shapes result in a distribution of stresses among interglobular bonds, and therefore degrade more slowly. In branched polymers the branches often rupture preferentially during milling: e.g., starch.

4. *Presence of oxygen and radical scavengers.* Reactions involving oxygen and radical scavengers during mastication are described by Eqs. (129) and (130). Relative to oxygen, at low oxygen concentration, e.g., 0.05 percent in argon, mastication generally produces branching and crosslinking in the polymer. Higher oxygen concentration results in degradation (scission), as in natural rubber. The mechanical degradation of styrene-butadiene in nitrogen results in crosslinking, whereas in presence of air degradation occurs. The Mooney viscosity of the styrene-butadiene rubber, initially at 96, increased slightly to 97 in the former case, but decreased markedly to 81 in the latter. In synthetic polymers, e.g., poly(vinyl chloride), poly(vinyl alcohol), the intrinsic viscosities decrease more when the polymers are masticated in air than in nitrogen.

Radical scavengers are variable in their behavior. Thus, for example, benzoquinone is less effective than phenol in stabilizing the mechanodegradation of natural rubber in an inert atmosphere. Furthermore, one percent of the radical scavenger thiophenol present in natural rubber will result in a rate of degradation on masticating in an inert atmosphere equal to that in air [50].

5. *Type of mechanical equipment.* Mechanochemical degradation can be influenced by the type of equipment used in processing. Thus, with roll mills, internal mixers, extruders, etc., shear and compression forces are primarily involved. For Hollander beaters, conical mills, etc., abrasion, crushing, and comminution generally occur. The consequence of impact action may be assessed by means of ball and hammer mills. Finally, high-frequency forces, e.g., jets of compressed air and ultrasonics, may be involved in the degradation. Ultrasonic degradation is treated in the following section, since more is known about this method.

6. *Miscellaneous.* Results obtained during mechanical degradation may be influenced by the presence of metallic contaminants from the equipment, which catalyze the oxidation (cf. Oxidative Degradation, page 4, and Chap. 5). A further complication is that shear degradation may be superimposed on a purely thermal oxidative degradation.

Ultrasonic Degradation [51-53]

When polymers in solution are subjected to ultrasonic irradiation of high intensity, degradation often results. This breaking of chemical bonds has been attributed to "cavitation" effects, i.e., the formation and violent collapse of small bubbles or voids in the liquid, as a result of pressure changes which occur in a sound wave. The violent collapses that occur can lead to shearing forces of sufficient magnitude to cause the rupture of bonds. Cavitation effects appears to be enhanced by the presence of gas or solid nuclei. These nuclei are less effective when a high external pressure is maintained above the solution. Thus, it was found that polystyrene in toluene solution underwent considerable bond breakage when irradiated under atmospheric pressure. Less breakage occurred under 8 atm of oxygen and still less under 15 atm of oxygen (cavitation effects can also occur in the presence of other gases, e.g., helium). In the absence of gas nuclei, i.e., when the solution is carefully degassed, little bond rupture occurred. It may be mentioned here that if a molecule is greater than a critical size, $D_{P,e}$ and becomes involved in a collapsing cavity, bond breakage is likely to occur. However, below $D_{P,e}$ the shearing forces which develop may be expended in a manner other than in bond rupture, e.g., in overcoming intermolecular forces, thereby causing greater mobility of the relatively short chains (cf. Machining Processess, page 44).

Considering the above, we may write for the rate of bond breakage,

$$\frac{dB_i}{dt} = k(D_{P,i} - D_{P,e})n_i \quad \text{for} \quad D_{P,i} > D_{P,e} \tag{150}$$

and

$$\frac{dB_i}{dt} = 0 \quad \text{for} \quad D_{P,i} \lessgtr D_{P,e} \tag{151}$$

where n_i is the number of molecules of degree of polymerization $D_{P,i}$. These expressions can lead to various other expressions, depending upon the assumptions made. A relatively simple case will be considered.

For all degrading species in a polymer solution containing originally n_0 molecules of degree of polymerization $D_{P,r}$ the rate of bond rupture is

$$\frac{dB}{dt} = k \sum_{e+1}^{r} (D_{P,i} - D_{P,e}) n_i \tag{152}$$

The limiting degree of polymerization $D\hat{P}$ below which molecules will not degrade, $D_{P,e}$, is not equal to the final value for the degraded polymer since a molecule of $D\hat{P}$ of $(D_{P,e} + 1)$ has a finite probability of degrading and will give rise to molecules with degrees of polymerization less than $D_{P,e}$. Therefore, the assumption was made that only fragments of $D\hat{P} > D_{P,e}/2$ will rupture from a degrading molecule and that all chain bonds are equally likely to rupture except those within $D_{P,e}/2$ monomer units from either chain end. Thus, those molecules which will not degrade further possess degrees of polymerization in the range of about $D_{P,e}/2$ to $D_{P,e}$ (average of $3D_{P,e}/4$, assuming an even distribution in this range).

Upon expanding the above equation,

$$\frac{dB}{dt} = k \left(\sum_{1}^{r} D_{P,i} n_i - D_{P,e} \sum_{1}^{r} n_i + D_{P,e} \sum_{1}^{e} n_i - \sum_{1}^{e} D_{P,i} n_i \right) \tag{153}$$

Furthermore, we may write

$$\sum_{1}^{r} D_{P,i} n_i = D_{P,r} n_0 ; \quad \sum_{1}^{r} n_i = n_0 + B \tag{154}$$

and therefore

$$\frac{dB}{K\,dt} = n_0 D_{P,r} - D_{P,e}(n_0 + B) + D_{P,e} \sum_{1}^{e} n_i - \sum_{1}^{e} D_{P,i} n_i \tag{155}$$

If we write $N_L \equiv \sum_{1}^{e} n_i$, and since the totally degraded molecules possess a number-average DP of $3D_{P,e}/4$,

$$\frac{dB}{K\,dt} = n_0 D_{P,r} - D_{P,e}(n_0 + B) + \frac{D_{P,e}N_L}{4} \tag{156}$$

For low degrees of conversion, B is small and the last term of Eq. (156) can be neglected (assuming there are no molecules in the original polymer with $D\hat{P} < (D_{P,e} + 1)$ to give

$$\frac{dB}{dt} + D_{P,e}kB = kn_0(D_{P,r} - D_{P,e}) \tag{157}$$

Upon integration,

$$B = n_0\left(\frac{D_{P,r}}{D_{P,e}} - 1\right)[1 - \exp(-D_{P,e}kt)] \tag{158}$$

This expression was found to hold well during the initial stages of the irradiation of a solution of polystyrene in toluene. Some values obtained (at ambient temperature) were: $k = 13 \times 10^{-6}$ min^{-1}, $D_{P,e} = 1740$, and the extrapolated ultimate value of $D\hat{P}$ was 1310 (0.71$D_{P,e}$ versus the theoretically estimated at 0.75 $D_{P,e}$). Other expressions were developed which were found to be valid for the initial and later stages of polystyrene degradation by ultrasonic irradiation.

CHEMICAL DEGRADATION

The literature on this aspect of degradation is extremely vast [54] since there are a large number of chemicals that can attack polymeric systems. (In this connection oxygen has been treated separately under Oxidative Degradation, page 4, and in Chap. 5.) The number of chemical agents considered in the following must, therefore, of necessity be restricted. Accordingly, the authors chose what they believe are agents that are important from a theoretical and practical viewpoint. Thus, atmospheric pollutants, such as nitrogen dioxide and sulfur dioxide, have been emphasized. Other chemical agents considered are ozone, acids, alkalies, halogens, and nitric acid.

Nitrogen Dioxide (NO_2)

In the following mathematical treatments, it will be assumed that at the pressures of NO_2 used, the formation of dimer N_2O_4 can be neglected.

In absence of oxygen. (NO_2) may react with suitable hydrocarbons by preferentially removing a tertiary hydrogen atom, namely,

$$RH + NO_2 \longrightarrow R\cdot + HNO_2 \tag{159}$$

$$R\cdot + NO_2 \longrightarrow RNO_2 \tag{160}$$

$$RNO_2 \xrightarrow[\substack{\text{and/or} \\ \text{crosslinking}}]{\text{scission}} \text{Products} \tag{161}$$

For example, polystyrene (PS) has a tertiary hydrogen atom available, and should therefore be subject to ready attack by NO_2. In this connection the degradation of PS films ($M_w = 3.72 \times 10^5$; $M_n = 1.62 \times 10^5$; 20 μ thick) by NO_2 appeared to be random [55]. Thus, the following expression was experimentally observed (cf. Chap. 4).

$$\frac{1}{D_P} - \frac{1}{D_{P,0}} = kt \tag{162}$$

This expression was found to hold for various NO_2 pressures at 55°C, e.g., 15 to 60 cm Hg. The k values, determined for each NO_2 pressure, were plotted as a function of NO_2 pressure. The plot was found to be linear. Furthermore, an Arrhenius plot of k afforded an activation energy of 16.2 kcal mole^{-1}. The following expression for k was found to hold:

$$k = P_{NO_2} (5.52 \times 10^4 e^{-16,200/RT}) \, \text{hr}^{-1} \tag{163}$$

where $P_{NO_2} \equiv$ pressure NO_2, in cm Hg.

The presence of stretching vibrations due to nitro groups on an aliphatic chain was detected by infrared spectroscopy. The following mechanism was proposed:

$$\begin{array}{cccc} H & H & H & H \\ \sim\!\!\sim\!\!\sim C & \!\!\!-C\!\!\!- & C\!\!\!- & C \sim\!\!\sim\!\!\sim \; + \; NO_2 \longrightarrow \\ H & | & H & \phi \\ & \phi & & \end{array}$$

$$(164)$$

$$\begin{array}{cccc} H & \cdot & H & H \\ \sim\!\!\sim\!\!\sim C & \!\!\!-\dot{C}\!\!\!- & C\!\!\!- & C \sim\!\!\sim\!\!\sim \; + \; HNO_2 \\ H & | & H & \phi \\ & \phi & & \end{array}$$

$$R\cdot + NO_2 \begin{cases} \longrightarrow RNO_2 \text{ (nitro)} & (165) \\ \\ \longrightarrow RONO \text{ (nitrite)} & (165a) \end{cases}$$

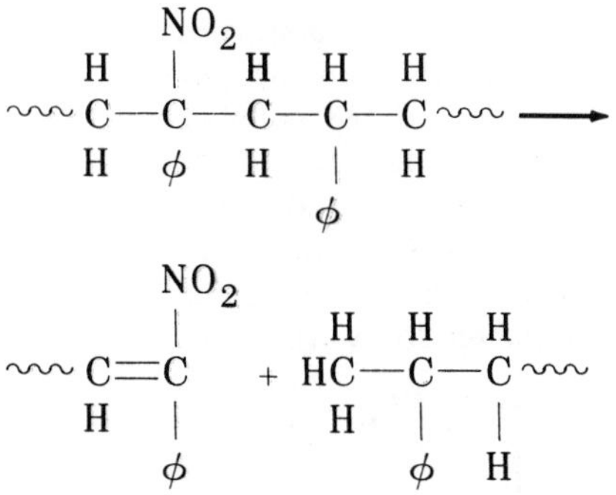

$$
\begin{array}{ccccccc}
 & \overset{\displaystyle NO_2}{|} & & & & \\
H & | & H & H & H & \\
\sim\!\!\sim\!\!\sim C & \!\!-C\!\!- & C & \!\!-C\!\!- & C \sim\!\!\sim\!\!\sim & \longrightarrow \\
H & \phi & H & | & H & \\
 & & & \phi & &
\end{array}
$$

$$
\begin{array}{cccccc}
 & \overset{\displaystyle NO_2}{|} & & & & \\
\sim\!\!\sim\!\!\sim C\!\!=\!\!C & & +\;HC\!\!- & H & H & H \\
H & | & & C\!\!- & C \sim\!\!\sim\!\!\sim & \\
 & \phi & & H & | & | \\
 & & & & \phi & H
\end{array}
\tag{166}
$$

or

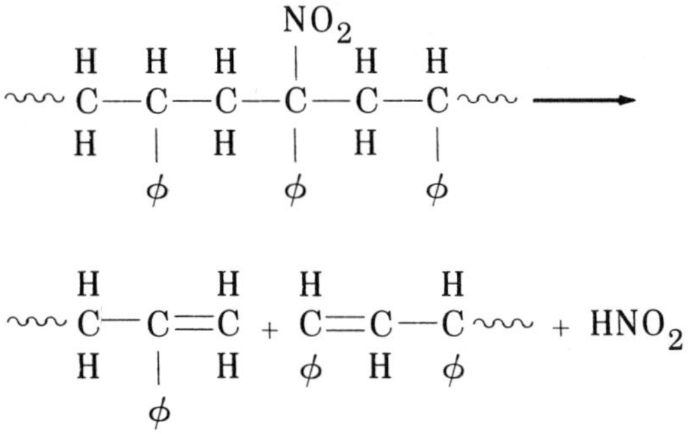

$$
\begin{array}{cccccc}
 & & & \overset{\displaystyle NO_2}{|} & & \\
H & H & H & | & H & H \\
\sim\!\!\sim\!\!\sim C\!\!- & C\!\!- & C\!\!- & C\!\!- & C\!\!- & C \sim\!\!\sim\!\!\sim \longrightarrow \\
H & | & H & | & H & | \\
 & \phi & & \phi & & \phi
\end{array}
$$

$$
\begin{array}{cccccc}
H & & H & H & & H \\
\sim\!\!\sim\!\!\sim C\!\!- & C\!\!=\!\!C & +\; & C\!\!=\!\!C\!\!- & C \sim\!\!\sim\!\!\sim & +\;HNO_2 \\
H & | & H & \phi & H & \phi \\
 & \phi & & & &
\end{array}
\tag{167}
$$

From the above scheme, Eq. (162) may be derived as follows. Letting $[n_0]$ and $[n]$ represent concentration of main chain links initially and at time t, respectively, and $[\alpha - H]$ denote concentration of tertiary hydrogens, we may write, assuming Eq. (164) to be rate controlling,

$$
-\frac{d[n]}{dt} = k'[\alpha - H][NO_2] = k[\alpha - H]P_{NO_2}
\tag{168}
$$

Assuming that $[\alpha - H] \approx \text{const} (= [n_0])$ we may write for the degree of degradation α

$$
\alpha = \frac{[n_0] - [n]}{[n_0]} \approx \frac{[n_0] - [n]}{[\alpha - H]}
\tag{169}
$$

Integrating Eq. (168) and combining with Eq. (169), we obtain [cf. Chap. 4, Eqs. (205), (206), (53')]

$$
\alpha = \frac{1}{D_P} - \frac{1}{D_{P,0}} = kP_{NO_2}t
\tag{162'}
$$

which is similar to Eq. (162), which was experimentally observed.

Other saturated polymers react differently with NO_2 and are in general less susceptible to NO_2 attack than unsaturated polymers, e.g., polyisoprene, polybutadiene, and butyl rubber (BR) (copolymer consisting of isoprene and isobutylene). In connection with BR the following mathematical treatment has been presented [56] for the chain scission of BR by NO_2. The mathematical treatment is based on the following reaction scheme (only isoprene units are considered to react).

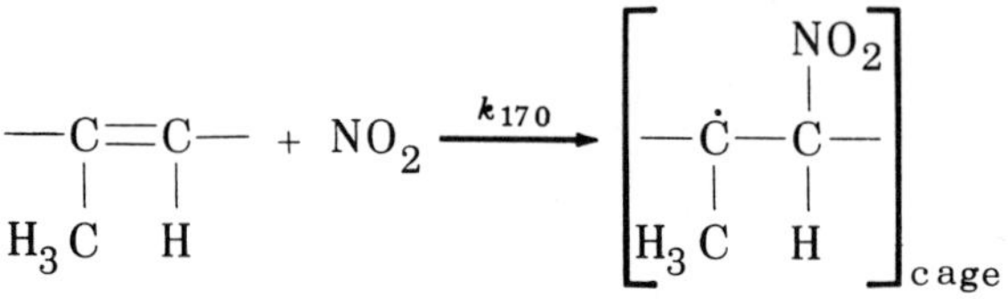

$$\begin{matrix} -C=C- \\ | \quad | \\ H_3C \quad H \end{matrix} + NO_2 \xrightarrow{k_{170}} \left[\begin{matrix} \quad \quad NO_2 \\ \quad \quad | \\ -\dot{C}-C- \\ | \quad \quad | \\ H_3C \quad H \end{matrix}\right]_{cage} \tag{170}$$

$$[Cage] \xrightarrow{k_{171}} \begin{matrix} -C=C- \\ | \quad | \\ H_3C \quad H \end{matrix} \tag{171}$$

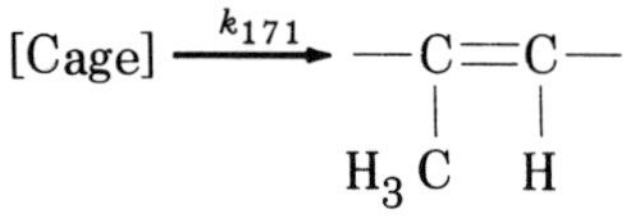

$$[Cage] + NO_2 \xrightarrow{k_{172}} \text{Chain scission} + NO_2 \tag{172}$$

From the above scheme the rate of chain scission is given by

$$-\frac{d[n']}{dt} = k_{172}[cage] \tag{173}$$

where $[n']$ denotes concentration of isoprene units at time t, and $[NO_2]$ is constant throughout the reaction. Assuming that $[n'] \approx [n'_0]$, since the average number of chain scissions per original chain molecule is small, and assuming steady-state concentrations for the cage, we may write

$$[Cage] = \frac{k_{170}[n'_0][NO_2]}{k_{171} + k_{172}[NO_2]} \tag{174}$$

Combining Eqs. (173) and (174), we obtain

$$-\frac{d[n']}{dt} = \frac{k_{170}k_{172}[n'_0][NO_2]}{k_{171} + k_{172}[NO_2]} \tag{175}$$

Upon combining the integrated form of Eq. (175) with the definition of α

[cf. Eq. (169)], i.e.,

$$\alpha = \frac{[n_0'] - [n']}{[n_0]} \qquad (169')$$

where $[n_0]$ = total number of isoprene and isobutylene links originally present in BR,

$$\alpha = \frac{1}{D_P} - \frac{1}{D_{P,0}} = \frac{k_{170} k_{172} [n_0'][NO_2] t}{[n_0] (k_{171} + k_{172}[NO_2])} \equiv K_e t \qquad (176)$$

Figure 1.9 shows plots of α versus time for various NO_2 concentrations. The plots are linear, after an initial period of time, in accordance with Eq. (176). During this initial time it was postulated that some structures degrade, which appear to be more susceptible to chain scission than the normal structures in the polymer molecule. However, these abnormal structures, which are present only in very small amounts, upon degradation become quickly exhausted; only "normal" chain scission proceeds after the kinks in the curves. This is a situation equivalent to that of "weak links" in a polymer chain, which are assumed to be some oxygenated structures in the isoprene, or some residual catalyst fragments. (See Weak Links and End Groups, Chap. 6.) Values of K_e were obtained (cf. Fig. 1.9) and plotted as a function of NO_2 pressure. The dependence of K_e on P_{NO_2} obeyed Eq. (176).

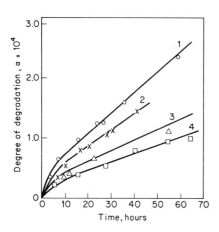

FIG. 1.9 Degree of degradation α as a function of time for various NO_2 pressures in absence of air at 35°C: (1) 1 mm Hg, (2) 0.5 mm Hg, (3) 0.25 mm Hg, (4) 0.20 mm Hg [56].

The mathematical treatment for the initial portions of the curves in Fig. 1.9 resembles the above treatment for the chain scission of normal links. Thus, if it is assumed that the degradation of weak links may be expressed by an equation similar to Eq. (175) [Eqs. (170) to (172) are also assumed to apply], then we may write

$$- \frac{d[n_w]}{dt} = \frac{k'_{170} k'_{172} [n_w] [NO_2]}{k'_{171} + k'_{172} [NO_2]} \tag{177}$$

where $[n_w]$ = concentration of weak links, and the primed k's with subscripts correspond to steps (170) to (172) except that now the hydrocarbon moieties being attacked by NO_2 involve weak links. Upon integration of Eq. (177),

$$- \ln \left(1 - \frac{S'}{D_{P,w}} \right) = \frac{k'_{170} k'_{172} [NO_2] t}{k'_{171} + k'_{172} [NO_2]} \equiv K'_e t \tag{178}$$

where S' = average number of broken weak links per original chain molecule, $D_{P,w}$ = average number of weak links per original chain molecule, and $(1 - S'/D_{P,w}) \equiv [n_w]/[n_w]_0$, where $[n_w]_0$ = weak-link concentration at zero time. Various values of K'_e were estimated for various values of P_{NO_2}, and the dependence of K'_e on P_{NO_2} could satisfactorily be accounted for by Eq. (178).

In presence of air. Studies of BR in presence of air and NO_2 were also conducted [56]. The conditions consisted of using NO_2 in the range of 0.01 to 1.0 mm Hg in the presence of 1 atm of air at 35°C. Plots of α versus time were different from those obtained in absence of air. The rates of chain scission were above five times those for the experiments conducted in presence of NO_2 only. It was further found that the rate of chain scission increased with temperature.

The above results could be satisfactorily accounted for by the mechanism below. It should be mentioned that weak links do not play a significant role here since the chain scission is now much faster than in presence of NO_2 alone. The first part of the proposed mechanism consists of the contribution of the reaction with NO_2 only:

$$RH + NO_2 \xrightarrow{k_{170a}} Cage_1 \tag{170a}$$

$$Cage_1 \xrightarrow{k_{171a}} RH \tag{171a}$$

$$Cage_1 + NO_2 \xrightarrow{k_{172a}} Chain\ scission + NO_2 \tag{172a}$$

The second part is the contribution by oxygen alone:

$$RH + O_2 \xrightarrow{k_{179}} R\cdot + HO_2\cdot \tag{179}$$

$$R\cdot + O_2 \xrightarrow{k_{130}} RO_2\cdot \tag{130}$$

$$RO_2\cdot + RH \xrightarrow{k_{131}} ROOH + R\cdot \tag{131}$$

$$ROOH \xrightarrow{k_{180}} \text{Inert products} \tag{180}$$

$$ROOH \xrightarrow{k_{181}} Cage_2 \tag{181}$$

$$Cage_2 \xrightarrow{k_{182}} ROOH \tag{182}$$

$$Cage_2 + O_2 \xrightarrow{k_{183}} \text{Chain scission} + O_2 \tag{183}$$

The third part consists of the synergistic reaction of NO_2 and O_2:

$$ROOH + NO_2 \xrightarrow{k_{184}} NO_2{-}ROOH \tag{184}$$

$$NO_2{-}ROOH \xrightarrow{k_{185}} Cage_3 \tag{185}$$

$$Cage_3 \xrightarrow{k_{186}} NO_2{-}ROOH \tag{186}$$

$$Cage_3 \xrightarrow{k_{187}} \text{Chain scission} \tag{187}$$

$$R\cdot + R\cdot \xrightarrow{k_{188}} Cage_4 \tag{188}$$

$$Cage_4 + O_2 \xrightarrow{k_{189}} R\cdot + R\cdot \tag{189}$$

$$Cage_4 \xrightarrow{k_{190}} R{-}R \tag{190}$$

From the scheme a general expression for the degree of degradation α was derived. This expression was simplified to yield

$$\alpha = K_e t + \frac{K_e''}{(K_1)^3} \left(1 - e^{-K_1 t}\right) \tag{191}$$

where K_e was defined in Eq. (176):

$$K_e'' = \frac{k_{184} k_{185} k_{187} [NO_2] K_2}{(k_{186} + k_{187})[n_0]}$$

$$K_1 = k_{180} + k_{181} + k_{184}[NO_2] - \frac{k_{181} k_{182}}{k_{182} + k_{183}[O_2]}$$

$$K_2 = \left\{ \frac{k_{179}(k_{189}[O_2] + k_{190})}{k_{188} k_{190}} \right\}^{1/2} k_{130}[n']^{1/2}[O_2]^{3/2}$$

The first term in Eq. (191) is due to chain scission by NO_2 alone [cf. Eq. (176)], and the second term represents the synergistic action of NO_2 and O_2. When α was plotted against $1 - e^{-K_1 t}$ (Fig. 1.10), linear relationships were obtained in accordance with Eq. (191) (the linear relationships indicate that the second term involving the synergistic action of NO_2 and O_2 is

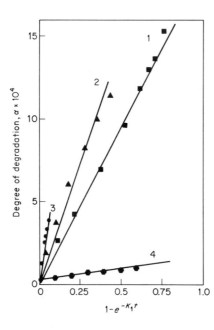

FIG. 1.10 Degree of degradation α plotted versus $(1 - e^{K_1 t})$ according to Eq. (191) for various NO_2 pressures at 35°C (1 atm of air): (1) 1 mm Hg; (2) 0.5 mm Hg; (3) 0.2 mm Hg; (4) 0.01 mm (at abscissa x 10^2) [56].

much greater than the first term involving chain scission by NO_2 alone). Other experimental data were found to be in satisfactory agreement with the proposed mechanism. The effects of NO_2 under various conditions for various polymer films (20 to 30 μ) are summarized in Table 7 [57]. These effects involve changes in intrinsic viscosity $[\eta]$ and crosslinking (Cl).

Sulfur Dioxide

The effects of SO_2 at 25°C on various polymer films (20 to 30 μ) are summarized in Table 8 for various polymers [57]. These effects involve changes in intrinsic viscosity $[\eta]$ and crosslinking (Cl).

The mechanism of the reaction of SO_2 with polymers in presence of ultraviolet light (> 3000 Å) is difficult to ascertain because of the many side reactions which the SO_2 can undergo. Thus,

$$nSO_2 \xrightarrow{h\nu} SO_3 + O + \cdots \tag{192}$$

In presence of oxygen, ozone may form, namely,

$$SO_2 + 2O_2 \xrightarrow{h\nu} SO_3 + O_3 \tag{193}$$

TABLE 7 Summary of Experimental Results on Exposure to NO$_2$ at 35°C [57]

Intrinsic Viscosity and Percent Crosslinking

Exposure	Intrinsic viscosity [η], dl g^{-1}										% crosslinking		
	At 30 hr						At 1/2 hr		At 5 min		At 30 hr	At 5 min	
	PE	PPR	PS	PMMA	PVC	PAN	Nylon 66	BR	PISO	PBD	PPR	PISO	PBD
Before exposure, dark	0.91	2.10	0.96	1.16	0.94	0.82	1.36	0.65	3.44	2.73			
NO$_2$ (15 cm Hg), dark	0.92	1.77	0.83	1.06	0.80	0.82	1.17*	0.62*	1.41*	- - -	- - -	- - -	53.5*
NO$_2$ (15 cm Hg), air (1 atm), dark	0.92	1.92	0.83	1.08	0.88	0.82	0.59*	0.51*	0.48*	- - -	- - -	0.6	61*†
UV, vacuum (10^{-5} mm Hg)	0.92	1.87	0.96	1.07	0.94	0.85	1.36*	0.64*	3.06*	2.58*			
UV, O$_2$ (15 cm Hg)	0.96	1.94	0.93	1.11	0.89	0.85	1.34*	0.65*	2.94*	2.73*	2.6		
UV, O$_2$ (15 cm Hg), NO$_2$ (1 cm Hg)	0.95	1.83	0.80	1.0	0.88	0.81	0.82*	0.61	0.46*	- - -	4.7	31.6*	39.4*

BR   Butyl rubber: 99% isobutylene, 1% isoprene.
PAN   Polyacrylonitrile.
PBD   Polybutadiene.
PE   Low-density polyethylene.
PISO   Polyisoprene.
PMMA   Poly(methyl methacrylate).
PPR   Polypropylene.
PS   Polystyrene.
PVC   Poly(vinyl chloride).
*These values for NO$_2$ pressure of 1 cm Hg.
†Percent crosslinking at 7 min = 76.

TABLE 8 Summary of Experimental Results on Exposure to SO$_2$ at 25°C [57]
Intrinsic Viscosity and Percent Crosslinking

Column groups — Intrinsic viscosity [η], dl g^{-1}: "At 10 hr" = PE, PPR, PS, PMMA, PVC, PAN, Nylon 66; "At 5 min" = BR, PISO, PBD. % crosslinking: "At 10 hr" = PE, PPR; "At 5 min" = PBD.

Exposure	PE	PPR	PS	PMMA	PVC	PAN	Nylon 66	BR	PISO	PBD	PE	PPR	PBD
Before exposure, dark	0.82	2.10	1.23	1.18	1.00	0.59	2.40	0.66	3.40	2.80			
SO$_2$ (1 cm Hg), dark	0.8	2.15	1.20	1.16	0.93	0.45	3.06	0.64	3.64	----	----	----	70
SO$_2$ (1 cm Hg), air (1 atm), dark	0.81	2.21	1.33	1.19	0.96	0.59	2.68	0.65	3.57	----	----	----	32
UV, vacuum 10^{-5} mm Hg	1.00	2.19	1.21	1.17	0.98	0.79	2.37	0.66	3.28	----	----	----	18
UV, O$_2$ (15 cm Hg)	1.03	2.10	1.14	1.10	0.97	0.50	2.35	0.64	3.03	----	----	----	53
UV, O$_2$ (15 cm Hg)	----	----	1.02	0.84	0.95	0.48	2.47	0.59	3.15	----	100	69	52

BR Butyl rubber: 99% isobutylene, 1% isoprene.
PAN Polyacrylonitrile.
PBD Polybutadiene.
PE Low-density polyethylene.
PISO Polyisoprene.
PMMA Poly(methyl methacrylate).
PPR Polypropylene.
PS Polystyrene.
PVC Poly(vinyl chloride).

A possible mechanism for the photoreaction of SO_2 with saturated polymers (RH) is [57]

$$SO_2 + h\nu \longrightarrow SO_2^* \tag{194}$$

$$SO_2^* + RH \longrightarrow RSO_2H \tag{195}$$

$$SO_2^* + RH \longrightarrow SO_2 + RH \tag{196}$$

$$SO_2^* + SO_2 \longrightarrow SO_2 + SO_2 \tag{197}$$

$$SO_2^* \longrightarrow SO_2 \tag{198}$$

An unsaturated hydrocarbon may react [57] as follows, where B is unsaturated hydrocarbon:

$$SO_2 + h\nu \longrightarrow SO_2^* \tag{194}$$

$$SO_2^* \longrightarrow SO_2 \tag{198}$$

$$SO_2^* + B \longrightarrow SO_2 + B \tag{199}$$

$$SO_2^* + B \longrightarrow BSO_2 \tag{200}$$

Ozone

In association with the air pollutant SO_2 it can be seen from Eq. (193) that ozone may be formed in presence of sunlight and oxygen. The deleterious effects of ozone on polymeric substances, e.g., natural rubber, may be enhanced by the additional ozone produced from the above reaction.

Chain scission and crosslinking can occur in polymers exposed to ozone. Thus, main-chain scission may occur as follows with unsaturated polymers:

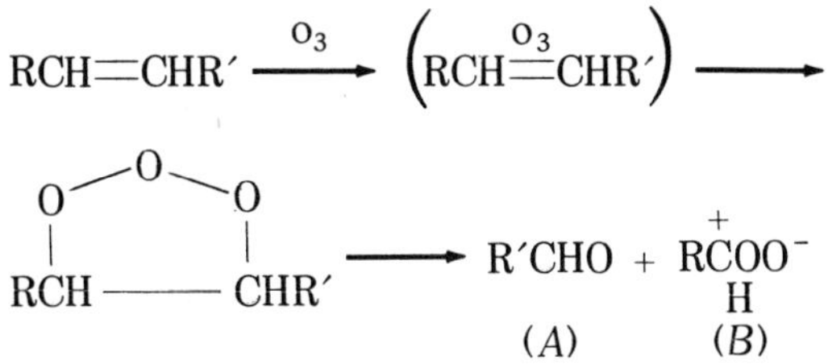

$$RCH{=}CHR' \xrightarrow{\;O_3\;} \left(RCH\overset{O_3}{=\!=}CHR'\right) \longrightarrow$$

$$\begin{matrix} O\!\!\overset{O}{\diagup}\!\!\diagdown\! O \\[2pt] | \qquad\quad | \\ RCH \!-\!\!-\!\!-\! CHR' \end{matrix} \longrightarrow R'CHO + R\overset{+}{C}OO^{-}_{\;\;H} \tag{201}$$

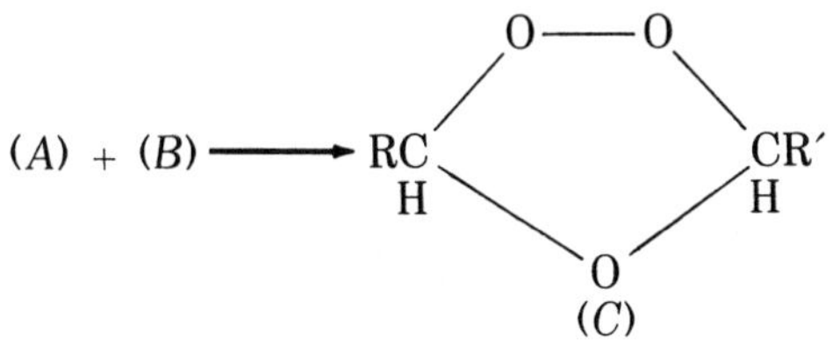

$$\qquad\qquad\qquad\qquad\qquad (A)\qquad\quad (B)$$

$$(A) + (B) \longrightarrow \begin{matrix} O\!-\!\!-\!O \\ \diagup \qquad\quad \diagdown \\ RC \qquad\qquad CR' \\ H \qquad\qquad H \\ \diagdown \qquad\quad \diagup \\ O \end{matrix} \tag{202}$$

$$\qquad\qquad\qquad\qquad\qquad\qquad (C)$$

Automobile tires often contain antiozonants to inhibit premature break-down of the rubber. An example of the effect of ozone on natural rubber of regular structure (*cis*-1,4-polyisoprene) is as follows:

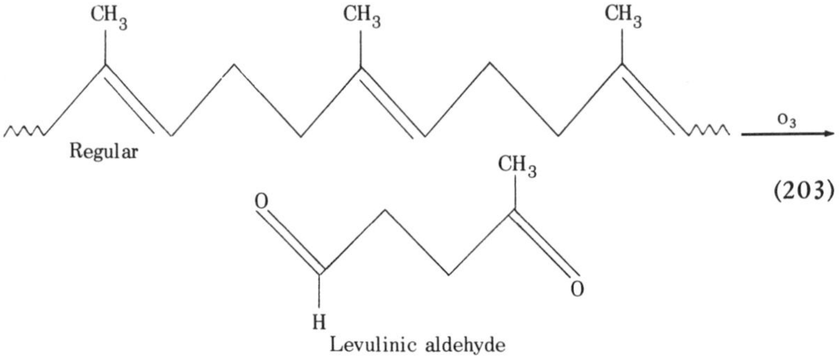

Levulinic aldehyde

(203)

Rubber of irregular structure can yield other products as follows:

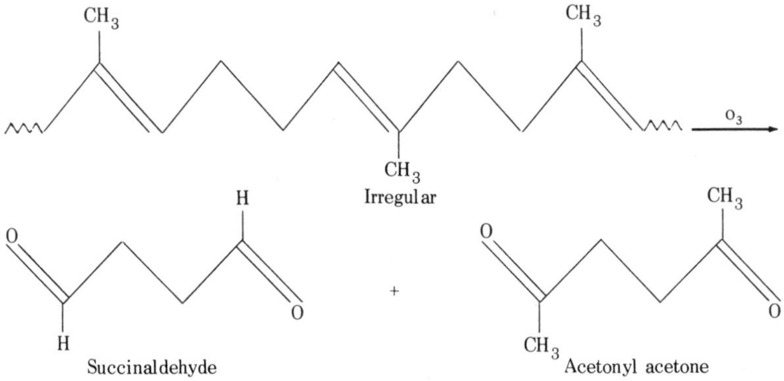

Succinaldehyde   +   Acetonyl acetone

(204)

In connection with unsaturated synthetic polymers may be mentioned the ozonization of polybutadiene (PB). Thus, two principal products form during ozonolysis: succinic acid and butane-1,2,4-tricarboxylic acid. These products were presumably derived from 1,4- and 1,2-addition moieties in the polybutadiene. Furthermore, Delman and coworkers [58a] studied the degradation of SBR (styrene-butadiene rubber) in presence of ozone and oxygen. The ozonization was carried out at room temperature, employing toluene solutions of SBR. It was observed that the intrinsic viscosity $[\eta]$ initially decreased rapidly with ozonization time, and then decreased with time at a slower rate until a limiting $[\eta]$ was attained. Several factors were examined to assess the cause of the observed $[\eta]$ changes. It was found that molecular-weight distribution and molecular chain length of the SBR did not influence appreciably the rate of degradation. Further, the rate of chain scission was found to be independent of the ozonization products formed by SBR and of the type of solvent used (toluene or orthodichlorobenzene). However, it was discovered that the concentration of active oxygen-containing groups in the SBR polymers during ozonization were related to the rate of $[\eta]$ changes. From the preceding and because pure oxygen does not produce any major structural changes in SBR at 25°C, it was postulated that severe SBR degradation is initiated by the random attack of O_3 on double bonds present. The resulting peroxidic-type structures [cf. Eq. (202), structure (C)] presumably decompose to yield free radicals which undergo the usual autoxidation steps in solution, i.e.,

$$R\cdot + O_2 \longrightarrow RO_2\cdot \tag{8a}$$

$$RO_2\cdot + RH \longrightarrow RO_2H + R\cdot \tag{8b}$$

$$\left.\begin{array}{r} 2R\cdot \\[6pt] R\cdot + RO_2\cdot \\[6pt] 2RO_2\cdot \end{array}\right\} \longrightarrow \text{Products} \qquad \begin{array}{l} (9a) \\[6pt] (9b) \\[6pt] (9c) \end{array}$$

The last termination step, Eq. $(9c)$, should be predominant in presence of oxygen under the experimental conditions used. Further, the formation of products containing —OH and $\diagdown C{=\!=}O$ structures was accounted for by

$$2R_2CHO_2\cdot \longrightarrow R_2C{=\!=}O + R_2CHOH + O_2 \tag{10}$$

The apparent approach of SBR to a limiting size was accounted for in the following: $R\cdot$ radicals no longer formed and/or oxygen was removed from the system.

Saturated polymers are relatively inert to ozone, e.g., polyethylene. However, ozone enhances the autoxidation of saturated polymers in presence of oxygen (as in the case of SBR). Thus, the rate of oxidation is increased and the induction period decreased. When atactic polypropylene was treated with ozone it was found to degrade less than when in the presence of pure oxygen. This was ascribed to decomposition of hydroperoxides to alcohols in presence of ozone. The synergistic effect of ozone during autoxidation in presence of oxygen may occur by virtue of the formation of free radicals during the initial phases of autoxidation [58]. Thus, for simple hydrocarbons, e.g., alkanes, the following was proposed:

$$RH + O_3 \longrightarrow RO\cdot + HO_2\cdot \tag{205}$$

$$RO\cdot + RH \longrightarrow ROH + R\cdot \tag{206}$$

$$R\cdot + O_2 \longrightarrow RO_2\cdot \tag{130}$$

. . . .

The ozonization of polyethylene (Alathon, molecular weight 18,000 to 19,000) in presence of oxygen was similar to atactic propylene in connection with enhancement of autoxidation. Any double bonds present in polyethylene are presumably attacked initially by ozone. Thus, it was found that infrared band intensity of double bonds in low-density polyethylene decreased on exposure to ozone [58b].

Derived polymers, e.g., methylcellulose (MC), are also susceptible to ozone attack in presence of oxygen. For example, the molecular weight of MC decreased as the ratio of consumed ozone to MC was increased [58c]. It appeared that various radical chain oxidations occurred on ozonization of MC (as in the case of α-methyl glucoside) and that a main result was the introduction of oxidized functions. However, the main chain degrading processes were attributed to the direct attack of ozone which induced hydrolysis.

Hydrolysis

Proteins and cellulose probably were the early polymers in which any extensive hydrolytic studies were made. In recent years investigations were extended to include many synthetic polymers: polyamides, polyesters, lactam polymers, polyacrylates, polyurethanes, poly(vinyl acetal), etc. The importance of hydrolytic studies stems from the breakdown of physical properties, e.g., tensile strength, in the polymer during either processing or in end use.

Hydrolytic studies can be conducted under neutral, acid, or alkaline conditions. Generally, neutral and acid hydrolyses are similar, whereas alkaline hydrolysis can be quite different. Most hydrolytic studies are heterogeneous, though in most neutral and acid hydrolyses the reaction within the polymer network is homogeneous. Hydrolytic studies may be conducted on polymers in the bulk phase (generally as films) over a wide range of relative humidity or in aqueous solutions over a wide range of pH. Hydrolytic degradation can involve main-chain scission because of functional interunits in condensation polymers (as, for example, the amide and ester linkages in polyamides and polyesters, respectively) and/or side groups.

Certain factors can affect the extent of hydrolysis, e.g., film thickness, morphology (crystallinity, orientation), relative humidity, concentration of acid catalyst, dielectric constant of the polymer, autocatalysis by functional groups within the polymer, molecular orientation without formation of crystallinity, type and number of ionizable or functional groups, electrostatic and steric effects, adsorption of water on the polymer, and chain conformation [59] (cf. Chap. 6). It should be mentioned here that there are some conflicting views of the effect of some of the above parameters. Examples of some of these factors affecting hydrolysis are given below.

McMahon [60] and coworkers in studying the hydrolysis of poly(ethylene terephthalate) found that thicker films tend to decrease rate of hydrolysis (owing to differences in diffusion rates among different thicknesses of films), and that the hydrolysis rate for a given film thickness is dependent on the relative humidity. The latter result, taken in conjunction with the adsorption isotherm reported by Ravens and Ward [59, 61], indicates that the rate of hydrolysis is directly proportional to the concentration of water in the polymer. Sharples [62] showed that the crystalline regions of cellulose are more resistant to hydrolysis than the amorphous regions. Glavis [63] in studying the alkaline hydrolyses in heterogeneous medium of poly(methyl methacrylate) of various tacticities found that the isotactic polymer hydrolyzes more rapidly and to a higher degree of conversion. Smets and Van Humbeeck [64] found that the isotactic copolymer of acrylic acid–methyl acrylate hydrolyzed three to five times more rapidly than the conventional copolymer, and its final degree of conversion was higher. In this vein might also be mentioned the work of Chapman [65] who made hydrolytic measurements on tactic and atactic poly-N,N-dimethyl acrylamide and found that the tactic species reacts about six to seven times more rapidly than the atactic polymer. In the acid hydrolysis of poly(ethylene terephthalate) Ravens [66] reported that the order of higher hydrolytic reactivity, based on morphology, was: unoriented amorphous > unoriented crystalline (48 percent) > oriented crystalline (30 percent). In the latter connection might be mentioned the hydrolyses of simple esters in neutral or acid media. The following mechanisms have been advanced:

$$
\underset{\substack{\| \\ O}}{R-C-OR'} + H^+ \;\rightleftharpoons\; \underset{\substack{\| \\ O}}{R-C-OHR'^+} \tag{207}
$$

This reaction is established quite rapidly after which one of two of the following reactions may occur:

$$
\underset{\substack{\| \\ O}}{R-C-O^+HR'} \;\rightleftharpoons\; \underset{\substack{\| \\ O}}{R-C^+} + R'OH \tag{208}
$$

or

$$
\underset{\substack{\| \\ O}}{R-C-O^+HR'} + H_2O \;\rightleftharpoons\; \underset{\substack{\| \\ O}}{R-C-OH_2^+} + R'OH \tag{209}
$$

In the acid hydrolysis of poly(ethylene terephthalate), using hydrochloric acid, it was found that the rate increased rapidly above $3N$ HCl [66]. This was explained by considering the concentration of the acid not in the aqueous phase but rather in the polymer itself; by assuming that, owing to the low dielectric constant of the polymer, only undissociated HCl diffused into the polymer but not the hydrogen ions [59]. Further, it was found that the hydrolysis rate can increase as the reaction proceeds, which was described as probably due to autocatalysts by the carboxyl end groups [61].

In the discussion to follow on acid and alkaline degradation other factors affecting hydrolysis will be mentioned where appropriate.

Acid. In presence of 85 percent phosphoric acid, cellulose underwent a random-type degradation in accordance with an equation similar to Eq. (162). Random degradation also applied to the hydrolysis of methylated cellulose with fuming hydrochloric acid at $0°C$ [54]. In the case of the ester, poly(ethylene terephthalate), acid hydrolysis was found to also proceed in a random manner ($70°C$, $5N$ HCl), which was also true with neutral hydrolysis. However, there appears to be some evidence to suggest that hydrolysis by concentrated sulfuric acid is not random since terephthalic acid can be isolated early in the reaction.

Recently, Cagliostro and coworkers [67] presented a kinetic model for the acid-catalyzed decomposition of Delrin (an acetal-formaldehyde resin). Citric acid was used as the catalyst, 2 to 20 percent by weight in the polymer mixture. Samples of these mixtures were degraded in a thermogravimetric analyzer (TGA) heated at $3°C$ min^{-1} under N_2 atmosphere (cf. Chap. 2). Formaldehyde was the main product liberated as determined from mass spectroscopy. TGA thermograms for 2, 5, and 20 percent by weight citric acid–Delrin mixtures are shown in Fig. 1.11.

As the starting concentration of citric acid increases, the temperature at which decomposition occurs is lowered. Also presented in Fig. 1.11 is a TGA for pure Delrin. As noted from the figure, all mixtures containing citric acid decompose at lower temperatures than does pure Delrin. For

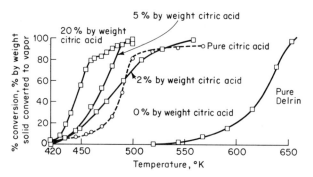

FIG. 1.11 Decomposition of Delrin-citric acid mixtures, using TGA [67].

the TGA representing an initial concentration of citric acid of 20 percent by weight, a plateau is reached at 80 percent conversion (80 percent weight loss). This represents the point at which all the Delrin has decomposed, with just the critic acid remaining. When the temperature of this mixture is increased, the remaining citric acid also decomposes. A TGA for the decomposition of pure citric acid is shown in Fig. 1.11. It can be seen from this figure that in the temperature range of interest the kinetic model must include the thermal decomposition of citric acid.

The following mechanism was proposed to account for the experimental data obtained:

Ionization of citric acid (HC)

$$HC \xrightleftharpoons{K} H^+ + C^-$$ (210)

Decomposition of citric acid

$$HC \xrightarrow{k_{HC}} X$$ (211)

Acid hydrolysis of Delrin (D)

$$D + H^+ \xrightarrow{k_h} P$$ (212)

Unzipping of polyformaldehyde (P)

$$P \xrightarrow{k_z} F$$ (213)

where F ≡ formaldehyde.

The net equation for polyformaldehyde production is

$$D + H^+ \xrightarrow{k_h} P \xrightarrow{k_z} F$$ (214)

From Eq. (210), the following may be written:

$$[H^+] = K^{1/2}[HC]^{1/2}$$ (215)

The rate of decomposition of citric acid may be written as

$$-\frac{d[HC]}{dt} = k_{HC}[HC]$$ (216)

Also the rate of disappearance of Delrin may be represented by

$$-\frac{d[D]}{dt} = k_h[D][H^+]$$ (217)

When Eq. (215) is substituted into Eq. (217), we obtain

$$-\frac{d[D]}{dt} = k_h K^{1/2}[D][HC]^{1/2}$$ (218)

Finally, the unzipping of P (cf. Weak Links and End Groups, Chap. 6) is considered to be of zero order, i.e.,

$$-\frac{d[P]}{dt} = k_z \tag{219}$$

From a total material balance we may write

$$W_o = W + F + X \tag{220}$$

where W_o denotes total weight of mixture before degradation; W represents weight remaining at any time t; F is weight of formaldehyde liberated; and X represents weight of decomposition products of citric acid.

Differentiating, it follows that

$$-\frac{dW}{dt} = \frac{dF}{dt} + \frac{dX}{dt} \tag{221}$$

From Eq. (216) it may be seen that

$$\frac{dX}{dt} = k_{HC}[HC]M_{HC} \tag{222}$$

where M_{HC} denotes molecular weight of citric acid.

The expression for dF/dt may be either one of two possible cases. From Eq. (214) when the hydrolysis of the Delrin proceeds much more rapidly than the unzipping of P ([P] > 0), we may write

$$\frac{dF}{dt} = k_z M_P \tag{223}$$

where M_P is molecular weight of polyformaldehyde.

When the unzipping of P is much more rapid than the hydrolysis of D in Eq. (214) we may write ([P] $\approx$ 0)

$$\frac{dF}{dt} = k_h[D][H^+] = k_h[D]K^{1/2}[HC]^{1/2}M_P \tag{224}$$

Employing the Arrhenius equation,

$$k = Ae^{-E/RT} \tag{225}$$

the following expressions may be written:

$$-\frac{dW}{dt} = A_z e^{-E_z/RT} M_P$$

$$+ A_{HC} e^{-E_{HC}/RT} [HC] M_{HC} \quad \text{for} \quad [P] > 0 \tag{226}$$

and

$$-\frac{dW}{dt} = A_T e^{-E_T/RT} [D][HC]^{1/2} M_P$$

$$+ A_{HC} e^{-E_{HC}/RT} [HC] M_{HC} \quad \text{for} \quad [P] \approx 0 \tag{227}$$

In Eq. (224) the $k_h K^{1/2} \equiv k_T$; the Arrhenius expression for k_T was used in Eq. (227). Rates of weight loss were converted into terms more appropriate for analysis of TGA data, i.e. (cf. Chap. 2),

$$\frac{dW}{dt} = (RH) \frac{dW}{dT} \tag{228}$$

where $(RH) \equiv$ constant heating rate used during TGA experiment, and T represents temperature.

Equations (227) and (228) were applied to the data as shown in Fig. 1.11. Where possible, kinetic parameters were determined from the individual materials; e.g., for citric acid alone A_{HC} and E_{HC} were independently determined. By using a nonlinear regression technique values of the parameters A_z, E_z, A_T, and E_T were determined. It was found that for compositions containing 2 percent citric acid Eq. (226) was applicable. For compositions above 2 percent Eq. (227) applied.

Alkaline. In the alkaline hydrolysis (20 percent KOH solution at 98°C) of poly(ethylene terephthalate) Waters [68] reported that the reaction was heterogeneous with respect to polymer, and not homogeneous as in neutral or acid hydrolysis. The alkaline attack was primarily a surface reaction resulting in weight loss as ethylene glycol and terephthalate ions were removed into the solution. This was explained on the basis that perhaps the OH⁻ ions cannot penetrate the polymer structure because of the low dielectric constant of the polymer, and that the resulting carboxylate ions might be repulsing the attacking hydroxyl ions [59].

In the alkaline hydrolysis of simple esters, the absence of acid catalysis is offset by having a more strongly nucleophilic reagent, OH⁻ instead of H_2O. Alkaline and acidic hydrolysis involves cleavage of the bond between the oxygen and acyl group, forming alcohol and acid salt in the former and acid in the latter.

The degradation of poly(ethylene terephthalate), of various morphological forms, with aqueous methylamine occurred in three stages [69]. The initial attack by methylamine occurred in the amorphous region with little low-molecular-weight material formed. Scission followed in the second stage, resulting in low-molecular-weight products but an increase in the crystallinity of the polymer. The third stage was characterized by a gradual decrease in the rate of the reaction attributed to slower attack on both the amorphous and crystalline moieties in the polymer.

Polyurethanes appear to be more resistant to hydrolysis than polyesters. The urethane linkage is more stable to hydrolytic attack than the ester group. Polyurethane foams made from polyethers were found to have excellent resistance to hydrolysis, whereas those based on polyesters underwent hydrolysis under similar conditions [70]. Polyurethanes prepared from aliphatic diols and diisocyanates were not appreciably affected by 1 percent NaOH at $50°C$, whereas the corresponding phenol analogues were attacked. The hydrolytic products of the latter were found to be mainly CO_2, a diamine, a phenol, and a polyurea (which was formed by the action of the diamine with some unreacted polyurethane [71]).

Miscellaneous

Polymer degradation was defined in the Introduction as a chemical modification of either the main chain or side groups which results in the breaking of primary valence bonds. The former involves chain scission, leading to a reduction in molecular weight, cross linking, etc. Since this book is essentially concerned with the degradative process that results in a decrease in polymer molecular weight, reactions other than main-chain scission, such as halogenation and sulfonation, are superficially treated in relation to their effect on properties. In this regard should be mentioned that there are a vast number of chemical reactions [54] carried out on polymeric systems in obtaining certain desired properties. For example, polyethylene film when exposed to chlorine vapors in the dark improves its dielectric heat sealability, dyeability, and printability while not affecting its original transparency [72].

However, when these chemical modifications are carried on to excess, or under certain conditions, e.g., high temperature and ultraviolet light, degradation can occur. For example, polyethylene film, exposed in the dark to fluorine vapor containing up to 10 percent N_2, will degrade unless the halogen is introduced slowly [73, 74], and polyethylene chlorinated below $100°C$ will degrade unless oxygen is removed. Further, polypropylene and polyisobutylene are more susceptible to degradation by chlorination than polyethylene [54].

Polystyrene can be nitrated to the dinitro derivative by reacting with nitric acid–sulfuric acid mixtures that dissolve the polymer. Time and temperature can markedly influence the reaction, ultimately leading to severe degradation [75].

As in the chlorination and sulfonation of various polymers, the use of fuming nitric acid leads initially to side-chain reactions which may be followed by main-chain scission. Temperatures can influence the type of attack. Thus it was found [76] that during the treatment of low-molecular-weight polyethylene between 25 and 83°C by fuming nitric acid, chain scission was predominant at the lower temperatures with nitration becoming dominant at the higher temperatures.

Crystalline polyethylene (PE) and crystalline polypropylene (PP) were subjected to attack by fuming nitric acid under isothermal conditions. It was found with polypropylene that as the duration of the acid treatment was increased, the molecular weight decreased rapidly initially, the crystallinity increased, and the melting point decreased slightly [77, 78] (see Fig. 1.12). Generally, for PE and PP it was observed that the nitric acid readily attached the amorphous region initially, thereby solubilizing and etching out this region. The highly crystalline region, which the acid cannot easily attack, remained nearly intact. The latter region presumably underwent some chain scission. In case of point, it may be mentioned that PP initially melted at 173°C, but the crystalline residue melted at about 160°C (cf. Fig. 1.12). This may be due to a lowering of the molecular weight of the crystalline region and to the depressing effect of a small number of nitro and carbonyl groups formed during the acid treatment. In this connection may be mentioned some aspects of the chemistry of the acid-treated polymer, as appraised by infrared spectroscopy. After an initial acid treatment a small percentage of NO_2 groups could be detected. This

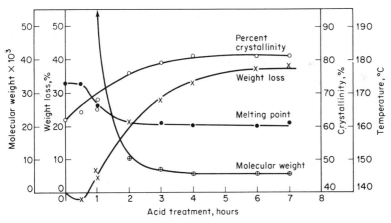

FIG. 1.12 **Isothermally (125°C) crystallized polypropylene treated with boiling (120°C) HNO$_3$ [78].**

percentage rose as the duration of the treatment increased. Furthermore, carbonyl moieties (presumably as —COOH) increased gradually until a constant value was reached. This leveling-off point was attained at the same time that the sample reached a constant weight as a result of removal of amorphous polymer. It may be noted from Fig. 1.12 that there is an initial weight increase during the early stages of the acid attack (about 1/2 hr). This was presumably brought about by the formation of NO_2 groups without formation of volatile fragments.

BIOLOGICAL DEGRADATION [79]

Biological degradation is chemical in nature. However, it is not treated in the section on Chemical Degradation since the source of the attacking chemicals are from microorganisms, such as fungi and bacteria. These chemicals are of a catalytic nature, e.g., enzymes. The susceptibility of a polymer to microbial attack generally depends on enzyme availability, availability of a site in the polymer for enzyme attack, enzyme specificity for the polymer, and coenzyme presence, if required.

Fungi, including bacteria, are widely distributed throughout the world in soil, water, and air. In the air the fungi occur mainly as spores, whereas in the soil (top few inches) they thrive usually in an active, growing condition. They generally require carbon, hydrogen, oxygen, nitrogen, among other elements, for nutrients. Accordingly, there are many substances that can be used as sources of these elements. Generally, water is required for the microorganisms to function properly. (Therefore, hydrolytic influences on degradation may occur with or without biological effects.)

In the following, for the sake of convenience, synthetic and natural polymers are treated separately.

Synthetic Polymers

Pure synthetic polymers are generally resistant to microbial attack because of several factors such as hardness, limited water absorption, and type of chemical structure. However, commercially used polymeric systems, e.g., plastics, fibers, elastomers, and adhesives, are usually formulated products containing such components as plasticizers, pigments, antioxidants, and lubricants. Often biological susceptibility is due to one or more of these nonpolymeric constituents, causing gross changes in the commerical product. Among these components, plasticizers are most vulnerable to microbial attack, especially those containing fatty acid residues, e.g., glycerol esters, laurates, oleates, phthalates, and stearates. However, the abietic acid derivatives appear to be more resistant. Some synthetic polymers (pure) with their susceptibility to biological attack are:

1. Poly(methyl methacrylate): resistant.
2. Epoxy resins: resistant.
3. Polyacrylonitrile: resistant.
4. Polystyrene: resistant.
5. Polyvinyl chloride: resistant.
6. Polyurethanes: resistant.
7. Synthetic rubber: resistant.
8. Cellulose derivatives: cellulose acetate, cyanocellulose, and ethylcellulose are resistant, whereas cellulose nitrate, methylcellulose, and sodium carboxy–methylcellulose are less resistant.

9. Polyethylene: the low-molecular-weight polymer supports biological growth, whereas the high-molecular-weight polyethylene is resistant.
10. Polyvinyl acetate: may show some attack.

The incorporation of additives, e.g., plasticizers, as stated earlier, can enhance biological attack in many of the above resistant polymers.

Natural Polymers

The celluloses and proteins are important natural polymers which undergo degradation by enzymatic action.

In proteins, all the bonds in the chain are peptide linkages, and under certain conditions they may be equally susceptible to attack. However, unlike many synthetic polymers, protein molecules contain a variety of side groups, and therefore only a few linkages may contain the same pair of side chains. Because of this, degradation is often specific, where reagents preferentially cleave peptide bond linking residues having particular kinds of side chains.

By way of illustration of the above may be considered the hydrolytic enzyme trypsin acting on peptide bonds represented by $-NH-CHR_1-CO-NH-CHR_2-CO-$. Trypsin will split those peptide linkages only if R_1 in the protein represents an arginine or lysine side chain, being specific to these amino acids. In oxidized ribonuclease, a protein containing 124 amino acid residues, only 12 of the 123 peptide bonds are specific to the enzyme trypsin [80, 81]. In the hydrolysis of this protein by trypsin, the reaction will proceed until the 12 bonds have been cleaved, after which there is essentially no further reaction. The products of the hydrolysis will consist of only 13 peptides with no other types of fragments. It should be mentioned here that even hydrolysis by means of strong acids may also tend to be specific initially. However, all peptide bonds are eventually broken upon prolonged hydrolysis.

TABLE 9   Some Natural Polymers Degraded by Various Microorganisms [82]

Polymer	Organism	Degradation product
Starch	*Bacillus macerans*	Dextrin
Xylan	*Sporocytophaga myxococcoides*	Xylose
Pectic acid	*Bacillus polymyxa*	Galacturonic acid
Pectin protein	*Erwinia carotovora*	≈100% methanol

Purified native proteins, on the other hand, are generally not readily attacked, even by proteolytic organisms. A nitrogen source other than the protein must be present for active proteolysis (splitting of proteins into simpler products) to occur. However, denatured proteins, e.g., casein and gelatin, are readily attacked.

Other natural products susceptible to biological attack are:

1. *Industrial gums.* Plant polysaccharides or their derivatives such as gum arbaic (slowly degraded by bacteria); gum guar (hydrolyzed by microorganisms); pectic substances (these polyuronides are attacked by numerous organisms, e.g., fungi); hemicelluloses (e.g., xylan) are hydrolyzed by bacteria and fungi.

2. *Natural rubber.* Such products are almost completely consumed by soil microorganisms.

3. *Starch.* Degraded by bacteria and fungi.

4. *Cellulose.* Most abundant organic polymer in nature; subject to attack by biological agents through enzymatic hydrolysis.

Some typical microorganisms responsible for biological degradation of natural polymers are summarized in Table 9 [82].

REFERENCES

1. Grassie, N.: "Chemistry of High Polymer Degradation Processes," p. 79, Interscience Publishers, Inc., New York, 1956.
2. Tatarenko, L. A., and V. S. Pudov: *Zh. Fiz. Khim.*, **41**:2951 (1967).
3. Reich, L., and S. S. Stivala: "Autoxidation of Hydrocarbons and Polyolefins," Marcel Dekker, Inc., New York, 1969.
4. Howard, J. A., K. U. Ingold, and M. Symonds: *Can. J. Chem.*, **46**:1017 (1968).
5. Adamic, K., J. A. Howard, and K. U. Ingold: *Can. J. Chem.*, **47**:3803 (1969).
6. Jellinek, H. H. G.: *Pure Appl. Chem.*, **4**:419 (1962).
7. Jellinek, H. H. G.: *J. Appl. Polymer Sci.* (Applied Polymer Symposia), **1967**(4):41.
8. Jellinek, H. H. G., and I. J. Bastien: *Can. J. Chem.*, **39**:2056 (1961).
9. Stokes, S., and R. B. Fox: *J. Polymer Sci.*, **56**:507 (1962).
10. Ershov, Y. A., S. I. Kuzina, and M. B. Neiman: *Russ. Chem. Rev.*, **38**:147 (1969).
11. Kamiya, Y., and K. U. Ingold: *Can. J. Chem.*, **42**:2424 (1964).
12. Grassie, N., and N. A. Weir: *J. Appl. Polymer Sci.*, **9**:987 (1965).
13. Kujirai, C., S. Hashiya, K. Shibuya, and K. Nishio: *Chem. High Polymer (Japan)*, **25**:193 (1968).
14. Betts, A. T., and N. Uri: *Chem. Ind.*, **1967**:512.
15. Ershov, Y. A., A. F. Lukovnikov, and A. A. Baturina: *Kinetika i Kataliz*, **7**:597 (1966).
16. Cicchetti, O.: *Advan. Polymer Sci.*, **7**:70 (1970).
17. Trozzolo, A. M., and F. H. Winslow: *Macromolecules*, **1**:98 (1968).
18. Carlsson, D. J., and D. M. Wiles: *Macromolecules*, **2**:587 (1969).
19. Carlsson, D. J., and D. M. Wiles: *Macromolecules*, **2**:597 (1969).
20. Hatton, J. R., J. B. Jackson, and R. G. J. Miller: *Polymer*, **8**:41 (1967).
21. Drogin, I.: *SPE J.*, **1965**:248.
22. FMC Corporation: U.S. Patent 3,296,191, 1967.
23. Bovey, F. A.: "The Effects of Ionizing Radiation on Polymers," Interscience Publishers, Inc., New York, 1958; A. Charlesby: "Atomic Radiation and Polymers," Pergamon Press, New York, 1960.
24. Shultz, A. R.: In E. M. Fettes (ed.), "Chemical Reactions of Polymers," Interscience Publishers, a division of John Wiley & Sons, Inc., New York, 1964.
25. Wall, L. A., and J. H. Flynn: *Rubber Chem. Technol.*, **35**:195 (1962).
26. Charlesby, A.: "Radiation Chemistry of Polymeric Systems," Interscience Publishers, a division of John Wiley & Sons, Inc., New York, 1962.
26a. Ohnishi, S., Y. Ikeda, M. Kashiwagi, and I. Nitta: *Polymer*, **2**:119 (1961).
27. Charlesby, A.: *Proc. Roy. Soc. (London)*, **A215**:187 (1952).
28. Lanza, V. L.: In R. A. V. Raff and K. W. Doak (eds.), "Crystalline Olefin Polymers," pt. II, pp. 301ff., Interscience Publishers, a division of John Wiley & Sons, Inc., New York, 1964.

29. Harlen, F., W. Simpson, F. B. Waddington, J. D. Waldron, and A. C. Baskett: *J. Polymer Sci.*, **18**:589 (1955).
30. Charlesby, A.: *Proc. Roy. Soc. (London)*, **A222**:60 (1954).
31. Charlesby, A., and S. H. Pinner: *Proc. Roy. Soc., (London)*, **A249**:367 (1959).
32. Dole, M., C. D. Keeling, and D. G. Rose: *J. Am. Chem. Soc.*, **76**:4304 (1954).
33. Miller, A. A., E. J. Lawton, and J. S. Balivit: *J. Phys. Chem.*, **60**:599 (1956).
34. Dole, M., D. C. Milner, and T. F. Williams: *J. Am. Chem. Soc.*, **80**:1580 (1958).
35. Charlesby, A.: *J. Polymer Sci.*, **11**:513, 521 (1953).
36. Wall, L. A., and D. W. Brown: *J. Phys. Chem.*, **61**:129 (1957).
37. Fischer, H.: *Kolloid-Z.*, **180**:64 (1962).
38. Shultz, A. R., P. I. Roth, and G. B. Rathmann: *J. Polymer Sci.*, **22**:495 (1956).
39. Todd, A.: *J. Polymer Sci.*, **42**:223 (1960).
40. Wall, L. A.: In W. A. Lundberg (ed.), "Autoxidation and Antioxidants," vol. 2, Interscience Publishers, a division of John Wiley & Sons, Inc., New York, 1962.
41. Hancock, T.: "Personal Narrative of the Origin and Progress of the Caoutchouc or India-Rubber Manufacture in England," Longmans Green & Co., Ltd., London, 1857.
42. Staudinger, H., and W. Heuer: *Ber.*, **67**:1159 (1934).
43. Staudinger, H., and E. Dreher: *Ber.*, **69**:1091 (1936).
44. Baramboim, N. K.: In W. F. Watson (ed.), "Mechanochemistry of Polymers," Maclaren and Sons, Ltd., London, 1964.
45. Tobolsky, A. V.: "Properties and Structure of Polymers," John Wiley & Sons, Inc., New York, 1960.
46. Baranwal, K.: *J. Appl. Polymer Sci.*, **12**:1459 (1968).
47. Bueche, F.: *J. Appl. Polymer Sci.*, **4**:101 (1960).
48. Karrer, E.: *Ind. Eng. Chem., Anal. Ed.*, **2**:96 (1930).
49. Ceresa, R. J., and W. F. Watson: *J. Appl. Polymer Sci.*, **1**:101 (1959).
50. Watson, W. F.: In E. M. Fettes (ed.), "Chemical Reactions of Polymers," p. 1089, Interscience Publishers, a division of John Wiley & Sons, Inc., New York, 1964.
51. Weissler, A.: *J. Appl. Phys.*, **21**:171 (1950).
52. Ovenall, D. W., G. W. Hastings, and P. E. M. Allen: *J. Polymer Sci.*, **33**:207 (1958).
53. Allen, P. E. M., G. M. Burnett, G. W. Hastings, H. W. Melville, and D. W. Ovenall: *J. Polymer Sci.*, **33**:213 (1958).
54. Fettes, E. M. (ed.): "Chemical Reactions of Polymers," Interscience Publishers, a division of John Wiley & Sons, Inc., New York, 1964.
55. Jellinek, H. H. G., and Y. Toyoshima: *J. Polymer Sci.*, **A-1, 5**:3214 (1967).
56. Jellinek, H. H. G., and F. F. Flajsman: *J. Polymer Sci.*, **A-1, 8**:711 (1970).
57. Jellinek, H. H. G., F. F. Flajsman, and F. J. Kryman: *J. Appl. Polymer Sci.*, **13**:107 (1969).
58. Bailey, P. S.: *Chem. Rev.*, **58**:925 (1958).
58a. Delman, A. D., B. B. Simms, and A. E. Ruff: *J. Polymer Sci.*, **45**:415 (1960).
58b. Cooper, G. D., and M. Prober: *J. Polymer Sci.*, **44**:397 (1960).
58c. Katai, A. A., and C. Schuerch: *J. Polymer Sci.*, **A-1, 4**:2683 (1966).
59. Ravens, D. A. S., and J. E. Sisley: In E. M. Fettes (ed.), "Chemical Reactions of Polymers," Interscience Publishers, a division of John Wiley & Sons, Inc., New York, 1964.
60. McMahon, W., H. A. Birdsall, G. R. Johnson, and C. T. Camilli: *J. Chem. Eng. Data,* **4**:57 (1959).
61. Ravens, D. A. S., and I. M. Ward: *Trans. Faraday Soc.*, **57**:150 (1961).
62. Sharples, A.: *Trans. Faraday Soc.*, **54**:913 (1958).
63. Glavis, F. J.: *J. Polymer Sci.*, **36**:547 (1959).
64. Smets, G., and W. Van Humbeeck: *J. Polymer Sci.*, **A-1**:1227 (1963).
65. Chapman, C. B.: *J. Polymer Sci.*, **45**:237 (1960).
66. Ravens, D. A. S.: *Polymer*, **1**:375 (1960).
67. Cagliostro, D. E., S. Riccitiello, and J. A. Parker: *J. Macromol. Sci. Chem.*, **A3**:1601 (1969).

68. Waters, E.: *J. Soc. Dyers Coluorists,* **66:**609 (1950).
69. Farrow, G., D. A. S. Ravens, and I. M. Ward: *Polymer,* **3:**17 (1962).
70. Saunders, J. H., S. Steingiser, P. G. Gemeinhardt, A. S. Morecroft, and E. E. Hardy: *J. Chem. Eng. Data Ser.,* **3:**153 (1958).
71. Dyer, E., and G. W. Bartels, Jr.: *J. Am. Chem. Soc.,* **76:**591 (1954).
72. Angier, D. J.: In E. M. Fettes (ed.), "Chemical Reactions of Polymers," Interscience Publishers, a division of John Wiley & Sons, Inc., New York, 1964.
73. Kropa, E. L.: U.S. Patent 2,497,046, 1950.
74. Joffre, S. P.: U.S. Patent 2,811,468, 1957.
75. Zenftman, H.: *J. Chem. Soc.,* **1950:**982.
76. Beachell, H. C., and S. P. Nemphos: *J. Polymer Sci.,* **21:**113 (1956).
77. Hock, C. W.: *J. Polymer Sci.,* **B-3:**573 (1965).
78. Hock, C. W.: *J. Polymer Sci.,* **A-2:4,** 227 (1966).
79. Rosato, D. V.: In D. V. Rosato and R. T. Schwartz (eds.), "Environmental Effects of Polymeric Materials," vol. 1, Interscience Publishers, a division of John Wiley & Sons, Inc., New York, 1968.
80. Tanford, C.: "Physical Chemistry of Macromolecules," John Wiley & Sons, Inc., New York, 1961.
81. Hirs, C. H. W., W. H. Stein, and S. Moore: *J. Biol. Chem,* **221:**151 (1956).
82. Wallen, L. L., F. H. Stodola, and R. W. Jackson: Type Reactions in Fermentation Chemistry, *U.S. Dept. Agr. Rept.* ARS-71-13, May, 1959.

2

DYNAMIC THERMOGRAVIMETRIC ANALYSIS (TGA) AND DIFFERENTIAL THERMAL ANALYSIS (DTA)

It has been reported [1] that instrument manufacturers estimate that sales of all thermal analysis equipment as of August, 1969, were growing at a rate of 30 percent per year and would approach 8 million dollars in 1961, and 15 million dollars in 1972. Furthermore, the number of research chemists' references to thermal analysis in basic chemistry journals has been increasing exponentially—from about 150 in 1960 to over 5,000 in 1968. These trends are undoubtedly due, in part, to the increasing search for new materials which will withstand high thermal and/or oxidative degradation. This chapter will discuss two important techniques of thermal analysis, dynamic thermogravimetric analysis (TGA) and differential thermal analysis (DTA). Theoretical aspects of TGA and DTA will be treated initially and will be followed by practical applications. Although polymer degradations will be discussed primarily, it should be realized that similar discussions are applicable to other types of materials, e.g., inorganic crystalline compounds and ceramics. Several monographs [2, 3] provide rather comprehensive descriptions of apparatus, techniques, and evaluations of TGA, DTA, and other methods of thermal analysis. There have also recently appeared reviews on the theoretical and practical aspects of TGA [4] and DTA [5].

TGA

Dynamic thermogravimetric analysis (TGA) may be defined as a continuous process which involves the measurement of sample weight as a function of temperature (or time) as the reaction temperature is changed by means of a programmed (usually constant) heating rate. TGA measurements can be made in either presence or absence (in vacuum) of gases, e.g., oxygen.

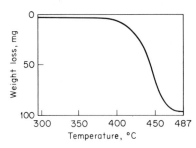

FIG. 2.1   Thermogravimetric curve for the degradation of polyethylene at 1 mm Hg pressure. Sample weight 100 mg, heating rate 5°C min^{-1} [6].

In Fig. 2.1 is shown a typical TGA curve for polyethylene at 1 torr and at a heating rate of 5°C min^{-1} [6].

In Fig. 2.2 is shown a schematic diagram of a Chevenard recording thermobalance, which can afford such a TGA curve. Simply, the TGA apparatus comprises a sample and its container, which are placed in an environment the temperature of which is regulated by some form of a temperature programmer. Changes in sample weight are monitored by an appropriate transducer (differential transformer) producing an electric output which is associated with the weight changes. This output is amplified electronically and applied to a readout device, usually a potentiometric XY recorder. The dependent variable (weight) is usually plotted on the Y axis against the sample temperature on the X axis (cf. Fig. 2.1). (In the isothermal operation of the apparatus, the X axis would be used to plot time.) There are several advantages in using TGA instead of isothermal methods [4]. These include the following:

1.   Considerably fewer data are required. The temperature dependence of the volatilization rate may be determined over various temperature ranges from the results of a single experiment, whereas several separate experiments are required for each temperature range when isothermal methods are employed.

2.   The continuous recording of weight loss versus temperature ensures that no features of the pyrolysis kinetics are overlooked.

3.   A single sample is used for the entire TGA trace, thereby avoiding a possible source of variation in the estimation of kinetic parameters.

4.   In the isothermal method a sample may undergo premature reaction, and this may make the subsequent kinetic data difficult, if not impossible, to analyze properly.

However, a major disadvantage is that precise temperature control for kinetic experiments is much more difficult with TGA than with isothermal methods. Furthermore, there are various variables which can affect the accuracy of the results obtained by TGA, as mentioned in the following.

Factors Affecting Accuracy of TGA Results

These factors are [7] :

1. The arrangement of the furnace to ensure uniform sample heating
2. The shape of the sample crucible
3. The location and arrangement of temperature-measuring elements
4. The physical characteristics of the sample such as particle size
5. The surrounding atmosphere and the prior thermal treatment of the sample.

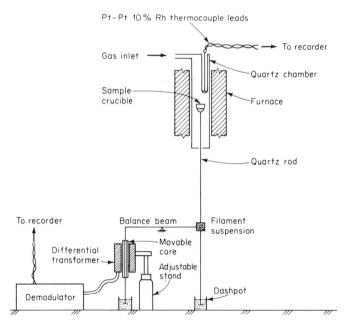

FIG. 2.2 Schematic diagram of Chevenard recording thermobalance.

The effects of these factors can be minimized by the following:

1. The use of small enough samples for temperature uniformity, and the use of direct sample temperature measurement (temperature sensor in or near the sample)
2. The use of samples of uniform size which are uniformly packed in the sample crucible

3. The adjustment of gas flow, pressure, and sample shape in order to reduce the effects of effluent gases emerging from the sample during degradation.

Through careful control of such factors, marked improvement can be made in the reproducibility of various kinetic parameters, i.e., activation energy E, reaction order, and frequency factor, which can be derived from TGA data. Nevertheless, there are two important topics which deserve separate mention. These are topochemical and heat-transfer factors and their influence on kinetic parameter values.

Effect of particle topochemistry on kinetic parameter values. In a consideration of crystal decomposition, it is often necessary to take into account the spatial and temporal coordinates (topochemistry) of the crystal. In such solid-state reactions, the reactive zone often comprises surfaces of growing nuclei of products. The number of nuclei and the surface area of each nucleus both increase with time, and either of these dependencies may determine the overall rate law [8]. Furthermore, reaction rate may be dependent upon transport of matter which invariably requires the presence of solid-phase defects. Such defects (point) can serve as reaction intermediates, e.g., the lattice vacancy defect. Point defects are often capable of rapid motion, and their interactions can give rise to simple first- or second-order rate laws closely analogous to most fluid-phase reactions. In the following will be obtained various kinetic expressions based upon a topochemical approach. Subsequently, such types of expressions will be assumed to be applicable to polymers, amorphous and/or crystalline.

Rate-law dependence upon shape of reaction interface. In the following will be presented examples which illustrate how rate laws may arise from the constant isotropic motion of a reaction interface at a constant linear rate. When a macroscopic interface is formed by coalescence of growing nuclei, zero time may be difficult to establish accurately and, accordingly, may hinder precise derivation of rate laws. These laws are, theoretically, as numerous as the possible initial shapes of the reaction interface. Only a few cases will be considered.

Eckhardt and Flanagan [9] studied the dehydration of manganous formate dihydrate (MFD), using a topochemical approach. A model of this dihydrate is portrayed in Fig. 2.3. Indicated in this figure are the directions of interfacial growth (only in two isotropic directions). The use of a three-dimensional isotropic model led to derived results which agreed poorly with experimental results. Letting C = percentage of dehydration (based on the initial reactant volume destroyed), and using the symbols given in Fig. 2.3b, we may write

$$C = 100 \frac{L_1 L_2 - (L_1 - 2k\tau \cos\theta)(L_2 - 2k\tau)}{L_1 L_2} \tag{1}$$

where k = rate constant for interfacial growth; and $\tau = t - t_0$, where t = total exposure time and t_0 = time for the interface to be instantaneously established at the end of the induction period. From Eq. (1), $k\tau$ can be calculated for various values of C, and then by plotting $k\tau$ versus t, values of k and t_0 could be estimated. Values of C may be calculated for various conversions from the values of k, t_0, θ (15.6°), L_1, and L_2. These were

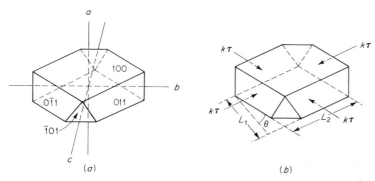

FIG. 2.3 (a) **Model of a monoclinic single crystal of manganous formate dihydrate;** (b) **model showing directions of interfacial growth** [9].

found to agree well with experimental values. In a typical dehydration run (Fig. 2.4), a linear region of dehydration existed between $C = 0.15$ and 0.45. Equation (1) can be rewritten as

$$C = \frac{200k\tau(L_1 + L_2 \cos\theta)}{L_1 L_2} - \frac{(400(k\tau)^2 \cos\theta}{L_1 L_2} \tag{2}$$

In the linear region (Fig. 2.4), it would appear that only the first term of Eq. (2) should be significant. However, this leads to a higher experimental linear region than predicted and to a higher value of k than observed. In order to overcome these difficulties, both terms in Eq. (2) were included, and it was found that the second term could be approximated as $C = f(\tau)^1$, thereby leading to a linear region similar to that observed. Further, upon differentiation, Eq. (2) yields

$$\frac{dC}{dt} = \frac{200k(L_1 + L_2 \cos\theta)}{L_1 L_2} - \frac{800k^2\tau' \cos\theta}{L_1 L_2} \equiv k' \tag{3}$$

where τ' = time interval over which k' = const. From Eq. (3) and the Arrhenius relationship,

$$k'(T) = C_1 e^{-E/RT} - C_2\tau' e^{-2E/RT} \tag{3a}$$

where $C_1 = 200(L_1 + L_2 \cos\theta)/L_1 L_2$, and $C_2 = 800 \cos\theta/L_1 L_2$. If the reasonable assumption is made that $\tau' = f(1/k)$ (which was experimentally confirmed), then Eq. (3a) becomes

$$k'(T) = C_3 e^{-E/RT} \tag{3b}$$

where $C_3 = [200 (L_1 + L_2 \cos \theta) - 800 \cos \theta]/L_1 L_2$.

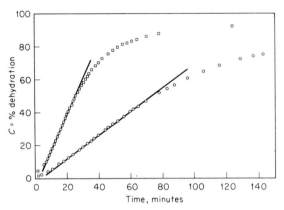

FIG. 2.4 Dehydration of crystalline power in vacuo [9]:
○ 22.9°C; □ 45.1°C.

It can be seen from Eq. (3b) that the value of E determined from the slope of the C versus t linear portion of the curve (Fig. 2.4) should be identical with that determined from values of k, using Eq. (1), which was found to be true. In this connection may be mentioned the results reported by Clarke and Thomas [10]. They determined a value of $E = 15.7 \pm 2.5$ kcal mole^{-1} based upon data obtained during a TGA run, as follows. During the linear region, if the Arrhenius relationship is used, then

$$\frac{dC}{dt} = k' = Z e^{-E/RT} \tag{4}$$

where Z = frequency factor. From Eq. (4), a plot of ln (dc/dt) versus $1/T$ should afford a linear relation whose slope should yield the value of E. From isothermal experiments a value of $E = 17.6 \pm 0.6$ kcal mole^{-1} had been previously reported [9]. Beyond the linear region, $C = 15$ to 45 percent (based on the theoretical percentage dehydration), an Arrhenius plot did not afford a linear relation. Thus, from Eq. (4), the activation energy for interfacial growth is associated with a zero-order reaction.

The topochemical approach used for single crystals may also be employed for powdered samples of MFD provided the average particle sizes of the various powdered samples used are nearly equal. That this should be so can be seen as follows [11]. From Eq. (2),

$$\frac{dC}{dt} = C_1 k - C_2 k^2 \tau \tag{3c}$$

If for the same crystal (or particle), we consider an arbitrary value of dehydration, C_c, for a series of hypothetical temperatures, T_1, T_2, etc., then from Eq. (2),

$$C_c = C_1 k_1 \tau_1 - \frac{C_2}{2} (k_1 \tau_1)^2$$

$$= C_1 k_2 \tau_2 - \frac{C_2}{2} (k_2 \tau_2)^2$$

$$= C_1 k_n \tau_n - \frac{C_2}{2} (k_n \tau_n)^2 \tag{5}$$

or

$$k_1 \tau_1 = \cdots = k_n \tau_n \equiv b \tag{5a}$$

Upon combining Eqs. (3c) and (5a) there is obtained

$$\frac{dC_c}{dt} = k(C_1 - bC_2) \tag{3d}$$

where C_1, C_2, and b are constants which are temperature-independent. Equation (3d) may be written as

$$\frac{dC_c}{dt} \equiv k'' = (C_1 - bC_2) Z e^{-E/RT} \tag{3e}$$

where $k'' = $ slope of the C versus t curve where $C = C_c$. In practice it is not possible to test Eq. (3c) by means of TGA. Instead, different samples of finely powdered MFD may be used at a series of temperatures, provided the average C_1 and C_2 values (and, therefore, the average particle size) of each of the powdered samples are reasonably equal. Such a condition can be met by passing the powdered samples through a standard mesh. Thus, from Eq. (3e), a plot of $\ln (dC_c/dt)$ versus $1/T$ for a series of isothermogravimetric runs on powdered MFD should yield values of E for the advance of the $\{011\}$ interface (cf. Fig. 2.3a). Since the isothermogravimetric curves obtained were linear for $C = 15$ to 50 percent as anticipated, the specified value of C_c may be chosen so that it lies within this range. Then dC_c/dt can be obtained with great accuracy. An Arrhenius plot of Eq. (3e) afforded a value of $E = 17.8 \pm 0.6$ kcal mole[-1], which agrees well with values obtained for the dehydration of individual single MFD crystals.

The topochemical approach can readily be applied to isotropic solids, e.g., cubes and spheres. Thus, for cubes,

$$C = \frac{a_0^3 - (a_0 - 2k\tau)^3}{a_0^3} \tag{6}$$

where a_0 = length of one side of the cube. Equation (6) may be readily converted into

$$(1 - C)^{1/3} = 1 - \frac{2k\tau}{a_0} \tag{6a}$$

which is often referred to as the *contracting formula*. It is applicable to isodimensional particles. The contracting formula has been found to be applicable to many substances, e.g., $Cd(OH)F$ [12]. Furthermore, other forms of expressions have been found to apply to various substances. For example, a first-order expression was found to be valid for the decomposition of $Zn(OH)F$ [12] and a second-order expression for the decomposition of $Cd(OH)Cl$ [12]. Also, from Eqs. (4) and (6), it can be shown that zero and two-thirds orders may be applicable to decompositions, respectively. From the preceding it appears that a general expression for the decomposition of solids may be written, based on a topochemical approach, as

$$\frac{dC}{dt} = k_e (1 - C)^n \tag{7}$$

where $n = 0, 2/3, 1,$ and 2. Equation (7) may also be written as

$$-\frac{dW}{dt} = k_e' W^n \tag{7a}$$

where W = weight or weight fraction of active material undergoing degradation. Equation (7a) will be used extensively in discussions of kinetic parameters obtained for polymers by means of TGA and DTA (in these cases n may assume any value in accordance with the experimental data).

Finally, it may be mentioned that the topochemical model used based on microscopic examination may not afford derived results which are in good agreement with experimental results. The use of a hypothetical model may be better in this respect. Engberg [13] investigated the dehydration kinetics of copper(II) chloride dihydrate in a vacuum and found that better agreement could be obtained between calculated and experimental results when a cubic model was assumed instead of the physically more realistic model of crystal needles (observed microscopically). This represents one example of a weakness of the topochemical approach.

Rate-law dependence upon the nucleation and development of the reaction interface. When nuclei are forming, developing, coalescing, etc., at the reaction interface, various expressions may be obtained [8] which involve an autoacceleration period. Thus, the following expressions have been derived, Eqs. (8) and (9), which involve a power law and an exponential law (self-catalyzed nucleation), respectively,

$$C = k_e t^n \tag{8}$$

and

$$C = k_1 e^{k_2 t} \tag{9}$$

A form similar to Eq. (8) has been applied to explosive-type materials, e.g., mercury fulminate ($n = 3$). Often, when Eq. (8) is not applicable, a form similar to Eq. (9) may be valid. Various other types of expressions can be derived, depending upon the conditions assumed for the topochemical model, e.g., the Avrami expression $[dC/dt = k(1 - C)t^2]$, and the Prout-Tompkins equation $[dC/dt = kC(1 - C)]$. However, it will be assumed that these expressions do not apply in polymer degradations. Instead, a form similar to Eq. (7a) will be assumed to be valid for these degradations (no autoaccelerated degradation occurs).

Effect of heat transfer on kinetic parameter values. Values of activation energy E for polymer degradations as determined by TGA measurements are often lower than those obtained by means of isothermal techniques. A case in point is Teflon [poly(tetrafluorethylene)] whose values of E, from TGA measurements, vary from 66 to 77 kcal mole^{-1} whereas when isothermal techniques were employed, E varied from 85 to 88 kcal mole^{-1} [4] (however, values of reaction order n were consistently close to unity for both techniques). The preceding suggested that there were factors affecting values of E during TGA experiments which were less operative in the case of isothermal experiments. One such factor may involve heat-transfer effects which would be expected to occur preferably during TGA experiments, owing to the dynamic nature of such experiments. The use of small, finely powdered samples or very thin-filmed samples usually used in TGA should minimize such effects. Reich and Stivala [14] investigated the influence of heat-transfer effects during Teflon degradation upon the value of E, using TGA techniques. Commercially available Teflon (Du Pont) cylindrical rods were lathed and cut into various cylindrical sample sizes. These samples were then embedded in a Transite thermal insulator so that only one face of each cylindrical sample was exposed.

Various average sample sizes (diameter and thickness, in inches) and their degradation codes are listed in the following: 3/16 x 1/16(1T); 1/8 x 3/32(3T); and, 3/16 x 1/37(5T). These samples were then placed in a 950 Du Pont Thermogravimetric Analyzer and heated at a rate of 2°C min⁻¹ in a helium atmosphere. A blank run indicated that the rate of weight loss of the Transite was relatively small above 500°C, at which temperature incipient degradation of Teflon becomes important. The TGA data obtained were analyzed with the aid of the following theoretical expressions.

The one-dimensional heat-conduction equation without chemical reaction or radiative cooling may be written for a solid as

$$\frac{C'\rho \partial T}{\partial t} = \frac{K\partial^2 T}{\partial x^2} \tag{10}$$

where T = temperature in a solid at position x at time t, K = thermal conductivity, C' = specific heat, and ρ = density. Values of K, C', and ρ were assumed to be constant during a TGA run. When chemical reaction is included, Eq. (10) becomes

$$\frac{C'\rho \partial T}{\partial t} = \frac{K\partial^2 T}{\partial x^2} + q \tag{11}$$

where q = amount of heat absorbed by the chemical reaction in a unit volume per unit time. Assuming that the Arrhenius law is applicable and that for Teflon decomposition $n = 1$, we may write

$$q = -QZ\rho e^{-E/RT} \tag{12}$$

where Q = thermal factor (e.g., cal g^{-1}), Z = frequency factor (e.g., sec⁻¹), and the minus sign denotes that heat is absorbed. Upon combining Eqs. (11) and (12),

$$\frac{C'\rho \partial T}{\partial t} = \frac{K\partial^2 T}{\partial x^2} - QZ\rho e^{-E/RT} \tag{13}$$

In order to obtain tractable expressions from Eq. (13), it is assumed that

$$\frac{\partial T}{\partial t} << \frac{QZe^{-E/RT}}{C'} \tag{13'}$$

Justification for the use of Eq. (13') can be obtained from various values reported for Teflon degradation [15]. Thus, if it is assumed that $E = 83$ kcal mole⁻¹, $Q = 400$ cal g^{-1} (enthalpy of depropagation), $Z = 3 \times 10^{19}$

sec^{-1}, $C' = 15 + 0.034T$ cal $mole^{-1}$ deg^{-1} (strictly applicable only up to 120°C), and $T = 800°K$, and if it is further assumed that $\partial T/\partial t$ may be well approximated by 0.033°C sec^{-1} (heating rate), then $0.03 \ll 1$ and Eq. (13′) would be justified. Equation (13) may now be written as

$$\frac{d^2 T}{dx^2} = \frac{QZ\rho e^{-E/RT}}{K} \tag{14}$$

Let us assume the following boundary conditions for the insulated Teflon samples: $x = 0$, $T = T_0$; $x = l$, $dT/dx \ll dT/dx|_{x=0}$, $T = T_l$; where $l =$ sample thickness which is involved in heat-transfer effects. Equation (14) may be solved as follows. It can be shown that

$$\int d\left(\frac{dT}{dx}\right)^2 = \int 2 \frac{d^2 T}{dx^2} \, dT \tag{15}$$

Upon substituting Eq. (14) into (15) and integrating,

$$\left(\frac{dT}{dx}\right)^2 \Bigg|_{x=0}^{l} = \frac{2QZ\rho}{K} \int_{T_0}^{T_l} e^{-E/RT} \, dT \tag{16}$$

The integral on the right-hand side of Eq. (16) may be solved by two approximate methods. Since these solutions will be used later on, the methods employed in solving this integral will now be described in detail.

In the integral $\int_0^T e^{-E/RT} \, dT$ (IT), let $u = E/RT$, and so $du = -(Ru^2/E) \, dT$. Then (IT) becomes, in terms of u,

$$-\frac{E}{R} \int_\infty^u \frac{e^{-u} \, du}{u^2} \tag{IU}$$

When the integral (IU) is integrated by parts, i.e., in the expression

$$\int z \, dv = vz \Bigg| - \int v \, dz \tag{17}$$

let $z \equiv 1/u^2$ and $dv \equiv e^{-u} \, du$; then (IU) becomes,

$$(IU) = -\frac{E}{R} \left(\frac{-e^{-u}}{u^2} - \int_\infty^u \frac{2e^{-u} \, du}{u^3} \right) \tag{18}$$

When the integral on the right-hand side of Eq. (18) is integrated by parts and this process repeated (asymptotic expansion), the following expressions results:

$$(IU) = \frac{Ee^{-u}}{Ru^2}\left(1 - \frac{2!}{u} + \frac{3!}{u^2} - \frac{4!}{u^3} + \cdots\right) \tag{19}$$

When E/RT is substituted for u, Eq. (19) becomes

$$(IT) \equiv \int_0^T e^{-E/RT}\,dT =$$

$$\frac{RT^2 e^{-E/RT}}{E}\left[1 - \frac{2RT}{E} + \frac{6(RT)^2}{E^2} + \cdots\right] \tag{20}$$

Another form of the solution of (IT) can be obtained as follows. If in Eq. (17), we let $z \equiv e^{-u}$ and $dv \equiv du/u^2$, then it can be readily shown that

$$(IU) = -\frac{E}{R}\left(\frac{-e^{-u}}{u} - \int_\infty^u \frac{e^{-u}\,du}{u}\right) \tag{21}$$

or

$$(IT) = \frac{E}{R/p(u)} \tag{22}$$

where

$$p(u) = \frac{e^{-u}}{u} - \int_u^\infty \frac{e^{-u}\,du}{u}$$

[In subsequent use of Eqs. (20) and (22), it will be assumed that $(IT) \approx \int_{T_0}^T e^{-E/RT}\,dt$ to a good approximation.] The integral in the expression for $p(u)$ is often denoted as the exponential integral $[E_i(u)]$, and values for this integral have been tabulated. Hence values of $p(u)$ [often denoted as $p(x)$] may be readily calculated. In this manner, Doyle [16] found that a good approximation for $p(u)$ for $20 \leqslant u \leqslant 60$ could be expressed as

$$\log p(u) = -2.315 - 0.457u \tag{23}$$

Returning to Eq. (16), it can now be seen that in conjunction with Eq. (20) (assuming that $1 \gg 2RT/E$), Eq. (16) becomes

$$\left(\frac{dT}{dx}\right)^2 \bigg|_0^l = -\left(\frac{dT}{dx}\right)^2_{x=0} = \frac{2QZ\rho R}{KE}\left(T_l^2 e^{-E/RT_l} - T_0^2 e^{-E/RT_0}\right) \tag{24}$$

Assuming that $T_l \ll T_0$, Eq. (24) becomes

$$\left(\frac{dT}{dx}\right)_{x=0} = \left(\frac{2QZ\rho R}{KE}\right)^{1/2} T_0 e^{-E/2RT_0} \tag{25}$$

Now, the amount of material which decomposes per unit time in a dx layer may be written as [cf. Eq. (12)]

$$dg = -Z\rho e^{-E/RT} S dx \tag{26}$$

where S = cross-sectional area of sample. The variation in the amount of material per unit time throughout the cylindrical rod becomes

$$-\frac{dW}{dt} \equiv \rho_t = k_e W = -\int_0^l Z\rho e^{-E/RT} S dx \tag{27}$$

where W = weight of active material remaining during decomposition. When Eq. (14) is substituted into Eq. (27) and the resulting expression integrated (S = const),

$$\rho_t = \frac{KS}{Q}\left(\frac{dT}{dx}\right)_{x=0} \tag{28}$$

When Eqs. (25) and (28) are combined,

$$\frac{\rho_t}{T_0} = \frac{2KZ\rho R}{QE}^{1/2} S e^{-E/2RT_0} \tag{29}$$

Since the thermal degradation of Teflon by TGA occurs over a relatively small temperature range and since these temperatures are high (about 800°K), $\rho_t/T_0 \approx$ const x ρ_t. Thus, Eq. (29) resembles a zero-order reaction. It has been indicated [17] that at relatively low conversions, the degree of conversion as a function of temperature is independent of reaction order. In view of this and Eq. (29), when a thermal barrier is controlling during the TGA degradation of Teflon, a first-order plot should afford a value of E which is approximately one-half the value that would be obtained in the absence of strong heat-transfer effects. For the first-order degradation of Teflon, we may write [4]

$$\log \frac{\rho_T}{W} = - \frac{E}{4.6T} + \log \frac{Z}{(RH)} \tag{30}$$

where $\rho_T = \rho_t/(RH) \equiv -dW/dT$, and (RH) = heating rate. When $\log (\rho_T/W)$ is plotted against $1/T$ [cf. Eq. (30)], the value of E should be about one-half that value of E obtained from a plot of $\log (\rho_T/T_0)$ versus $1/T$ [cf. Eq. (29)], when heat-transfer effects are important. (Derived data for the

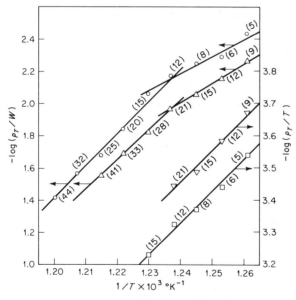

FIG. 2.5 Plots of $-\log (\rho_T/W)$ and $-\log (\rho_T/T)$ against $1/T$, ○ and □, run $1T$; △ and ▽, run $3T$. (Values in parentheses on plots denote percentage conversion.) [14].

Teflon degradation were obtained only up to about 50 percent conversion, since beyond this conversion the slopes of the primary thermograms became too steep to be utilized satisfactorily.) In Fig. 2.5 are depicted plots of Eqs. (29) and (30) for runs using samples $1T$ and $3T$.

From this figure, a first-order plot gave a value of $E \approx 48$ kcal mole^{-1} up to about 15 percent conversion for run $1T$. However, after about 20 percent conversion, the value of E changed drastically, giving a value of about 88 kcal mole^{-1}. Similarly, up to about 21 percent conversion a value of

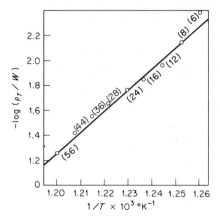

FIG. 2.6 Plot of $-\log(\rho_T/W)$ against $1/T$ (run $5T$). (Values in parentheses on plot denote percentage conversion.) [14].

$E \approx 53$ kcal mole^{-1} was obtained for run $3T$, whereas after about 28 percent conversion the value rose to about 85 kcal mole^{-1}. When Eq. (29) was employed for the data of runs $1T$ and $3T$ at low conversions (where $E \approx 48$ and 53 kcal mole^{-1}), linear relations are obtained which yielded values of $E \approx 88$ kcal mole^{-1} (run $1T$) and $E \approx 86$ kcal mole^{-1} (run $3T$). Thus, when heat-transfer effects are taken into account, values of E can be obtained which are of the correct order of magnitude. Otherwise, low values of E are obtained.

From Fig. 2.5 and sample sizes, it was observed that heat-transfer effects diminish considerably at sample thicknesses below 50 to 65 mils. To test this observation a cylindrical Teflon sample thickness of 27 mils (run $5T$) was employed in a TGA degradation. In Fig. 2.6 can be seen a plot of Eq. (30) for this sample size. A fairly good linear relationship was obtained, commencing at the relatively low conversion of about 8 percent, which gave a value of $E \approx 77$ kcal mole^{-1}. In this connection, it may be noted that Siegle and coworkers [15] degraded a 60-mil Teflon film isothermally at 480°C and observed that during the early conversions the first-order rate constant was of much smaller magnitude than at later conversions. These workers also observed that when thin films (about 2 mils) of Teflon of low melt index were degraded isothermally at 480°C, the first-order plot assumed a sigmoid-type curve and that the value of the rate constant at higher conversions became similar to that obtained for the degradation of thin Teflon films (about 2 mils) of higher melt index, at 480°C. These effects were attributed to the diffusion barrier which resulted from sample thickness.

From Eq. (29), it can be seen that when heat-transfer effects are important, the ratio ρ_t/T should be proportional to the surface area of the exposed cylindrical sample face. Thus, for runs $1T$ and $3T$ at low conversions, the ratio $\rho_t(3T)/\rho_t(1T)$ should be about 0.5. The observed ratio was found to be about 0.7. Although this agreement is not too good, it should be remembered that Eq. (29) is considered to be only approximate, that the surface area S may not remain constant during Teflon degradation, and that sample dimensions are not exact.

If we extend this work, it would appear that when heat-transfer effects become important during TGA degradations (owing to conglomeration, melting, etc.), and when linear first-order plots are constructed, lower average values of E should obtain than when heat-transfer effects are absent. This may be true of Teflon. In the isothermal degradation of Teflon these heat-transfer effects may be minimal owing to better control of experimental conditions, e.g., isothermal degradation of a sample may be carried out well below the sintering point of the sample, and better temperature equilibrium may be established.

Estimation of Kinetic Parameters (E, n, and Z) during Degradation by TGA

TGA traces generally represent relatively complex decompositions which occur by multistep mechanisms. When two or more volatilization reactions are occurring and values of the kinetic parameters E, n, and Z are such that the regions of weight loss occur at separated temperature ranges, then a stepwise TGA trace (or thermogram) results. In such cases each curve between successive horizontal segments may be separately analyzed. However, more complex cases arise when two or more decomposition reactions are represented by a single TGA trace. In the following section will be treated two types of composite cases [18].

Composite cases represented by a single TGA trace. *Two competitive first-order reactions.* For this case, we may write [cf. Eq. (7)],

$$\frac{dC}{dT} = \frac{k_1}{(RH)}(1 - C) + \frac{k_2}{(RH)}(1 - C) \tag{31}$$

or, when the Arrhenius equation is used,

$$\frac{dC}{dT} = \frac{(1 - C)}{(RH)}\left(Z_1 e^{-E_1/RT} + Z_2 e^{-E_2/RT}\right) \tag{32}$$

where the subscripts for Z and E denote the two competitive reactions occurring. Upon integration, Eq. (32) becomes, employing Eq. (22) $[T \gg T_0]$,

$$1 - C = \exp\left[-\frac{Z_1 E_1 p(u_1)}{R(RH)} - \frac{Z_2 E_2 p(u_2)}{R(RH)}\right] \tag{33}$$

Two independent first-order reactions. In this case a fraction of the material volatilizes by a first-order reaction with Arrhenius parameters, Z_1 and E_1, and the remaining material volatilizes by an independent first-order reaction with kinetic parameters Z_2 and E_2. Then we may write

$$\frac{dC_1}{dT} = \frac{Z_1}{(RH)} e^{-E_1/RT} (1 - C_1) \tag{34}$$

and

$$\frac{dC_2}{dT} = \frac{Z_2}{(RH)} e^{-E_2/RT} (1 - C_2) \tag{35}$$

When Eqs. (34) and (35) are integrated, using Eq. (22), then we obtain respectively

$$1 - C_1 = \frac{W_1}{W_{1,0}} = \exp\left[-\frac{Z_1 E_1 p(u_1)}{R(RH)} \right] \tag{34a}$$

and

$$1 - C_2 = \frac{W_2}{W_{2,0}} = \exp\left[-\frac{Z_2 E_2 p(u_2)}{R(RH)} \right] \tag{35a}$$

where $W_{1,0}$ and $W_{2,0}$ denote the initial weights involving two independent processes, and W_1 and W_2 refer to the weights remaining at time t in respect to the independent processes 1 and 2. Upon the addition of Eqs. (34a) and (35a), it can be seen that

$$1 - C = a \exp\left[-\frac{Z_1 E_1 p(u_1)}{R(RH)} \right] + (1 - a) \exp\left[-\frac{Z_2 E_2 p(u_2)}{R(RH)} \right] \tag{36}$$

where $a \equiv W_{1,0}/W_0$; and $1 - a \equiv W_{2,0}/W_0$ where W_0 = total sample weight.

From Eq. (33) it can be seen that when $E_1 \approx E_2$, then the two competitive first-order reactions appear as a simple first-order reaction regardless of the difference in the values of Z_1 and Z_2. However, for different orders, competitive reactions are resolvable, in theory. On the other hand, the differentiation between two independent first-order reactions with $E_1 \approx$

E_2 depends upon the relative difference in magnitude between Z_1 and Z_2 [cf. Eq. (36)]. In this case the resolution is relatively unaffected by changing the heating rate (RH). However, (RH) can affect the resolution when $E_1 \neq E_2$. In order to observe the effect of variation of (RH) on the two competitive and the two independent first-order reactions, let $a = 0.5$, $E_1 = 30$ kcal mole^{-1}, $E_2 = 60$ kcal mole^{-1}, $Z_1 = 4.458 \times 10^6$ sec^{-1}, and $Z_2 = 10^{15}$ sec^{-1} for values of (RH) of 0.001, 0.01, 0.1, and 1.0°K sec^{-1} [18]. From these data and Eqs. (32) to (36) were constructed integral (Fig. 2.7) and differential (Fig. 2.8) curves for the competitive and independent first-order reactions. It is apparent from Figs. 2.7 and 2.8 that for the competitive and independent first-order reactions, wherein E_1 and E_2 differ greatly, the corresponding thermograms exhibit different variations in shape upon changing (RH). Thus, from Fig. 2.7, as the value of (RH) decreases from 1.0 to 0.001°K sec^{-1}, the two independent first-order reactions become more and more distinguishable from their thermograms. However, for the competitive reactions, the low-activation energy reaction predominates at low temperature and heating rates. At the higher heating rate, the low-activation energy reaction dominates initially, with the high energy reaction modifying considerably the later stages of the thermogram.

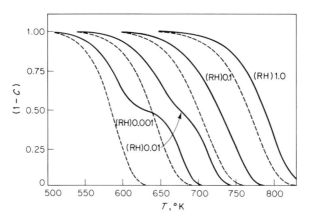

FIG. 2.7 Effect of heating rate on residual fraction versus temperature for composite cases [18]: - - - competitive reactions; —— independent reactions.

Factors affecting the shape of the TGA thermogram. *Thermogravimetric rate versus conversion curves for various reaction orders n.* From Eq. (7) and the Arrhenius expression, we may write

$$\frac{dC}{dT} = Ze^{-E/RT}\frac{f(C)}{(RH)} \tag{37}$$

where $f(C) \equiv (1 - C)^n$. Upon differentiation, Eq. (37) becomes

$$\frac{d^2 C}{dC \, dT} = \frac{E}{RT^2} + \frac{f'(C)}{f(C)} \frac{dC}{dT}$$

(38)

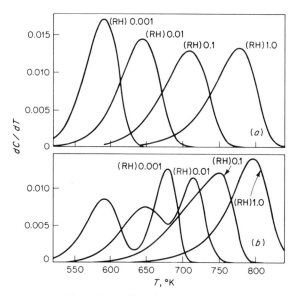

FIG. 2.8 Effect of rate of heating on the thermogravimetric rate versus temperature for composite cases [18]: (a) competitive reactions; (b) independent reactions.

where $f'(C) = -n(1 - C)^{n-1}$. Equation (38) may also be written as

$$\frac{d^2 C}{dC \, dT} = \frac{E}{RT^2} - \frac{n}{1 - C} \frac{dC}{dT}$$

(38a)

From Eq. (38a), the slope of curve arising from the plot of dc/dT versus C for $n = 0$ should be E/RT^2. At ordinary values of E, T is a slowly changing variable over the reaction temperature range so that E/RT^2 should be essentially constant. Furthermore, as $C \to 0$, dC/dT is generally of low magnitude, and so other cases where $n \neq 0$ should approach that for $n = 0$ at low conversions. The preceding is illustrated in Fig. 2.9 for $n = 0, 0.5, 1.0, 2.0, 3.0,$ and 4.0; $Z/(RH) = 10^{16}$ °K^{-1}; $E = 60$ kcal mole^{-1}; and case A involves random degradation for $L = 2$ and $N \gg L$ (see Chap. 4). (Cases B and C need not concern us here.) The effect of reaction order on the shape of primary TGA traces may be seen from Fig. 2.10 for $n = 0, 0.5, 1.0, 2.0, 3.0,$ and 4.0; $Z/(RH) = 10^{16}$°K^{-1}; and, $E = 60$ kcal mole^{-1}. Furthermore,

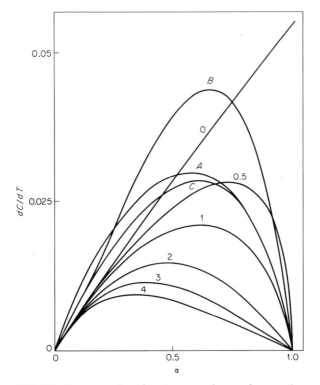

FIG. 2.9 Thermogravimetric rate versus degree of conversion [18]. n = 0, 0.5, 1.0, 2.0, 3.0 and 4.0.

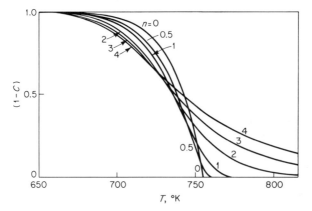

FIG. 2.10 Effect of order on residual fraction versus temperature [18].

when Eq. (38*a*) is equated to zero, we obtain

$$\left(\frac{dC}{dT}\right)_{\max} = \frac{E(1 - C)_{\max}}{nRT^2_{\max}} \tag{39}$$

From Eq. (39), it can be seen that when values of n change much more than corresponding values of $(1 - C)_{\max}$ and $T^2_{\max}$, then the maximum thermogravimetric rate, $(dC/dT)_{\max}$, should decrease as n increases, as may be seen in Fig. 2.9. Similar results obtain when dC/dT is plotted against T for various values of n (cf. Fig. 2.10).

Effects of activation energy E. The effect of changing E upon the character and shape of thermograms may be ascertained by considering three cases: I, II, and III. The following values were assumed: $n = 1$ for all cases; and $E(I) = 60$, $E(II) = 40$, and $E(III) = 80$ kcal mole^{-1}. In order that the effects of E be reasonably comparable in all cases by means of Eq. (39), it is necessary to maintain C and T constant in all cases. This may be attained by varying $Z/(RH)$, which can be understood from the following. From Eq. (37) with $n = 1$,

$$\frac{dC}{dT} = \frac{Ze^{-E/RT}(1 - C)}{(RH)} \tag{37a}$$

When Eq. (37*a*) is integrated using Eq. (22), we may write [cf. Eqs. (34*a*) and (35*a*)],

$$-\ln(1 - C) = \frac{Z}{(RH)} \frac{E}{R} p(u) \tag{37b}$$

From Eq. (37*b*), it can be seen that for an assumed constant value of $C_{\max}$, $p(u)$ can be maintained constant ($T_{\max} = $ const) by selecting appropriate values of $Z/(RH)$. Then, $(dC/dT)_{\max}$ can be calculated from Eq. (37*a*), and such values can be seen in Fig. 2.11 for cases I, II, and III. (Case IV need not concern us.) From this figure it can be seen that $(dC/dT)_{\max}$ decreases as E decreases, in accord with Eq. (39).

Effect of heating rate (RH). The effect of (RH) upon the shape and character of the primary TGA trace may be seen by means of Eq. (7). Thus, when C is maintained constant $[\int dC/(1 - C)^n = $ const$]$ then, at this particular value, for runs 1 and 2,

$$\frac{p(u_1)}{(RH)_1} = \frac{p(u_2)}{(RH)_2} = \cdots \tag{37c}$$

Upon taking the logarithm of both sides of Eq. (37*c*) and employing

Eq. (23),

$$\log \frac{(RH)_2}{(RH)_1} = \frac{0.457E}{R}\left(\frac{1}{T_1} - \frac{1}{T_2}\right) \tag{40}$$

It can be seen from Eq. (40) that when several TGA runs are carried out for a substance and several heating rates are employed, the TGA trace will

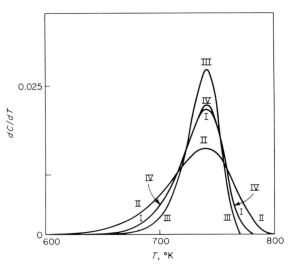

FIG. 2.11 Effects of activation energy on thermogravimetric rate versus temperature [18].

shift toward higher temperatures as (RH) is increased. This effect of (RH) may be observed in Fig. 2.12.

Specific methods for determining kinetic parameters from TGA traces. There are numerous methods available for determining kinetic parameters from TGA traces [4, 18]. However, owing to limitations imposed by the scope of this book, we have been necessarily selective. Therefore, only a relatively few such methods are described in the following.

Method of Freeman and Carroll [19]. The difference-differential method of Freeman and Carroll is one of the most widely used for the kinetic analysis of TGA data. From Eq. (7a) and the Arrhenius expression,

$$-\frac{dW}{dT} \equiv R_T = \frac{Z}{(RH)} e^{-E/RT} W^n \tag{41}$$

The difference form of Eq. (41) is, for a single trace,

$$\Delta \log R_T = n\Delta \log W - \left(\frac{E}{2.3R}\right)\Delta\left(\frac{1}{T}\right) \qquad (42)$$

From Eq. (42), when $\Delta(1/T)$ is maintained constant, then a plot of $\Delta \log R_T$ versus $\Delta \log W$ should afford a linear relationship whose slope yields the value of n and its intercept the value of E. Once n and E have been determined, the value of Z may be calculated from Eq. (41). Another

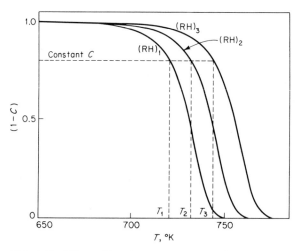

FIG. 2.12 Effect of heating rate on residual fraction versus temperature [18]: $(RH)_3 > (RH)_2 > (RH)_1$.

form which Eq. (42) may take is

$$\frac{\Delta \log R_T}{\Delta \log W} = n - \frac{E}{2.3R}\left[\frac{\Delta(1/T)}{\Delta \log W}\right] \qquad (42a)$$

From Eq. (42a), a plot of the left-hand side of Eq. (42a) versus $\Delta(1/T)/\Delta \log W$ should provide a linear relationship whose slope will afford the value of E and whose intercept the value for n. An example of the use of Eq. (42) follows.

Doyle [20] presented data for the decomposition of Teflon in nitrogen by TGA at $(RH) = 3°C$ min^{-1}. He analyzed these data by a rather tedious method, assuming $n = 1$, and obtained a value of $E \approx 67$ kcal mole^{-1}. From the data presented, a smooth primary TGA thermogram was constructed. From slopes of this thermogram at various temperatures, plots of R_T and of weight fraction remaining W versus reciprocal temperature $1/T$ were constructed, as shown in Fig. 2.13. Then, an arbitary constant value of

$\Delta(1/T) = 0.02 \times 10^{-3} \, °K^{-1}$ was chosen. Values of R_T and W which corresponded to the temperature at the ends of the $\Delta(1/T)$ intervals were then obtained from Fig. 2.13. Then, corresponding values of $\Delta \log R_T$ and $\Delta \log W$ were calculated, and a plot was made of $\Delta \log R_T$ versus $\Delta \log W$, as depicted in Fig. 2.14, in accordance with Eq. (42). From this plot, values of $n = 1.0$ and $E = 65$ kcal mole^{-1} were obtained which agree well with values reported by Doyle [20].

A major advantage of the Freeman and Carroll method is that it allows the simultaneous determination of E and n. However, in using this method, it is necessary to measure steep slopes accurately. Moreover, much scatter is generally encountered at low conversions in plots of $\Delta \log R_T$ versus $\Delta \log W$ (cf. Fig. 2.14). This is probably due to the magnification of experimental scatter by the use of the difference of a derivative.

Methods involving multiple heating rates. It can be observed from Eq. (40) that when multiple heating rates are used and a constant value of conversion C is employed (cf. Fig. 2.12), then

$$\frac{-d \log(RH)}{d(1/T)} = \frac{0.457E}{R} \tag{40a}$$

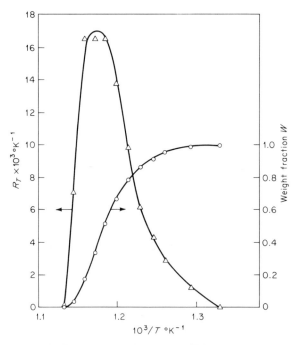

FIG. 2.13 Plots of R_T and W versus $1/T$ for the decomposition of Teflon in nitrogen at $(RH) = 3°C$ min^{-1}.

Thus, a plot of log (RH) versus $1/T$ (at a constant conversion and for several heating rates) should yield a linear relationship whose slope affords the value of E. A similar method which involves another approximate integral solution [cf. Eq. (20)] is as follows. From Eqs. (20) and (41),

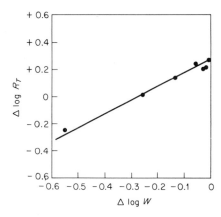

FIG. 2.14 Plot of $\Delta \log R_T$ versus $\Delta \log W$ for Teflon from data obtained in Fig. 2.13.

maintaining α constant $[-\int dW/W^n = \text{const}]$, it can be shown that $(1 \gg 2RT/E)$

$$\frac{ZRT_1^2}{(RH)_1 E} e^{-E/RT_1} =$$

$$\frac{ZRT_2^2}{(RH)_2 E} e^{-E/RT_2} = \cdots = \text{const} \tag{43}$$

or

$$\log \frac{(RH)}{T^2} = -\frac{E}{2.3RT} + \text{const} \tag{43a}$$

Thus, from Eq. (43a), a plot of log $[(RH)/T^2]$ versus $1/T$ should afford a linear relationship whose slope yields the value of E [cf. Eq. (40a)]. Advantages of using Eqs. (40a) and/or (43a) are: no prior knowledge of n is required, there is no need to measure steep slopes, and changes in degradation mechanisms may be detected when values of E become different at different conversions. However, some disadvantages are: n is indeterminate, several TGA thermograms are required, and the method is sensitive to errors in temperature. It may be interesting to note here that Eq. (43a) may be rewritten as

$$\frac{-d \log (RH)}{d(1/T)} = \frac{E}{2.3R} + \frac{2T}{2.3}$$ (43b)

From a comparison between Eqs. (40a) and (43b), both equations become almost identical when $(E/R) \times 0.06 \cong 2T$, e.g., when $T = 800°K$ and $E = 50$ kcal mole^{-1}. In the following will be described an application of Eq. (43a).

Ozawa [21] has studied the degradation of nylon 6 under high vacuum by means of TGA, using four different heating rates (four thermograms). From the thermograms the data in Table 1 were extracted. From these data plots of $-\log [(RH)/T^2]$ versus $1/T$ could be made for the various values of W listed in the table and above the appropriate straight lines in Fig. 2.15. From the slopes of these lines, $E = 29.7 \pm 1.5$ kcal mole^{-1}.

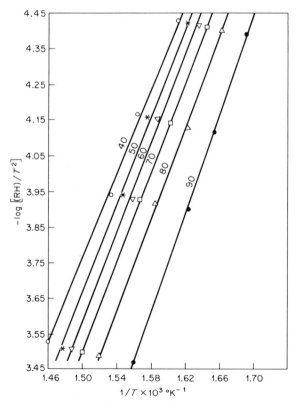

FIG. 2.15 Plot of $-\log [(RH)/T^2]$ versus $1/T$ for nylon 6 at various values of W.

Another method involving multiple heating rates from which values of E, n, and Z may be estimated is as follows. From Eq. (41),

$$\log R_t = \log Z - \frac{E}{2.3RT} + n \log W \qquad (44)$$

where $R_t = (RH)R_T$. When W is maintained constant, then a plot of $\log R_t$ versus $1/T$ should give linear relationships whose slopes yield the value of

TABLE 1   TGA Data for Decomposition of Nylon 6

W, %	Temperature T, $^\circ$K for (RH), $^\circ$C hr^{-1}			
	T for 140	T for 48	T for 28	T for 14.3
90	641	616	605	592
80	658	631	617	602
70	667	638	624	608
60	672	642	630	612
50	678	647	635	617
40	685	652	640	621

E. When the values of the intercepts of such plots for various values of W are plotted against $\log W$ [cf. Eq. (44)], then a linear relationship should obtain whose slope is equal to n and whose intercept to $\log Z$. A major disadvantage of this method is the necessity to determine slope accurately from the primary thermogram.

Methods involving the use of master curves. Ozawa [21] utilized master curves to estimate the values of n and Z during the degradation of nylon 6, under high vacuum, by means of TGA. The value of E was estimated by the use of Eq. (40a) employing several thermograms with different values (RH). Once the value of E had been determined (about 30 kcal mole^{-1}), the values of n and Z could be estimated as follows. For a random degradation, we may write (cf. Special Case of Random Degradation, Chap. 4)

$$\frac{d\alpha}{dt} = k_r(1 - \alpha) \qquad (45)$$

where α = fraction of the chain linkages which were broken by the random degradation. Upon integration and the use of the Arrhenius relationship, Eq. (45) becomes

$$-\ln(1 - \alpha) = \frac{ZE}{(RH)R} p(u) \qquad (46)$$

or

$$\log - [\ln(1 - \alpha)] = \log \frac{ZE}{(RH)R} - 2.315 - 0.457u \qquad (46a)$$

where $u \equiv E/RT$ [cf. Eq. (23)]. From Eq. (46a), it can be seen that for various values of the right-hand side, values of α can be estimated. Furthermore, for a random degradation, α is related to conversion C by the expression

$$1 - C = (1 - \alpha)^{L-1}\left[1 + \frac{\alpha(N - L)(L - 1)}{N}\right] \qquad (47)$$

where N = initial degree of polymerization, and L = least number of repeating polymer units not volatilized. (Cf. Special Case of Random Degradation, Chap. 4.)

When $N \gg L$ (often the case), Eq. (47) becomes

$$1 - C = (1 - \alpha)^{L-1}[1 + \alpha(L - 1)] \qquad (47a)$$

Thus, when values of α and L are known or assumed, then values of $1 - C$ (or W) may be estimated. In this manner, master plots of W versus log $[ZEp(u)/(RH)R]$ were constructed, as shown in Fig. 2.16, for assumed values of $L = 2$, 3, and 4. When W versus log $[Ep(u)/(RH)R]$ was plotted for nylon 6 [$(RH) = 28°C$ hr^{-1} and $E = 30$ kcal mole^{-1}; cf. Table 1] as shown in Fig. 2.16, it could be ascertained that the resulting curve resembled the $L = 2$ curve most closely. From a lateral shift of the resulting curve to the $L = 2$ curve, the value of log $Z \approx 10^9$ min^{-1} could be calculated. From Fig. 2.16, it can be seen that unless extreme precautions are taken, it is difficult to distinguish $L = 2$ from $L = 3$ from $L = 4$, etc. In this connection, Papkov and Slonimskii [22] have criticized Ozawa's method. Thus, they have indicated that the master curves for $L > 3$ practically coincide, and that the same TGA curve can be described by various values of L and Z.

Master curves according to the method of Ozawa may also be plotted for various values of reaction order n, as shown in Fig. 2.17. As indicated previously, once E has been determined, then a plot of W versus log $[Ep(u)/(RH)R]$ is constructed for the substance in question, and the reaction order is estimated from a comparison of this curve with those in Fig. 2.17. Then, a lateral shift to the curve of appropriate reaction order n allows for the estimation of log Z.

Reich [23] has presented another master-curve method. When Eq. (41) is integrated using the approximate integral solution given in Eq. (20),

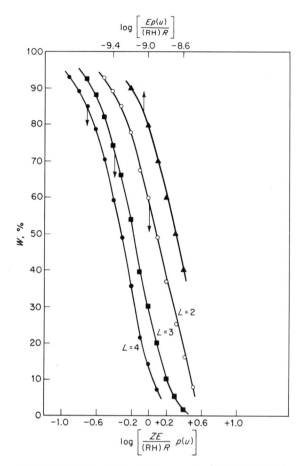

FIG. 2.16 Master plots of W versus log $[ZEp(u)/(RH)R]$ for random degradation (L = 2, 3, and 4): $\triangle$ = nylon 6; (RH) = 28°C hr⁻¹ ; E = 30 kcal mole⁻¹ .

$2RT/E \ll 1$, we obtain

$$-\int_{W_0}^{W} \frac{dW}{W^n} = \frac{Z}{(RH)} \frac{RT^2}{E} e^{-E/RT} \tag{48}$$

When Eqs. (41) and (48) are combined, eliminating the $Z/(RH)$ term, and generalizing by allowing for an inactive residue, there are obtained

$$\frac{E}{R} = \frac{S}{W_c \ln(W_{0,c}/W_c)} \qquad \text{for } n = 1 \qquad (49a)$$

$$\frac{E}{R} = \frac{S(1-n)}{W_c^{\,n}(W_{0,c}^{\,1-n} - W_c^{\,1-n})} \qquad \text{for } n \neq 1 \qquad (49b)$$

where

$$S = \frac{dW_c}{d(1/T)}$$

$$W_{0,c} = W_0 - W_R$$

$$W_c = W - W_R$$

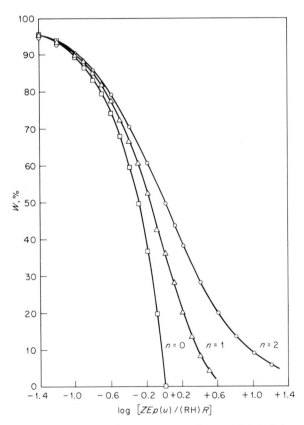

FIG. 2.17 Master plots of W **versus log** $[ZEp(u)/(RH)R]$ **for various reaction orders;** $\square \, n = 0$; $\triangle \, n = 1$; $\bigcirc \, n = 2$.

where W is weight fraction of material remaining at time t, and W_R is weight fraction of inactive material remaining after a pyrolysis. From Eqs. (49a) and (49b) the following expressions may be obtained respectively:

$$\frac{S_1}{S_2} = \frac{W_{1,c}}{W_{2,c}} \frac{\log(W_{0,c}/W_{1,c})}{\log(W_{0,c}/W_{2,c})} \qquad \text{for } n = 1 \qquad (49c)$$

and

$$\frac{S_1}{S_2} = \left(\frac{W_{1,c}}{W_{2,c}}\right)^n \left[\frac{1 - (W_{1,c}/W_{0,c})^{1-n}}{1 - (W_{2,c}/W_{0,c})^{1-n}}\right] \qquad \text{for } n \neq 1 \qquad (49d)$$

Before determining the overall activation energy of a pyrolysis, it is necessary to estimate n. From Eqs. (49c) and (49d) it can be seen that for a particular, arbitrarily selected ratio of $W_{1,c}/W_{2,c}$ and value of W_R, the corresponding ratio S_1/S_2 may be calculated for various values of n. Such calculated values were used to construct the theoretical curves shown in Fig. 2.18. In this figure, various values of W_1/W_2 or $(W_{1,c}/W_{2,c})$ and of W_R are given, from which the ratio S_1/S_2 could be calculated. The curves in Fig. 2.18 can be used to estimate n from experimental TGA curves. After the value of n has been determined, it may be substituted into Eqs. (49a) and (49b) to obtain the value of E. After E and n have been estimated, the value of Z may then be obtained by means of an expression such as Eq. (41). When this method was applied to TGA curves for Teflon and polyethylene, the kinetic parameters obtained were in satisfactory agreement with values given in the literature.

It may be of interest to note here that Eqs. (49c) and (49d) may be employed to obtain a simple expression involving n. Thus, at values of W equal to 0.8 and 0.1, values of S_1/S_2 (designated as $S_{0.8}/S_{0.1}$) were calculated for various values of n, with $W_R = 0$. From these values the following approximate expression may be written for values of n from about 1/4 to 2:

$$\frac{S_{0.8}}{S_{0.1}} \leftrightarrow D\left(\frac{T_{0.8}}{T_{0.1}}\right)^2 \approx 0.8n \qquad (49e)$$

where, $D \equiv (dW/dT)_{0.8}/(dW/dT)_{0.1}$. Since the value of $T_{0.8}/T_{0.1}$ is generally a little less than unity, Eq. (49e) may be written as

$$D \approx n \qquad (49f)$$

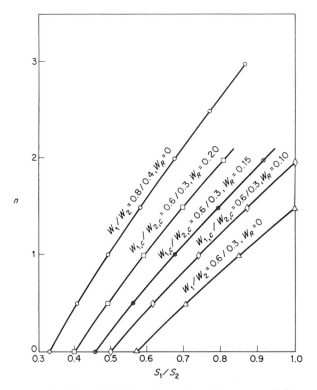

FIG. 2.18 Theoretical plots of reaction order, n versus S_1/S_2, for various values of W_1/W_2 or $W_{1,c}/W_{2,c}$ and W_R [23].

Equation (49f) may be used when it is desired to obtain a rapid, approximate estimation of n from a primary thermogram which possesses a value of $W_R = 0$. From experimental thermograms obtained in the author's (LR) laboratory for Teflon degradation in vacuum at different heating rates, values of D were determined which were consistently near unity, as anticipated. An illustration of the use of the preceding method is as follows. From Fig. 2.13 for Teflon, $W_R = 0$; for $W = 0.6$: $R_T = 16 \times 10^{-3}$ and $1/T = 1.19 \times 10^{-3}$; for $W = 0.3$: $R_T = 17 \times 10^{-3}$ and $1/T = 1.17 \times 10^{-3}$. Thus, $S_{0.6}/S_{0.3} = 0.91$, which affords a value of $n = 1.2$ from Fig. 2.18. Again, for $W = 0.8$: $R_T = 8.0 \times 10^{-3}$ and $1/T = 1.22 \times 10^{-3}$; for $W = 0.4$: $R_T = 17 \times 10^{-3}$ and $1/T = 1.18 \times 10^{-3}$. Thus, $S_{0.8}/S_{0.14} = 0.44$, which affords a value of $n = 0.8$ from Fig. 2.18. The average value of $n = 1.0$ as anticipated. We may now employ Eq. (49a) to calculate E. From the various values corresponding to W in the preceding, values of S may be calculated and, consequently, the value of E. In this manner, $E = 67 \pm 3$ kcal mole^{-1} for values of $W = 0.3, 0.4, 0.6$, and 0.8.

Some other methods. Under this heading will be described some other methods which are less quantitative than those described previously. Thus,

Reich and Levi [4] used an approximate and relatively rapid method for determining the value of E. This method involves the measurement of areas as depicted in Fig. 2.19. Two arbitrary temperatures are selected on a TGA trace, and E is evaluated from the equation

$$E = \frac{\log(A_{I+II}/A_I)\, 4.6 T_1 T_2}{T_2 - T_1} \tag{50}$$

where A = shaded area in Fig. 2.19 with subscripts indicating appropriate areas. The reaction order was arbitrarily assumed as zero in order to calculate E from Eq. (50). This is justified from the fact that for many polymeric degradations it is often difficult to distinguish clearly between various orders during early stages of conversion (cf. Fig. 2.17). Then, expression (50) may be derived as follows. From Eq. (41), we may write ($n = 0$)

$$\int_1^W -\frac{dW}{T^2} = \left[\frac{Z}{(RH)}\right] \int_0^T \frac{e^{-E/RT}}{T^2}\, dT \tag{51}$$

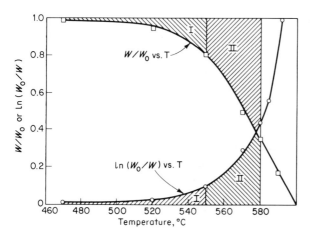

FIG. 2.19   Zero- and first-order rate plots [4].

or

$$\left(\frac{1}{T_{avg}}\right)^2 (1 - W) = \frac{ZR}{(RH)E}\, e^{-E/RT} \tag{52}$$

where

$$\left(\frac{1}{T_{avg}}\right)^2 = \frac{\int dW/T^2}{\int dW} \approx \text{const}$$

Equation (52) may further be converted into the expression

$$\left(\frac{1}{T_{avg}}\right)^2 (T'_{avg})^2 \int_0^T (1 - W)\, dT = \frac{Z}{(RH)}\left(\frac{R}{E}\right)^2 e^{-E/RT} \qquad (53)$$

where

$$\left(\frac{1}{T'_{avg}}\right)^2 = \frac{\int \dfrac{1 - W}{T^2}\, dT}{\int (1 - W)\, dT} \approx \text{const}$$

In Eq. (53), the integral represents the area of the upper portion of the TGA trace bounded by the zero-reaction horizontal line and extending from temperature T to the temperature at which the trace and the zero-reaction horizontal line meet (cf. Fig. 2.19). If this area is denoted by A, Eq. (53) may be transformed, for two different temperatures, T_1 and T_2, into Eq. (50). Similar considerations may be applied to first-order kinetics. In this respect, the method yielded values of E for Teflon (using data of Fig. 2.13) of 68 and 69 kcal mole^{-1} when zero- and first-order expressions were used respectively.

Another method which is limited to low conversions is one presented by Coats and Redfern [17]. When Eq. (37) is integrated using the approximation of Eq. (20), $2RT/E \ll 1$,

$$\int_0^C \frac{dC}{(1 - C)^n} = \frac{ZRT^2}{(RH)E} e^{-E/RT} \qquad (54)$$

when $C \ll 1$, $(1 - C)^n \approx 1 - nC$; thus, for $n = 0$ or low values of C, $(1 - C)^n \to 1$. As can be seen, this approximation to unity becomes less reliable as C and/or n values become larger. Applying the approximation to unity, Eq. (54) becomes

$$\log \frac{C}{T^2} = -\frac{E}{2.3RT} + \log \frac{ZR}{(RH)E} \tag{55}$$

Equation (55) was applied to the decomposition of Teflon and octamethyl-cyclotetrasiloxane (OMS), using data of Doyle [20]. Values of E of 61 and 11.5 kcal mole^{-1} were obtained, respectively. In the latter case, the agreement between calculated and reported values was excellent (reported, 11.6 kcal mole^{-1}), as might be expected since $n \approx 0$ for the volatilization of OMS by means of TGA.

DTA

Differential thermal analysis (DTA) is a thermal technique in which heat effects associated with physical changes (e.g., phase transitions such as fusion) or chemical changes (e.g., decomposition reactions) are recorded as a function of temperature (or time) as a sample is heated at a uniform

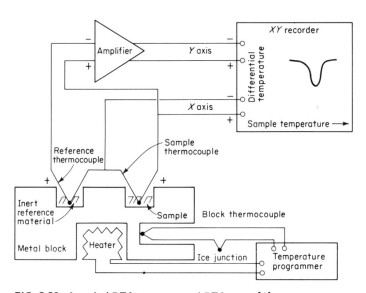

FIG. 2.20   A typical DTA apparatus and DTA curve [1].

rate. In DTA, the heat effects are measured by a differential method; i.e., the sample temperature is continuously compared with a reference-material temperature, and the difference in temperature ΔT is recorded with respect to sample temperature (or time) T. In Fig. 2.20 is shown a typical DTA apparatus and a typical XY recording of a DTA curve. The inert reference material used is often α-alumina. Further, the various factors which affect the accuracy of TGA results may also have an important influence upon DTA results (cf. TGA, page 83). For more details on DTA apparatus and the effect of various factors upon the accuracy of DTA results, the reader is referred to recent monographs on thermal analysis [2, 3].

Theoretical Aspects

In the following will be described a derivation of expressions which are of considerable value in the interpretation of DTA data [24]. Although these expressions were originally applied to stirred reaction systems and are, in theory, more applicable to liquids than solids, they have found increasing usage in the latter case [5]. In Fig. 2.21 is shown a simple DTA apparatus (with a differential thermocouple, DTC). It consists of two cells mounted in a bath of temperature T_3. One cell contains a solution of reactants at temperature T_1, and the other pure solvent or other inert liquid with temperature T_2. The bath temperature is uniformly increased by a heater assembly. We may first write a heat balance for the cell containing reactants as follows:

$$C_r dT_1 = dH + K_r(T_3 - T_1) dt \qquad (56)$$

where C_r = total heat capacity of the reactant solution, K_r = heat-transfer

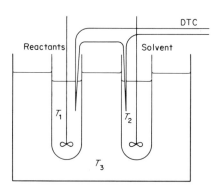

FIG. 2.21 DTA apparatus for obtaining kinetic data for reactions occurring in solution [24].

coefficient of the reactant cell, dH = heat evolved by the reaction, and dt = the time interval. Similarly, for the reference or solvent cell,

$$C_s dT_2 = K_s(T_3 - T_2) dt \qquad (57)$$

where the subscript s refers to the solvent cell. If identical cells are used, $K_r = K_s \equiv K$; also, assume that $C_r = C_s \equiv C_P$. Then, when Eq. (57) is subtracted from (56),

$$dH = C_p d(\Delta T) + K \Delta T dt \qquad (58)$$

where $\Delta T \equiv T_2 - T_1$. Upon integration of Eq. (58) between 0 and ∞,

$$\Delta H = C_P(\Delta T_\infty - T_0) + K \int_0^\infty \Delta T \, dt = KA_t \tag{59}$$

where $A_t = \int_0^\infty \Delta T \, dt$; since ΔT at infinite time (ΔT_∞), and ΔT initially (ΔT_0), are both equal to zero. Although Eq. (59) has been derived for a liquid medium, a similar expression may be derived in a similar manner for a solid system [25]. Note that a prerequisite for obtaining Eq. (59) is that the thermal characteristics of the sample and reference cells be as similar as possible. This equation states that the area under the DTA curve is directly proportional to the heat transfer due to reaction. Now, assume that $dH = f(dW_c)$, where dW_c = weight, or weight fraction of material undergoing reaction, and hence we may write

$$dH = -\frac{KA_t}{W_{0,c}} dW_c \tag{60}$$

where $W_{0,c}$ = initial weight or weight fraction of active material, and the minus sign denotes evolution of heat. Upon substituting Eq. (58) into (60) and upon rearranging,

$$-\frac{dW_c}{dt} = \frac{W_{0,c}}{KA_t}\left[C_p \frac{d(\Delta T)}{dt} + K\Delta T\right] \tag{61}$$

Upon integration of Eq. (61) between t and ∞,

$$W_c = \frac{W_{0,c}}{KA_t}(K\tilde{a}_t - C_p \Delta T) \tag{62}$$

where $\tilde{a}_t = \int_t^\infty \Delta T \, dt$. If it is assumed that the K terms $\gg C_p$ terms in Eqs. (61) and (62), and T is substituted for t,

$$-\frac{dW_c}{dT} \approx \frac{W_{0,c}}{A_T} \Delta T \tag{63}$$

and

$$W_c \approx \frac{W_{0,c}}{A_T} \tilde{a}_T \tag{64}$$

where $A_T = \int_{T_0}^{T}\Delta T \, dT$ and $\tilde{a}_T = \int_T^{T_0} \Delta T \, dT$. Equations (63) and (64) are of prime importance and will be employed subsequently. A very rigorous mathematical treatment of DTA has been carried out by Akita and Kase [26], but after using many simplifying assumptions, expressions were obtained which resembled Eqs. (63) and (64), after some mathematical manipulation. Much evidence has been presented in support of Eqs. (63) and (64) [5]. Thus, for instance, Eq. (63) predicts that the maximum reaction rate should occur approximately at the DTA peak. Recently, simultaneous measurements of DTA and derivative thermogravimetric analysis (DTG) (the derivative of the primary TGA trace) were carried out for the dehydration of calcium oxalate monohydrate [27]. From Fig. 2.22 can be seen the strong relation between the DTG and DTA peak shapes.

Besides assuming the validity of Eqs. (63) and (64) to bulk-phase reactions, it will be further assumed that the Arrhenius equation, $k = Ze^{-E/RT}$, is valid. By utilizing Eqs. (41), (63), and (64) it is possible to convert TGA expressions into DTA expressions. This interconversion will be carried out under Methods for the Estimation of Kinetic Parameters, page 119.

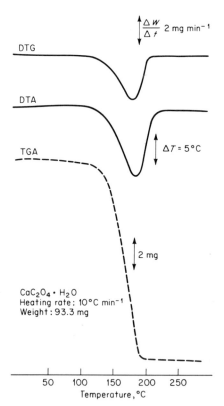

FIG. 2.22 Simultaneous TGA/ DTG/DTA of the dehydration of calcium oxalate monohydrate, showing the strong relation between the DTG and DTA peak shapes (from [27]).

Examples will be given of kinetic parameters calculated by some of the DTA expressions listed for polymer decompositions. These values will be compared with those obtained by TGA. In determining kinetic parameters in thermal pyrolysis by DTA techniques, it must be ascertained that over the apparent decomposition peak no other changes are occurring which are endothermic or exothermic. Furthermore, in order to maintain the thermal characteristics of the reference and sample cells as identical as possible, the sample is often considerably diluted by admixture with inert reference material.

Methods for the Estimation of Kinetic Parameters

1. When Eqs. (63) and (64) are applied to Eq. (42) (Freeman and Carroll expression), we may write

$$\Delta \log(\Delta T) = n \Delta \log \tilde{a} - \frac{E}{2.3R} \Delta \left(\frac{1}{T}\right) \tag{65}$$

(Henceforth, $\tilde{a}_T \equiv \tilde{a}$ and $A_T \equiv A$.)

From Eq. (65), it can be seen that when $\Delta(1/T)$ is held constant, a plot of $\Delta \log (\Delta T)$ versus $\Delta \log \tilde{a}_T$ should provide a linear relationship whose slope will afford the value of n and whose intercept will give the value of E. Such a plot is depicted in Fig. 2.23 for poly(methyl methacrylate) (PMMA), polyethylene (PE), poly(propylene oxide) (PPrO), and Teflon [28]. These polymers afforded the following respective values of n and E: 0.75, 46; 0.82, 76; 0.8, 49; and 0.73, 64. These values are in reasonably good agreement with reported values based on TGA. However, a major disadvantage of Eq. (65) (cf. TGA, page 104) is that plots of this equation often lead to considerable scatter in the derived data.

2. In estimating kinetic parameters from TGA traces, it was previously mentioned that Eqs. (49a) to (49d) were applicable. By inserting Eqs. (63) and (64) into these expressions, the following equations may be obtained:

$$F(T) = \frac{\tilde{a}_1}{\tilde{a}_2} \frac{\log(A/\tilde{a})_1}{\log(A/\tilde{a})_2} \qquad \text{for } n = 1 \tag{66}$$

$$F(T) = \left(\frac{\tilde{a}_1}{\tilde{a}_2}\right)^n \frac{1 - (\tilde{a}/A)_1^{(1-n)}}{1 - (\tilde{a}/A)_2^{(1-n)}} \qquad \text{for } n \neq 1 \tag{67}$$

where $F(T) \equiv (T_1/T_2)^2 (\Delta T_1/\Delta T_2)$.

$$\frac{E}{R} = \frac{T^2 \Delta T}{\tilde{a} \ln(A/\tilde{a})} \qquad \text{for } n = 1 \tag{68}$$

$$\frac{E}{R} = \frac{[(1 - n)T^2 \, \Delta T]/A}{(\tilde{a}/A)^n [1 - (\tilde{a}/A)^{(1 - n)}]} \qquad \text{for } n \neq 1 \qquad (69)$$

Before obtaining E from Eqs. (68) and (69), it is necessary to estimate n. From Eqs. (66) and (67), it can be seen that $F(T)$ may be plotted against n for various value of $\tilde{a}_1/A$ and $\tilde{a}_2/A$. Such plots have been constructed (Fig. 2.24) with various values of $(\tilde{a}_1/A)/(\tilde{a}_2/A)$ listed alongside each curve [29]. Values of the sample temperature T and the height of the thermogram ΔT were obtained where $\tilde{a}/A$ possessed the desired value on the thermogram. In obtaining values of areas, a compensating polar planimeter may be utilized. After a few rapid trial-and-error tracings of the thermogram with the planimeter, the sample temperature and peak height corresponding to the ratio $\tilde{a}/A$ can be readily obtained. For the values of ΔT and T, $F(T)$ can be calculated, and from the curves in Fig. 2.24, values of n can be estimated for various DTA thermograms. After the value of n has been determined, a value of E can be calculated for each ratio $\tilde{a}/A$, by means of Eq. (68) or (69). Finally, values of Z may be estimated from

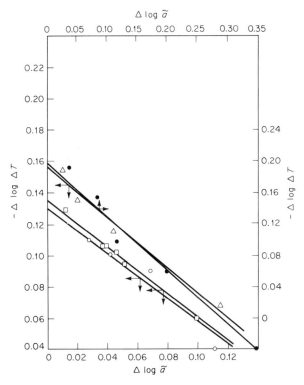

FIG. 2.23 Plots for determination of kinetic parameters from Eq. (65): O PMMA; △ PE; □ PPrO; (●) Teflon [28].

Eq. (41) after it has been converted into its respective DTA form [29]. This method was applied to various polymers. In the following are listed the polymers, corresponding values of n, E in kcal mole^{-1}, and Z in min^{-1} (reported values are listed in parentheses): polypropylene, 0.89 (1), 58 (55, 58), 5.9 × 10^{17} (2.2 × 10^{17}); polyethylene, 0.88 (1), 73 (67, 72), −; polystyrene, 0.97 (1), 78 (60, 74, 77), −. Calculated and reported values of

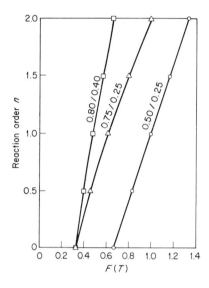

FIG. 2.24 Reaction order n as a function of $F(T)$ for various values of the ratio $(\tilde{a}_{T,1}/A_T)/(\tilde{a}_{T,2}/A_T)$ [29].

kinetic parameters appear to be in good agreement. Although this method appears to be very reliable, some disadvantages are that it involves trial-and-error procedures for estimating $F(T)$ (or n) (however, with a little practice, not much time and effort are required), and that overall values of kinetic parameters are obtained; i.e., it must be ascertained by other procedures whether changes in mechanism are occurring during polymer degradation.

3. When Eq. (64) is inserted into Eq. (55), there is obtained

$$\log \frac{a}{At^2} = -\frac{E}{2.3RT} + \log \frac{ZR}{(RH)E} \tag{70}$$

where $a \equiv \int_{T_0}^{T} \Delta T \, dT$. Equation (70) has been applied to various polymers [28]. Based upon calculated and reported values of E, it appears that the use of Eq. (70) tends to lead to lower values of E than reported. Nevertheless, the agreement between calculated and observed values of E appears to be reasonably good. Major disadvantages of this method are that it is strictly valid only at low conversions (unless $n = 0$), it does not provide for values of n, and it often leads to considerable scatter in the derived data.

Various other TGA expressions have been converted into corresponding DTA expressions by Eqs. (63) and (64), and the latter have been applied to polymer degradations [5, 28]. However, these are not described here as they are not within the scope of this book.

OTHER APPLICATIONS OF TGA AND DTA TO POLYMER DEGRADATIONS

Polyethylene

Anderson and Freeman [30] found that during the thermal decomposition of high-pressure polyethylene (DYNH) in vacuum by means of TGA there were three decomposition stages (cf. Fig. 2.25). In the first stage, up to 3 percent conversion, zero-order kinetics obtained with a value of $E = 48$ kcal mole^{-1}. These and subsequent values were obtained from a plot of Eq. (42) [$\Delta(1/T)$ being held constant]. During the second stage, 3 to 15 percent conversion, $n = 0$ and $E = 6$ kcal mole^{-1}. From 15 to 35 percent reaction, there appeared to be a transition from $n = 0$ to $n = 1$, while above 35 percent conversion, $n = 1$ and $E = 67$ kcal mole^{-1}. The initial stage was attributed to the degradation of short-lengthed branched chains which underwent bond rupture initially. The second stage was ascribed to end-chain cleavage, and the final stage to random rupture of carbon-carbon bonds, which requires an $E \approx 66$ kcal mole^{-1}.

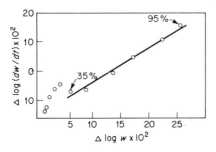

FIG. 2.25 Kinetics of the thermal degradation of polyethylene in vacuum [30].

Igarashi and Kambe [31] studied the degradation of polyethylene in air and nitrogen by means of DTA and TGA techniques. The polyethylenes used comprised two high-pressure samples (Sumikathene MF-40 and F-70-6) and two low-pressure samples (Hizex 5000 and 7000). In Fig. 2.26 are shown DTA curves for Hizex 5000 in air and in nitrogen. The initial large peaks denote melting, whereas the final peaks represent thermal decomposition. The upper curve also exhibits two small exothermic peaks (arrows), which are presumably due to oxidation since these peaks are absent in a nitrogen atmosphere. Thus, TGA techniques should reveal three stages during low-pressure polyethylene oxidation if each such stage is accompanied by weight loss. In Fig. 2.27 are shown derivative thermogravimetric curves (DTG) for the oxidation of H-7000 and F-70-6. As anticipated, there exist three stages of degradation: the first, up to 350°C; the second, from 350 to 390°C; and the third, from 390 to 455°C. Furthermore, from Fig. 2.26, there should only be one stage for the degradation of low-pressure polyethylene in nitrogen. This was verified by means of DTG

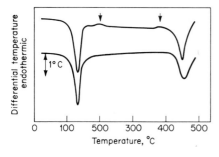

FIG. 2.26 Differential thermal analysis curves for low-pressure polyethylene Hizex 5000. Upper curve, in air; lower curve, in a nitrogen atmosphere [31].

curves. (Note that these results contrast with those obtained by Anderson and Freeman for high-pressure polyethylene, wherein three degradation stages were observed from TGA results.)

Polypropylene

Kaesche-Krischer [32] studied the decomposition of isotactic polypropylene (PPr) by means of TGA and DTA techniques. In Fig. 2.28 are shown DTA traces for the decomposition of PPr in vacuum and in presence of oxygen at various pressures. For the former, a strong endothermic peak appears at about 178°C. However, as the oxygen pressure is increased, this peak lessens and disappears at oxygen pressures above 200 torrs. But at oxygen pressures of 300 and 500 torrs, a sharp exothermic peak appears at 155°C, followed by a second broader exothermic peak at about 250°C. Based upon TGA traces (cf. Fig. 2.29), a strong endothermic peak should appear for PPr decomposition in vacuum between 400 and 455°C. The absence of such a peak in the DTA thermogram was ascribed to instrumental difficulties.

In Fig. 2.29 are shown TGA traces for PPr decomposition in vacuum and in presence of oxygen at various pressures [32]. As previously indicating, the PPr underwent extensive decomposition between 400 and 455°C. When oxygen was present, the PPr gained weight between 150 and 180°C and then decomposed rapidly, depending upon the oxygen pressure. As the latter increased, the decomposition rate increased, and the presence of hydroxyl- and carbonyl-containing compounds could be detected by infrared spectroscopy.

The transition from an endothermic to an exothermic peak at 150°C as the oxygen pressure was increased was mainly attributed to the following exothermic reaction:

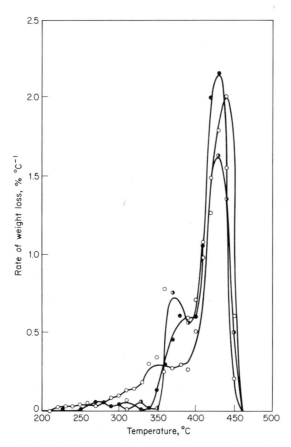

FIG. 2.27 Thermogravimetric analysis curves for polyethylenes in air: ● Hizex 7000, ◑ Sumikathene F-70-6 (unpurified), ○ Sumikathene F-70-6 (purified) [31].

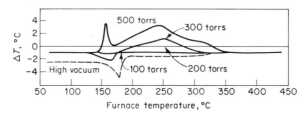

FIG. 2.28 DTA traces for PPr in vacuum and in presence of oxygen [32].

$$
\begin{array}{ccc}
\underset{\displaystyle H}{\overset{\displaystyle CH_3}{\underset{|}{\overset{|}{-C}}}}-\underset{\displaystyle H}{\overset{\displaystyle H}{\underset{|}{\overset{|}{C}}}}- + O_2 & \longrightarrow & \underset{\displaystyle \underset{OH}{\overset{|}{O}}}{\overset{\displaystyle CH_3}{\underset{|}{\overset{|}{-C}}}}-\underset{\displaystyle H}{\overset{\displaystyle H}{\underset{|}{\overset{|}{C}}}}-
\end{array}
\tag{71}
$$

As the oxygen pressure is increased, reaction (71) should be favored, and this exothermic reaction should counteract the endothermic melting phase. The exothermic reaction which appeared in the DTA trace at about 250°C

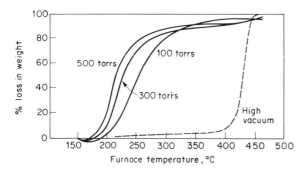

FIG. 2.29   TGA traces for PPr in vacuum and in presence of air [32].

as the oxygen pressure was increased was attributed to an exothermic decomposition which results in the formation of oxygen-containing compounds. In this respect, it was surmised that not only was the hydroperoxide decomposition [cf. Eq. (71)] of importance, but also the direct participation of oxygen in this decomposition step.

Upon comparing the oxidation behavior between PPr and polyethylene, it was noted that whereas an exothermic process commenced at about 150°C for the former, an exothermic reaction occurred for the latter at about 220°C. This difference was ascribed to the presence of a tertiary carbon atom in the PPr structure [cf. Eq. (71)].

Polyacrylonitrile

Kaesche-Krische [32] also studied the decomposition of polyacrylonitrile (PAN) both in presence and absence of oxygen, by means of DTA and TGA methods.

Decomposition in absence of oxygen. In Fig. 2.30 are shown TGA and DTA traces for PAN in high vacuum. From this figure, no instrumental response could be detected until a temperature of about 250°C was reached. Then a sharp exothermic peak resulted, and the slope of the TGA trace began to rise sharply so that within a 5°C interval there was a weight loss of about 20 percent and the PAN became orange-brown. The weight loss was associated with the formation of a colorless, unsaturated liquid nitrile (based on infrared analysis). After the initial steep TGA curve, there appeared a relatively flat portion with an inflection point at about 400°C and an end point at about 480°C. During this stage nitrogen-containing materials are evolving; there was a total weight loss of about 75 percent, and the PAN sample was now a black powder.

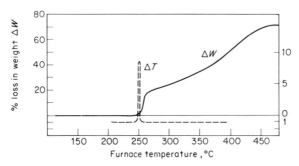

FIG. 2.30 TGA and DTA traces for PAN in high vacuum [32] .

From Fig. 2.30, it can be observed that despite the strong exothermic peak at 250°C, relatively little decomposition occurred. Apparently, at this temperature, an exothermic rearrangement is taking place, which involves color formation. However, the resulting product is not thermally stable and begins to lose weight (20 percent) rapidly. Grassie and Hay [33] have attributed heat discoloration of PAN to the conversion of C≡N groups to C=N groups without any weight loss. An initial rearrangement can occur owing to the reactivity of the tertiary C—H groups in PAN, namely,

$$\tag{72}$$

The initial step [Eq. (72)] could then be followed by further rearrangements involving the imino-hydrogen atoms,

$$
\begin{array}{c}
\underset{}{\text{CH}_2}\quad\underset{}{\text{CH}_2}\quad\underset{}{\text{CH}_2}\quad\underset{}{\text{CH}_2}\quad\underset{}{\text{CH}_2}\\
\text{NC}-\text{CH}\quad\text{CH}\quad\text{CH}\quad\text{CH}\quad\text{CH}\\
\quad|\qquad\quad|\qquad\ \ |\qquad\quad|\qquad\quad|\\
\quad\text{CH}_2\quad\text{C}{=}\text{NH}\ \ \text{CN}\qquad\text{CN}\qquad\text{CN}\\
\qquad\text{C}\\
\qquad\ \ \text{CN}
\end{array}
\longrightarrow
$$

$$
\begin{array}{c}
\underset{}{\text{CH}_2}\quad\underset{}{\text{CH}_2}\quad\underset{}{\text{CH}_2}\quad\underset{}{\text{CH}_2}\quad\underset{}{\text{CH}_2}\\
\text{NC}-\text{CH}\quad\text{CH}\quad\text{CH}\quad\text{CH}\quad\text{CH}\\
\quad|\qquad\quad\quad\qquad\qquad\qquad\\
\quad\text{CH}_2\quad\text{C}\quad\text{C}\quad\text{C}\quad\text{C}{=}\text{NH}\\
\qquad\text{C}\ \ {\diagdown}\text{N}\ {\diagdown}\text{N}\ {\diagdown}\text{N}\\
\qquad\ \ \text{CN}
\end{array}
\qquad(73)
$$

In order to account for the lack of a DTA peak above 250°C, it was postulated that the exothermic reaction (73) can proceed simultaneously with an endothermic pyrolysis reaction in such a manner that an overall thermally "neutral" reaction occurs, namely,

$$
\begin{array}{c}
\qquad\qquad\qquad\qquad\qquad\text{CN}\quad\text{CN}\quad\text{CN}\\
\qquad\qquad\qquad\qquad\qquad|\qquad\ |\qquad\ |\\
\text{CH}_2\ \ \text{CH}_2\ \ \text{CH}_2\ \ \text{CH}_2\ \ \text{CH}\quad\text{CH}\quad\text{CH}\\
\text{CH}\quad\text{CH}\quad\text{CH}\quad\text{CH}\quad\text{CH}_2\ \ \text{CH}_2\ \ \text{CH}_2\ \ \text{CH}_2\\
\text{C}\ \ \text{C}\ \ \text{C}\quad\text{C}{=}\text{NH}\\
\ \ {\diagdown}\text{N}\ \ {\diagdown}\text{N}\ \ {\diagdown}\text{N}
\end{array}
$$

$$
\begin{array}{c}
\quad\text{CH}_2\ \ \text{CH}_2\ \ \text{CH}_2\\
\text{CH}\quad\text{CH}\quad\text{CH}\quad\text{CH}_2\\
|\qquad\ |\qquad\ |\\
\text{C}\ \ \text{C}\ \ \text{C}\quad\text{C}{=}\text{NH}\\
\ {\diagdown}\text{N}\ \ {\diagdown}\text{N}\ \ {\diagdown}\text{N}\\
\\
\qquad\qquad\text{CN}\qquad\qquad\text{CN}\quad\text{CN}\\
\qquad\qquad|\qquad\qquad\quad\ |\qquad\ |\\
+\ \text{CH}_2{=}\text{CH}{-}\text{C}{=}\text{CH}_2\ +\ \text{CH}_2\quad\text{CH}\\
\qquad\qquad\qquad\qquad\qquad\qquad{\diagdown}\text{CH}_2{\diagup}
\end{array}
\qquad(74)
$$

Decomposition also sets in owing to nonuniformities in PAN chain structure which prevent further cyclization from occurring [Eq. (74)]. In this equation is shown the formation of 2-cyano-1,3-butadiene due to pyrolysis. The presence of this material was ascertained by means of refractive index and infrared spectroscopy measurements. The open-chain product could presumably undergo further cyclization. Others [34, 35] have investigated the decomposition of PAN in absence of oxygen by DTA techniques and have found similar trends, as shown in Fig. 2.30, and have observed that the temperature of the sharp exothermic peak varied, depending upon the PAN molecular weight.

Decomposition in presence of oxygen. In Fig. 2.31 can be seen DTA and TGA traces for PAN in presence of various oxygen pressures. At relatively low oxygen pressure (20 torrs), the first decomposition stage observed in vacuum is virtually gone. Instead, a broad decomposition curve is observed with an inflection point at 280°C, and a weight loss of 13 percent occurs at 450°C (cf. Fig. 2.31a). The decomposition behavior becomes different at higher oxygen pressures (cf. Fig. 2.31b and c). Thus, there is a slight weight gain between 180 and 260°C, followed by a rapid decomposition at 265°C, the weight loss being dependent upon the oxygen pressure. After this weight loss, there is a small weight gain (2 to 4 percent), and finally another weight loss occurs above 300°C; at 450°C about 30 percent of the original sample weight has been lost. The DTA traces

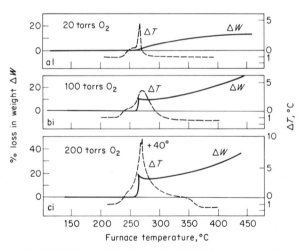

FIG. 2.31 TGA and DTA traces for PAN in presence of oxygen [32].

exhibit a shoulder at 250°C and a strong peak at about 270°C. As the oxygen pressure is increased, the breadth and height of the second peak increase.

The small weight gain between 200 and 260°C above 100 torrs was attributed to the formation of peroxy groups on the PAN tertiary carbon atoms. The ensuing rapid weight loss was then the result of the decomposition of these peroxides. The small weight increase between 270 and 300°C was attributed to the oxygen uptake by residual decomposition products. At the lower oxygen pressures (20 torrs), no weight gain could be observed between 200 and 265°C, as was observed at higher oxygen pressures. This was ascribed to two overlapping processes. In one there should be a weight increase due to oxygen uptake; however, in the other there should be a weight loss due to pyrolysis.

The DTA traces also show that the first peak does not vary much with oxygen pressure. However, as this pressure increases, the area of the second peak increases. The first exothermic peak was attributed to the heat of reaction from hydroperoxy formation, while the second exothermic peak was attributed to the heat of reaction due to the oxygen uptake of residual decomposition products. From the DTA traces, it is also evident that the cyclization stage postulated for PAN in vacuum is suppressed during oxidation. This was attributed to the formation of hydroperoxide groups which cause the disappearance of hydrogen atoms on the tertiary carbon atoms, thereby hindering the cyclization process.

In addition to the effect of oxygen pressure on DTA traces for PAN, it has been observed [35] that molecular weight also influenced the DTA thermogram. Thus, it was found that in presence of air, either broad shoulders or a second exothermic peak could be seen in all DTA traces obtained except for those which involved PAN of low molecular weight. At relatively high molecular weights the second exotherm peak temperature varied between 310 and 313°C whereas at low molecular weights this peak temperature possessed a value of about 245°C.

For descriptions on the utilization of DTA and TGA techniques for other polymer degradations, the reader is referred to Refs. 4 and 5.

REFERENCES

1. *Chem. Eng. News,* **46,** (Aug. 18, 1969).
2. Wendlandt, W. W.: "Thermal Methods of Analysis," Interscience Publishers, a division of John Wiley & Sons, Inc., New York, 1964.
3. Garn, P. D.: "Thermoanalytical Methods of Investigation," Academic Press, Inc., New York, 1965.
4. Reich, L., and D. W. Levi: In A. Peterlin, M. Goodman, S. Okamura, B. H. Zimm, and H. F. Marks (eds.), "Macromolecular Reviews," vol. 1, pp. 173ff, Interscience Publishers, a division of John Wiley & Sons, Inc., New York, 1966.
5. Reich, L.: In A. Peterlin, M. Goodman, S. Okamura, B. H. Zimm, and H. F. Marks (eds.), "Macromolecular Reviews," vol. 3, pp. 49ff., Interscience Publishers, a division of John Wiley & Sons, Inc., New York, 1968.
6. Anderson, D. A., and E. S. Freeman: *J. Polymer Sci.,* **54:**253 (1961).
7. Wiedemann, H. G., A. V. Tets, and H. P. Vaughan: Paper presented at Pittsburgh Conference on Analytical Chemistry and Applied Spectroscopy, Feb. 21, 1966.
8. Harrison, L. G.: In C. H. Bamford and C. F. H. Tipper (eds.), "Comprehensive Chemical Kinetics," vol. 1, pp. 377ff., American Elsevier Publishing Company, Inc., New York, 1969.
9. Eckhardt, R. C., and T. B. Flanagan: *Trans. Faraday Soc.,* **60:**1289 (1964).
10. Clarke, T. A., and J. M. Thomas: *Nature,* **219:**1149 (1968).

11. Clarke, T. A., and J. M. Thomas: *J. Chem. Soc.*, **1969**(pt. A):2227, 2230.
12. Ramamurthy, P., and E. A. Secco: *Can. J. Chem.*, **47**:2181 (1969).
13. Engberg, A.: *Acta Chem. Scand.*, **23**:557 (1969).
14. Reich, L., and S. S. Stivala: *Thermochim. Acta*, **1**:65 (1970).
15. Siegle, J. C., L. T. Muus, T. -P. Lin, and H. A. Larsen: *J. Polymer Sci.*, **A2**:391 (1964).
16. Doyle, C. D.: *J. Appl. Polymer Sci.*, **6**:639 (1962).
17. Coats, A. W., and J. P. Redfern: *J. Polymer Sci.*, **B-3**:917 (1965).
18. Flynn, J. H., and L. A. Wall: *J. Res. Natl. Bur. Std.*, **70A**:487 (1966).
19. Freeman, E. S., and B. Carroll: *J. Phys. Chem.*, **62**:394 (1958).
20. Doyle, C. D.: *J. Appl. Polymer Sci.*, **5**:285 (1961).
21. Ozawa, T.: *Bull. Chem. Soc. Japan*, **38**:1881 (1965).
22. Papkov, V. S., and G. L. Slonimskii: *Vysokomolekul. Soedin.*, **A10**:1204 (1968).
23. Reich, L.: *J. Appl. Polymer Sci.*, **9**:3033 (1965).
24. Borchardt, H. J., and F. Daniels: *J. Am. Chem. Soc.*, **79**:41 (1957).
25. Ramachandran, V. S.: "Applications of DTA in Cement Chemistry," Chemical Publishing Company, Inc., New York, 1969.
26. Akita, K., and M. Kase: *J. Phys. Chem.*, **72**:906 (1968).
27. *Mettler Instr. Corp. Tech. Bull. T-102*, Princeton, N. J.
28. Reich, L.: *J. Appl. Polymer Sci.*, **10**:813 (1966).
29. Reich, L.: *J. Appl. Polymer Sci.*, **10**:465 (1966).
30. Anderson, D. A., and E. S. Freeman: *J. Polymer Sci.*, **54**:253 (1961).
31. Igarashi, S., and H. Kambe: *Bull. Chem. Soc. Japan*, **37**:176 (1964).
32. Kaesche-Krischer, B.: *Chem.-Ingr.-Tech.*, **37**:944 (1965).
33. Grassie, N., and J. N. Hay: *Soc. Chem. Ind. Monograph*, **13**:184 (1961).
34. Gillham, J. K., and R. F. Schwenker: *Appl. Polymer Symp.*, **2**:59 (1966).
35. Thompson, E. V.: *J. Polymer Sci.*, **B-4**:361 (1966).

3

ANCILLARY METHODS
TO TGA AND DTA

This chapter will briefly discuss those methods used in polymer degradation studies which are ancillary to TGA and DTA. For more detailed treatments, the appropriate references cited should be consulted.

GEL PERMEATION CHROMATOGRAPHY (GPC) [1]

Gel permeation is generally used in estimating molecular-weight distribution (MWD) and number- and weight-average molecular weights of polymers, $\bar{M}_n$ and $\bar{M}_w$, respectively. Accordingly, one can obtain, for example, the effect of MWD on the thermal degradation of polymers.

In the operation of GPC, a solution of a polymer sample is passed through a gel consisting of rigid, crosslinked polystyrene. The separation of molecular species according to size presumably occurs as a result of differences in the extent to which different species permeate the gel particles. Whereas smaller molecules permeate the gel depending upon their size and the distribution of gel pore sizes, the larger molecules pass through the gel mainly by way of the interstitial volume (the space occupied by the solvent outside the porous gel beads). Thus, the larger molecules follow a less circuitous and tortuous path than the smaller molecules and emerge from the

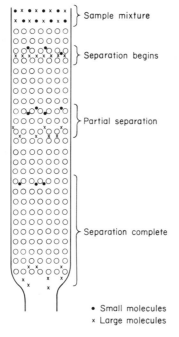

Sample mixture

Separation begins

Partial separation

Separation complete

• Small molecules
× Large molecules

FIG. 3.1 Separation of molecules according to size by GPC [1].

column first while the smaller molecules emerge last (cf. Fig. 3.1). Increased resolution (decrease in peak width rather than an increase in the differences between the elution volumes of the individual components) can be affected by temperature and eluent flow rate [2].

The GPC elution volume of a polymer, generally expressed as 5-ml "counts," has been correlated with its hydrodynamic volume. Grubisic and coworkers [3] found that a plot of log ($[\eta]M$) versus elution volume gave a curve upon which fell experimental values of various polymers, e.g., polystyrene, polymethyl methacrylate (cf. Fig. 3.2). This was referred to as a "universal" calibration curve which may be partially justified as follows.

From the Einstein viscosity law,

$$[\eta] = K\frac{V}{M} \tag{1}$$

where $[\eta]$ = limiting viscosity index (intrinsic viscosity), V = hydrodynamic volume of the particles, M = molecular weight of the particles, and K = const. From this equation, it can be seen that the product $[\eta]M$ is related to hydrodynamic volume which, from Fig. 3.2, is related to the elution volume. It can be further shown that $[\eta]M$ is related to the radius of gyration for branched polymers.

The set of columns selected is generally calibrated (log of angstrom length versus counts) by selecting commercially available polystyrene standards which are relatively monodisperse and determining their elution volume (cf. Fig. 3.3 [4]). After passing through the columns, the effluent flows into a 5-ml syphon tube (which is filled every 5 min). After the syphon empties the effluent flows into a differential refractometer which measures the difference between the refractive index of sample and of solvent.

The recorder marks each time the syphon is emptied (a "count"). The number of counts, calculated from the point of sample injection into the column, is automatically recorded along with the corresponding difference in refractive index (which is related to the weight of polymer in the 5 ml of effluent in the syphon tube) (cf. Fig. 3.4).

From a chromatogram as shown in Fig. 3.4, it is possible to calculate weight-average and number-average molecular weights $\overline{M}_w$ and $\overline{M}_n$, assuming that certain constants are known. Accordingly, it was previously indicated that

$$H_i = fM_iN_i = fQA_iN_i \tag{2}$$

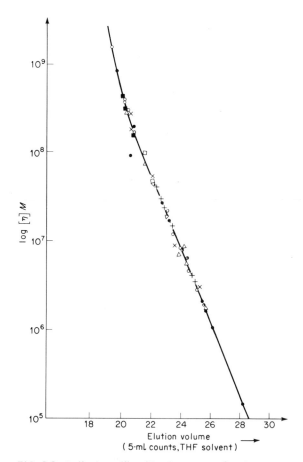

FIG. 3.2 A "universal" calibration curve: ● polystyrene,
○ polystyrene comb, + polystyrene star, △ heterograft copolymer, x polymethyl methacrylate, ◑ polyvinyl chloride,
▽ graft copolymer polystyrene-polymethyl methacrylate,
■ polyphenylsiloxane, □ polybutadiene [3].

where H_i = height of a given count (or polymer fraction) (arbitary units),
f = a proportionality constant, M_i = average molecular weight of polymer
fraction (or segment), N_i = number of polymer molecules represented by
the segment, A_i = average angstrom size of the segment, and Q = molecular-
weight units per angstrom. Further, from the definition of $\bar{M}_n$ and $\bar{M}_w$,

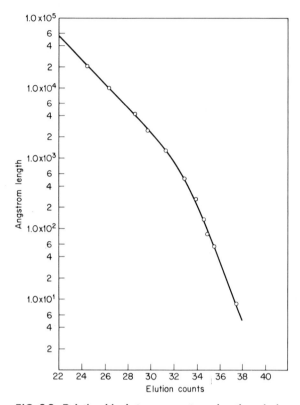

FIG. 3.3 Relationship between angstrom length and elution counts for standard polystyrene and poly(propylene oxide) [4].

$$\overline{M}_n = \overline{A}_n Q = \frac{W}{\displaystyle\sum_{i=1}^{\infty} N_i} = \frac{\displaystyle\sum_{i=1}^{\infty} M_i N_i}{\displaystyle\sum_{i=1}^{\infty} N_i} \qquad (3)$$

and

$$\overline{M}_w = \overline{A}_w Q = \sum_{i=1}^{\infty} W_i M_i = \frac{\displaystyle\sum_{i=1}^{\infty} M_i^2 N_i}{\displaystyle\sum_{i=1}^{\infty} M_i N_i} \qquad (4)$$

where W = total weight of polymer, W_i = weight fraction of a given polymer species i, and $\bar{A}_n$ and $\bar{A}_w$ denote number- and weight-average angstrom size of the polymer, respectively. From Eqs. (2) and (3),

$$\bar{A}_n = \frac{\sum_i H_i}{\sum_i (H_i/A_i)} \tag{5}$$

From Eqs. (2) and (4),

$$\bar{A}_w = \sum_i W_i A_i = \frac{\sum_i H_i A_i}{\sum_i H_i} \tag{6}$$

where $W_i = N_i M_i / \Sigma N_i M_i = H_i / \Sigma H_i$.

Thus, from Eqs. (3) to (6), it can be seen that once Q is known, values of $\bar{M}_n$ and $\bar{M}_w$ can be estimated. In Table 1 are listed data derived from Figs. 3.3 and 3.4. From the values in this table, $\bar{A}_w$ = 82,420 and $\bar{A}_n$ = 51,927 (the value of the ratio $\bar{M}_w/\bar{M}_n$ = 1.6). Also, from this table, values of the cumulative weight percent for corresponding values of A_i can be calculated. However, the estimation of reliable values of $\bar{M}_n$ and $\bar{M}_w$ depends on reliable estimates of the value of Q. When polystyrene standards are used, values of $\bar{M}_w$, $\bar{M}_n$, and molecular sizes are provided (Q = 41.3 for polystyrene). To obtain reliable values of Q for other polymers it would appear advisable to use experimental methods such as light scattering ($\bar{M}_w$) and osmometry ($\bar{M}_n$). The calculation of Q from bond lengths and angles

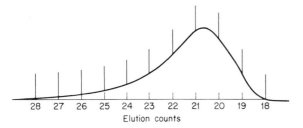

FIG. 3.4 Gel permeation chromatography curve for a poly-(isobutyl methacrylate) sample in tetrahydrofuran at 25°C [4].

can lead to erroneous results [4]. The experimentally obtained value of Q may then be employed for other polymer samples. Q values have been reported for various polymers [5, 6] such as polyethylene (Q = 11) and cis-1,4-polybutadiene (Q = 12.4).

Some limitations of GPC are [6] : (1) the amount of sample is limited since the column must not be overloaded, (2) air must be kept out of the system, and (3) the operating temperature should not exceed 150°C since above this temperature the polystyrene gel degrades. Some advantages of

TABLE 1   Gel Permeation Chromatography Data for
Poly(isobutyl Methacrylate) in Tetrahydrofuran at
25°C [4]

Counts	H_i	Cumulative height	Cumulative weight, %	A_i	H_i/A_i	H_iA_i
18	0.4	56.6	100	262,000	0.00000	104,800
19	5.8	56.2	99.3	175,000	0.00003	1,015,000
20	13.3	50.4	89.0	117,000	0.00011	1,556,100
21	14.7	37.1	65.5	78,500	0.00018	1,153,950
22	9.8	22.4	39.6	52,500	0.00018	514,500
23	5.6	12.6	22.3	35,100	0.00015	196,560
24	3.2	7.0	12.4	23,600	0.00013	75,520
25	2.0	3.8	6.7	15,800	0.00012	31,600
26	1.2	1.8	3.2	10,600	0.00011	12,720
27	0.6	0.6	1.1	7,100	0.00008	4,260
Σ	56.6	- - - -	- - - - -	- - - - - - -	0.00109	4,665,010

GPC are [6] : (1) a very small sample can be used, (2) the chromatogram can be rapidly obtained, (3) the system used is flexible in that it is possible to run several different types of samples at different temperatures and with widely varying solvents on the same system, and (4) reproducible results can be obtained, even though other types of runs are carried out between the two runs to be duplicated.

Finally, it should be remembered that polystyrene standards are generally used as the basis for the estimation of the linear molecular dimensions of other polymers. Since elution volume is related to the hydrodynamic volume of polymers, the estimation of linear sizes of polymers may only be very approximate. For reasons such as this, GPC is not an absolute method for the determination of molecular size or molecular weight.

GAS CHROMATOGRAPHY [7, 8]

Gas chromatography involves the separation of substances by repeated distribution between a moving gas phase and a solid or liquid stationary phase that are in equilibrium. Gas-liquid chromatography (GLC) is the term used when the stationary phase is liquid. This phase is generally spread on an inert solid support, e.g., Carbowax 400 on a 60- to 80-mesh Chromosorb solid support. The liquid-solid support mixture is packed in a column placed in an oven where the temperature can be closely controlled.

The mobile gas carrier of the solute is usually helium or nitrogen. The sample is injected into the gas stream which carries it into the column. Figure 3.5 is a simplified schematic diagram of a GLC column showing two partially separated substances, O and X (substance X possesses a greater

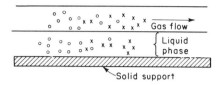

FIG. 3.5 Simplified chromatographic column showing the partial separation of compounds O and X.

relative volatility). Substance X emerges initially from the column and may be detected by devices that measure thermal conductivity (katharometers). Since absolute measurements are difficult, a differential procedure is generally used. Thus, two gas channels and wires which are as identical as possible are employed. The wires are heated by a constant electric current (these wires have a high-temperature coefficient of resistance). The temperature of the wires depends on the thermal conductivity of the surrounding gas, which in turn affects the wire resistance, the latter being measured. Thus, when pure carrier gas flows through one gas channel, and the carrier gas with solute (which has passed through the column) through the second gas channel, the resulting differences in resistance of the two wires, due to the effects of volatile components in the effluent, are recorded by such devices as a Wheatstone bridge. In this manner, an elution

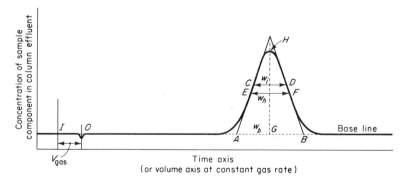

FIG. 3.6 Quantities in elution diagram [7]. (*Courtesy of Reinhold Publishing Co.*)

diagram may be obtained for any particular solute, as shown in Fig. 3.6. In this figure, the ordinate represents solute concentration, and the abscissa the gas volume. The term I represents zero time (when the sample is introduced); O denotes the time it takes (from I) for a nonabsorbing substance to emerge from the column (gas holdup of the column); IG = retention volume (RV); IA = initial retention volume (RV_i); IB = final retention volume (RV_f); OG = apparent retention volume (ARV); OA and OB denote initial and final apparent retention volumes, respectively $(ARV_i$ and $ARV_f)$; AB = peak width at the base w_b; CD = peak width at the inflection points w_i; EF = peak width at half height w_h; and GH = peak height. In the following will be presented some of the theoretical aspects of GLC [7, 8]. The theoretical plate concept will be stressed.

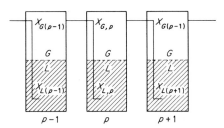

FIG. 3.7 Three successive absorption vessels [7]. (*Courtesy of Reinhold Publishing Co.*)

In the theoretical plate concept it is assumed that the volume of the inert gas phase V_G and the volume of the nonvolatile liquid V_L are the same in each vessel of Fig. 3.7, that these volumes remain constant during the stripping process, and that each vessel acts as an ideal plate (two phases are in equilibrium at any time). In Fig. 3.7 are shown three successive absorption vessels out of a series of $n + 1$, with ordinal numbers $0, 1, \ldots,$ n. Volatile solute is introduced into vessel zero where it dissolves in a nonvolatile liquid exerting vapor pressure above it. A constant flow of carrier gas is passed through the vessels carrying the vapors above the liquid from one vessel to another. Assuming a linear isotherm, i.e.,

$$X_{L,p} = kX_{G,p} \tag{7}$$

where $X_{L,p}$ = concentration of solute in the liquid phase of vessel p, $X_{G,p}$ = concentration of solute in the inert-gas phase of vessel p, and k = an equilibrium constant. A volume dV of gas that moves from vessel $p - 1$ into p carries with it the amount of solute equal to $X_{G,(p-1)} \, dV$ while an equal volume carries the amount $X_{G,p} \, dV$ of solute from p into $p + 1$. In this manner the concentration of solute in the liquid phase has changed by $dX_{L,p}$, and that in the gas phase by $dX_{G,p}$. From the conservation of mass principle,

$$[X_{G,(p-1)} - X_{G,p}] dV = V_G \, dX_{G,p} + V_L \, dX_{L,p} \tag{8}$$

From Eqs. (7) and (8),

$$\frac{dX_{G,p}}{dV} = \frac{X_{G,(p-1)} - X_{G,p}}{V_G + kV_L} \tag{9}$$

If solute is initially present in the first vessel only with concentration $X_{G,i}$, then from Eq. (9),

$$X_{G,0} = X_{G,i} e^{-V/\phi} \tag{10}$$

where $\phi = 1/(V_G + kV_L)$. When $p = 1$, Eq. (9) becomes

$$\frac{dX_{G,1}}{dV} = \frac{X_{G,0} - X_{G,1}}{\phi} \tag{11}$$

Combining Eqs. (10) and (11) leads to

$$X_{G,1} = X_{G,i} \frac{V}{\phi} e^{-V/\phi} \tag{12}$$

In a similar manner,

$$X_{G,2} = \frac{X_{G,i}}{2} \left(\frac{V}{\phi}\right)^2 e^{-V/\phi} \tag{13}$$

and the general expression becomes

$$X_{G,p} = \frac{X_{G,i}}{p!} \left(\frac{V}{\phi}\right)^p e^{-V/\phi} \tag{14}$$

or

$$X_{G,p} = X_{G,i} \frac{e^{-v} v^p}{p!} \tag{14a}$$

where $v = V/\phi$ (ϕ is often termed the "effective plate volume"). From Eq. (14a) can be plotted the quantity $x_p \equiv X_{G,p}/X_{G,i}$ versus v for various values of p, thus describing the situation in each vessel after an arbitrary "volume" v has passed through (see Fig. 3.8). Except for $p = 0$, a family of curves is obtained which all possess maxima and two inflection points and are of the Poisson distribution type. From Eq. (14a), it can be readily shown that the peak maximum occurs at $p = v$ (cf. Fig. 3.8). Further, the inflection points occur at $v = p - p^{1/2}$ and $v = p + p^{1/2}$. The points where the inflection tangent intersects the horizontal axis (cf. Fig. 3.8) may be determined as follows. The slope of the inflection tangent with positive slope is

$$\text{Slope}(+) = \frac{e^{-v_+} v_+^{p}}{p!} \frac{p - v_+}{v_+} \tag{15}$$

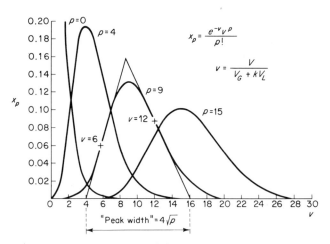

FIG. 3.8   Poisson distributions [7]. (Courtesy of Reinhold Publishing Co.)

where v_+ = inflection point of the inflection tangent with positive slope (+). From Fig. 3.8,

$$\text{Slope}(+) = \frac{e^{-v_+} v_+^{p}}{p!} \frac{1}{v_+ - I_+} \tag{16}$$

where I_+ = intercept of inflection tangent with positive slope on the horizontal axis. From Eqs. (15) and (16),

$$v_+ - I_+ = \frac{v_+}{p - v_+} \tag{17}$$

where $v_+ = p - p^{1/2}$. Solving for I_+

$$I_+ = p + 1 - 2p^{1/2} \tag{18a}$$

In a similar manner,

$$I_- = p + 1 + 2p^{1/2} \tag{18b}$$

where I_- = intercept of inflection tangent with negative slope on the horizontal axis. From Eqs. (18a) and (18b),

$$w_b \equiv I_- - I_+ = 4p^{1/2} \tag{19}$$

Also, at the peak maximum,

$$x_{p,m} = \frac{e^{-p} p^p}{p!} \tag{20}$$

The preceding formulas, derived for a solute which is introduced into a vessel zero and subsequently into a series of absorption vessels, can be applied to chromatographic columns. These columns are considered to consist of a *large number of discrete plates* (equivalent to absorption vessels) in which the amount of solute introduced is considered to be so small that it can be introduced initially into the first plate. If the total number of plates of such a column is denoted as $n + 1$, then from Eqs. (14a) and (20),

$$x_n = \frac{e^{-v} v^n}{n!} \tag{14b}$$

and

$$x_{n,m} = \frac{e^{-n} n^n}{n!} \tag{20a}$$

From Eqs. (14b) and (20a),

$$v_R = n \quad \text{or} \quad RV = n(V_G + kV_L) \tag{21}$$

From Eq. (21),

$$ARV = RV - nV_G = knV_L \tag{22}$$

The term ARV may be expressed in terms of activity coefficients and vapor pressures as follows. For nonideal solutions, the partial pressure of a solute p, may be expressed in terms of the mole fraction of solute in solution x as

$$p = \gamma \times p^\circ \tag{23}$$

where γ = activity coefficient, and p° = vapor pressure of the pure solute at the corresponding temperature. Above the solution, the mole fraction of solute in the gas phase y is related to p as

$$y = \frac{p}{P} \tag{24}$$

where P = total gas pressure, and

$$yP = y \times p^{\circ} \tag{25}$$

For dilute solutions and from the definition of k (amount of solute per unit volume of stationary liquid phase to the amount of solute per unit volume of moving phase),

$$k = \frac{xN_L}{yN_G} \tag{26}$$

where N_L and N_G denote moles of stationary liquid and gas respectively per unit volume. From Eqs. (25) and (26),

$$k = \frac{P}{yp^{\circ}} \frac{N_L}{N_G} \tag{27}$$

Applying the ideal gas law to Eq. (27),

$$k = \frac{RTN_L}{yp^{\circ}} \tag{28}$$

Thus, from Eqs. (22) and (28), for two solutes, 1 and 2,

$$\frac{(ARV)_1}{(ARV)_2} = \frac{(yp^{\circ})_2}{(yp^{\circ})_1} \equiv \alpha \tag{29}$$

where α = relative volatility (or separation factor). Thus, the larger the value of α, the greater should be the ease with which substances 1 and 2 may be separated. The number of plates may be estimated as follows. If the distance between the injection point and the projection of the peak maximum on the base line is denoted by d, and if d and w_b are expressed in similar dimensional units,

$$\frac{d}{w_b} = \frac{c \cdot n}{c \cdot 4n^{1/2}} = \frac{n^{1/2}}{4} \tag{30}$$

or

$$n = \left(4\,\frac{d}{w_b}\right)^2 \qquad\qquad (30a)$$

where c = proportionality factor.

The preceding equations are subject to various limitations, some of which are briefly discussed below (for detailed discussions, see Keulemans [7] and Purnell [8]).

It was assumed that the pressure gradient in the column could be neglected. When this is observed not to be experimentally true, then corrections must be applied, as in the case of retention volumes. Also, it will generally not be possible to introduce the total amount of sample into the first column plate. This can result in the formation of asymmetric bands and in such a case the plate formulas will not apply. Another factor which can cause asymmetric bands is the nonlinearity of the distribution isotherm (particularly at higher solute concentrations). Various other factors can affect column efficiency. These are, to mention a few, column temperature and flow rate. To obtain a good compromise between high separation efficiency and rapid chromatographic separations, a high flow rate and a low column temperature are used. When high-boiling volatiles are involved, a programmed heating rate may be used to reduce the retention volume times. Thus, for low-boiling volatiles, relatively low constant temperatures are used, and after these volatiles are detected, the column is subjected to a programmed heating rate. Thus, retention times for the higher-boiling materials are shortened (too high column temperatures can lead to less efficient separation).

Application of GLC to Polymer Pyrolysis

When the bands arising in GLC are to be identified, the effluent(s) corresponding to the band(s) may be passed into an infrared-absorption spectrometer and/or a mass spectrometer. In the latter case, the mass spectrum of a component can be directly observed on the cathode screen, thereby providing evidence for the identity of the component in question. Other methods of identification could involve the isolation of the effluents in a cold trap, which may then be identified by physical and/or chemical means. If, based on retention volumes, a certain component is suspect, this substance may be added to the sample to be injected into the GLC column. If the material represented by a band is identical with the added substance, then the area of the band in question should increase (w_b should be constant).

After the sample components have been identified, quantitative estimations of these substances can then be carried out. Several methods are: (1) the use of external standards, (2) the use of internal standards, and (3) the method of internal normalization. In method (3) it is assumed that the area under each peak (approximately the peak height multiplied by w_h, Fig. 3.6) is directly proportional to the corresponding amount of component and is independent of its nature. The composition of each component may then be estimated by dividing the particular peak area by the sum of the areas of all the peaks. For increased accuracy, a synthetic mixture corresponding to the composition estimated may be prepared and analyzed. The new peak areas are compared with those for the original sample, and the compositions of each component are adjusted accordingly. In method (1) a known amount of a pure substance (external standard) which is not one of the components originally present is added to the sample. The ratio of peak areas (or peak heights) of the sample component to the external standard is then measured. From a calibration curve of this ratio versus quantity of sample component, the amount of sample component may be determined. This ratio method tends to eliminate errors due to variations in column performance from run to run. Method 2 also is a ratio method, but the standard may be one of the bands which arise from the sample itself.

Some examples of the use of GLC in qualitative and quantitative aspects of polymer degradation follow. Tsuchiya and Sumi [9] investigated the thermal decomposition products of polyvinyl alcohol (PVOH) by means of GLC. The thermal decomposition of PVOH occurs in two stages; the first stage commences at about 200°C, and the second at about 500°C. The sample of PVOH was heated in a Pyrex tube connected to a liquid nitrogen trap and a vacuum pump. After evacuation of the system, the sample was heated at 240°C for 4 hr, and most of the volatile material was collected in the trap. When the liquid nitrogen was removed, two layers formed in the trap: a water layer and an oil layer. The fraction volatile at room temperature was transferred into a gas sampling bottle of known volume. The resulting three fractions were separately analyzed, and the pyrolysis tube containing residue was heated at 450°C for 4 hr. The products from the second-stage decomposition were analyzed in a similar manner to those from the first stage. The identification of the peaks obtained in the GLC apparatus were carried out by (1) comparison of retention times with those of known compounds, (2) collection of products with certain peaks, followed by chemical identification, and (3) determination of carbon structure of products by catalytic hydrogenation. For quantitative analysis, peak areas were determined and appropriate calibration factors were applied.

In Fig. 3.9 are shown products formed during the first-stage decomposition. The retention index of each component was determined, using n-alkanes as standards. The major peaks, A_0, A_1, A_2, and A_3 of Fig. 3.9 are uniformly spaced on the retention-index scale indicating that they belong to a homologous series of aldehydes (identified by a comparison of retention indices with those of known compounds and from melting points of the 2,4-dinitrophenylhydrazone derivatives), i.e., acetaldehyde, crotonaldehyde, 2,4-hexadiene-1-al, and 2,4,6-octatriene-1-al. Peaks K_0, K_1, K_2, and K_3 were also uniformly spaced and were identified as ketones, i.e., acetone, 3-pentene-2-one, 3,5-heptadiene-2-one, and 3,5,7-nonatriene-2-one. Thus, the general formula of the A compounds can be represented as

$$CH_3-(CH=CH)_n-\overset{\overset{\displaystyle O}{\displaystyle \|}}{C}-H$$

(I)

whereas the general formula of the B compounds can be denoted as

$$CH_3-(CH=CH)_n-\overset{\overset{\displaystyle O}{\displaystyle \|}}{C}-CH_3$$

(II)

As may be observed from Fig. 3.9, some of the other products isolated during the first-stage decomposition were ethanol, benzene, benzaldehyde, and acetophenone. About 37.2 weight percent of the original PVOH sample

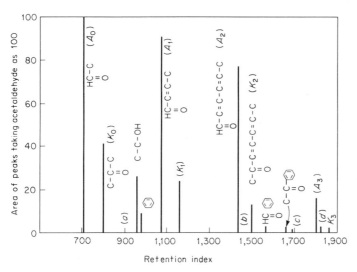

FIG. 3.9 First-stage decomposition products of poly(vinyl alcohol). Area of peaks versus retention index [9].

could be detected as decomposition products by GLC for the first-stage pyrolysis; water appeared as 33.4 percent, and aldehydes (A compounds) and ketones (K compounds) appeared as 3.3 percent. In Table 2 can be seen material balances for the first and second decomposition stages. A large portion of the oil layer was not analyzed because of insufficient volatility for use in GLC. However, conversion of the carbonyl compounds in the oil layer to 2,4-dinitrophenyl hydrazones indicated that a large portion of this layer could be represented by carbonyl compounds. The

TABLE 2 Decomposition Products of Poly(vinyl Alcohol) (Material Balance, in Weight Percentage of Original Polymer)

1st stage (240°C, 4 hr)			2d stage (450°C, 4 hr)	
Volatiles 47.9	Water layer	Water . . . . . .33.4 Organic compounds . . . 1.56	Volatiles 27.7	Water layer . . . 0.60
	Oil layer	Organic compounds: Analyzed . . . . 1.19 Not analyzed		Oil layer 22.30
	Gas . . .0.92 Loss . . .5.81			Gas . . .2.46 Loss . . .2.34
Residue 52.1			Residue 24.4	

volatile decomposition products of the second stage were mainly hydrocarbons consisting of n-alkanes, n-alkenes, and aromatic hydrocarbons. Many more aldehydes than ketones were found in the products. The following mechanistic equations were postulated.

$$\sim\sim \left(\underset{\underset{OH}{|}}{CH}-CH_2\right)_n -\underset{\underset{OH}{|}}{CH}-CH_2 \sim\sim \longrightarrow$$

$$\sim\sim (CH=CH)_n -\underset{\underset{OH}{|}}{CH}-CH_2 \sim\sim \quad + nH_2O \qquad (31)$$

$$\sim\sim \underset{\underset{OH}{|}}{CH}-CH_2-(CH=CH)_n -\underset{\underset{OH}{|}}{CH}\!\!-\!\!CH_2 \sim\sim \longrightarrow$$

(Continued on next page)

$$\sim\sim CH-CH_2-(CH=CH)_n-\overset{\overset{\displaystyle O}{\|}}{C}-H \; + \; CH_3-\underset{\underset{\displaystyle OH}{|}}{CH} \sim\sim \qquad (32)$$

where the left CH bears OH below.

$$\sim\sim \underset{\underset{\displaystyle OH}{|}}{CH}-CH_2-(CH=CH)_n-\underset{\underset{\displaystyle OH}{|}}{\overset{\overset{\displaystyle \cdot}{}}{C}}-CH_2-CH \sim\sim \; \longrightarrow$$

$$\sim\sim \underset{\underset{\displaystyle OH}{|}}{CH}-CH_2-(CH=CH)_n-\underset{\underset{\displaystyle OH}{|}}{C}=CH_2 \; + \; \sim\sim CH_2\overset{\overset{\displaystyle O}{\|}}{C}-H \qquad (33)$$

$$\sim\sim \underset{\underset{\displaystyle OH}{|}}{CH}-CH_2-(CH=CH)_n-\underset{\underset{\displaystyle OH}{|}}{C}=CH_2 \sim\sim \; \longrightarrow$$

$$\sim\sim \underset{\underset{\displaystyle OH}{|}}{CH}-CH_2-(CH=CH)_n-\overset{\overset{\displaystyle O}{\|}}{C}-CH_3 \qquad (34)$$

The preceding equations can account for the formation of more aldehyde than ketone since the former can arise from Eqs. (32) and (33) whereas the latter arises only from Eq. (34).

As indicated previously, GLC may be used for the qualitative and quantitative analysis of products obtained from the controlled degradation of polymers under various conditions. Besides giving an insight into the degradation mechanism, the volatile products liberated can serve as "fingerprints" for the identification of polymers. Thus, Cox and Ellis [10] pyrolyzed 0.1-mg polymer samples at 700°C and expressed elution times of the peaks as percentage fractions of the elution time of a reference peak, which was arbitrarily selected as the peak of the last product eluted during a 15-min period after pyrolysis. Retention data for several polymers are shown in Fig. 3.10. GLC techniques may also be employed in differentiating among random copolymers, block copolymers, and polymer mixtures [11, 12] as well as in estimating the degree of tacticity of polymers [13]. Quantitative data for copolymer compositions may also be obtained by the use of an internal standard (method 2). Thus, Groten [13] obtained pyrograms of typical SBR-polybutadiene (PBD) vulcanizates as well as a calibration curve relating peak height ratio (peak height at about 20 min to that at about 6 min (internal standard) to weight percent of styrene (cf. Fig. 3.11).

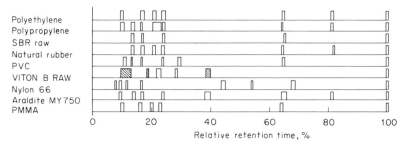

Polyethylene
Polypropylene
SBR raw
Natural rubber
PVC
VITON B RAW
Nylon 66
Araldite MY 750
PMMA

0   20   40   60   80   100

Relative retention time, %

FIG. 3.10 Chart of reduced retention time data for some polymers. Negative peaks are indicated by cross-hatching. Only peaks that have well-defined relative retention times have been included [10].

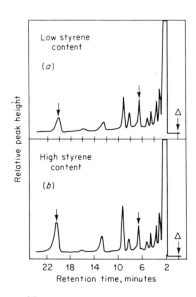

Low styrene content

(a)

High styrene content

(b)

Relative peak height

22   18   14   10   6   2

Retention time, minutes

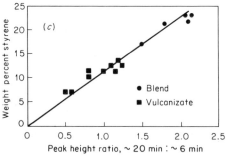

(c)

Weight percent styrene

25

20

15

10

5

0

0   0.5   1.0   1.5   2.0   2.5

Peak height ratio, ~ 20 min : ~ 6 min

● Blend

■ Vulcanizate

FIG. 3.11 Pyrograms of (a, b) typical SBR-polybutadiene vulcanizates (arrows indicate peaks used for quantitative styrene analysis) and (c) calibration curve for styrene content of SBR-PBD blends and vulcanizates [13].

INFRARED-ABSORPTION SPECTROSCOPY

Besides its use as an aid in the identification of volatile products from polymer pyrolysis, infrared absorption spectroscopy (IR) has been extensively employed in the analysis of nonvolatile products formed during polymer degradation, e.g., oxidative degradation [14]. The IR spectra is a manifestation of molecules undergoing transitions between quantum states corresponding to different internal energies. The energy difference is related to the frequency of the emitted or absorbed radiation, which is associated with molecular vibrations and rotations. Molecular vibrations and rotations are characteristic of chemical groups, e.g., CH_3—, —OH, and >C=O. Thus, IR spectra can be used for the qualitative and quantitative analysis of such groups in polymer chains [15–17]. For this purpose, a cell may be constructed which will allow constant monitoring of the polymer sample undergoing degradation. Such a cell has been designed for use in studies of oxidative degradation of polyolefin film samples such as polypropylene and polybutene-1 [14, 16, 17] and is depicted in Fig. 3.12, which is self-explanatory. In this studies, polymer films of about 2 mils thickness were oxidized, and the amount of carbonyl-containing compound formed was estimated as a function of oxygen concentration, temperature, and time. The rate of formation of nonvolatile carbonyl groups due to oxidation (at relatively low conversions) could be estimated by determining the area under the absorption bands between 5.4 and 5.9 μ, as shown in

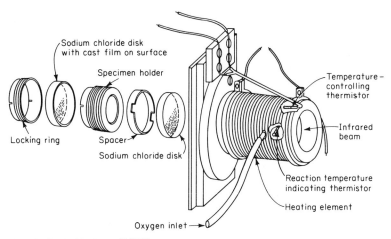

FIG. 3.12   Oxidation cell [16].

Fig. 3.13. These bands arise from the presence of carboxylic acids, aldehydes, esters, and ketones. Conversion of transmittance to absorbance can be accomplished by the use of the expression

$$A = -\log T \tag{35}$$

where A = absorbance and T = transmittance (generally only recorded by most IR instruments). In the following will be discussed the detection and estimation of structural changes in polymers undergoing autoxidation.

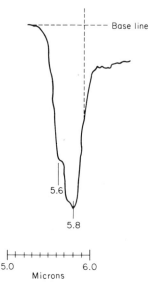

FIG. 3.13 Area limits of transmittance used to replot carbonyl content in terms of absorbance [16].

Polymer Tacticity

IR spectra may be used to estimate the degree of tacticity in polymers, e.g., polypropylene. Luongo [18] found that in a 100 percent isotactic polypropylene (IPP) sample the 974 and 995 cm^{-1} bands possess similar intensities whereas in the 100 percent atactic polypropylene (APP) sample the 995 cm^{-1} band appears only as a shoulder while the intensity of the 974 cm^{-1} band remains the same. By mixing IPP with APP in the molten phase and then cooling, samples may be prepared containing various weight percentages of APP. The IR spectra of these standards were recorded, and the intensity ratios of the 995 and 974 cm^{-1} bands of each standard. In this manner a working curve for APP content from 0 to 100 percent was obtained (Fig. 3.14). The accuracy of the method was estimated to be ±2 percent over the 0 to 100 percent range. In this connection may be mentioned the effect of crystallinity upon the results obtained. The degree of crystallinity will affect the results of the isotacticity measurements since the same spectral changes occur with crystallinity variations as with variations in isotacticity. However, with the proper experimental conditions, crystallinity will not vary significantly with sample preparation.

Polymer Film Thickness

Polymer film thickness may be estimated from IR spectra. Balaban and co-workers [19] studied the dependence of optical density (absorbance) of the 1460 cm^{-1} band upon the film thickness of atactic polypropylene (APP) (the absorbance is independent of polymer tacticity). A linear relationship was obtained which passed through the origin of the coordinates. This confirmed the validity of the Lambert-Beer law for film thicknesses varying from 0.2 to 2 mils. Recently, Stivala and coworkers [20] found that a similar linear relationship could be obtained between the film thickness of APP and the absorbance of the band at 1380 cm^{-1}. This relationship was used to investigate the effect of film thickness on the oxidation rate of APP (diffusion control). Thus, four film thicknesses of APP were prepared by casting from solution, i.e., 1.1, 2.5, 3.4, and 7.5 mils. These samples were oxidized at 130°C in pure oxygen in a cell similar to that in

Fig. 3.12. The IR spectra in the carbonyl region (cf. Fig. 3.13) were obtained for each film at various reaction times. The area of the band shown in Fig. 3.13 (in terms of absorbance) was estimated in units of cm^2. From the Lambert-Beer law, for a particular absorbing group,

$$A = adc \qquad (36)$$

where a = constant which depends upon wavelength λ, d = film thickness,

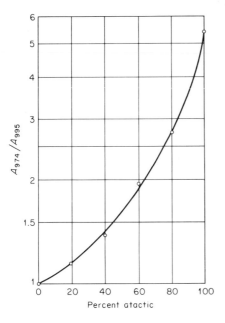

FIG. 3.14 Analytical curve for determining atactic content of polypropylene [18].

and c = concentration of absorbing specie, it can be shown that absorbance area (AA) for a particular specie may be expressed as

$$\frac{(AA)}{d} = K \cdot c \qquad (37)$$

where $K = \int_{\lambda_1}^{\lambda_2} a(\lambda) \, d\lambda$, and λ_1 and λ_2 denote the extremities of the base line in terms of reciprocal wavelengths (cf. Fig. 3.13). In order for K to remain a constant during a run, the base-line limits λ_1 and λ_2 should not be exceeded. From Eq. (37), it can be seen that if the rate of oxidation of APP is independent of film thickness (no diffusion control), the ratio $(AA)/d$ for any particular reaction time should be constant for various film thicknesses at relatively low degrees of oxidation. (As will be shown later, *AA* was found to be directly proportional to the total carbonyl content which accrued during the autoxidation at any time *t*.) Such a relationship was obtained for films of thickness, 1.1, 2.5, and 3.4 mils (no diffusion control), whereas diffusional effects commenced to become important for the 7.5-mil film (see Fig. 3.15).

Nonvolatile Oxidation Products

During the oxidation of polymers, the formation of oxygen-containing products may be followed by means of IR [14]. Thus, in Fig. 3.16 is shown the OH stretching region to 3000 to 4000 cm^{-1} for oxidized

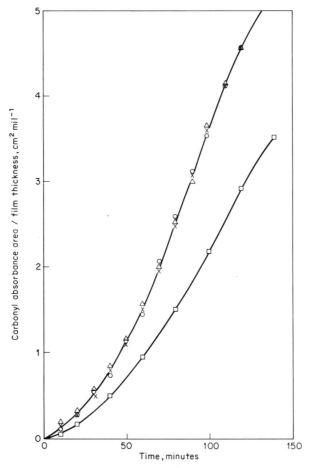

FIG. 3.15 Effect of APP film thickness upon oxidation rate at 130°C and 100 percent oxygen: △ 1.1 mils; × 2.5 mils; ○ 3.4 mils; □ 7.5 mils.

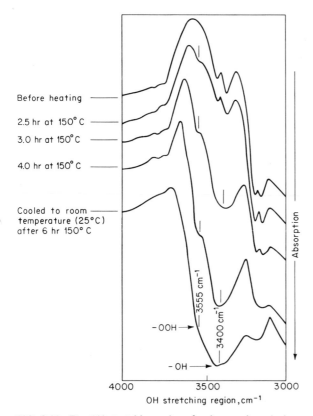

Before heating

2.5 hr at 150°C

3.0 hr at 150°C

4.0 hr at 150°C

Cooled to room
temperature (25°C)
after 6 hr 150°C

Absorption

3555 cm⁻¹

3400 cm⁻¹

–OOH→

–OH→

4000   3500   3000

OH stretching region, cm⁻¹

**FIG. 3.16   The OH stretching region of polypropylene during
and after accelerated oxidation [18].**

isotactic polypropylene [18]. After 2.5 hr oxidation, the band near 3400 cm^{-1} began to broaden toward 3550 cm^{-1}, owing to the formation of hydroperoxide groups (—OOH) which absorb at 3555 cm^{-1}. Upon further heating, both bands continued to increase in intensity, finally merging into a broad and strong absorption band. Rugg and coworkers [21] have indicated that during the oxidation of polyethylene the formation of a 3550 cm^{-1} band is related to the presence of free hydroperoxide groups whereas the band at 3360 cm^{-1} is related to associated hydroperoxide groups.

Besides the formation of hydroxyl and hydroperoxy bands in the IR spectrum during polymer oxidation, bands associated with nonvolatile carbonyl-containing compounds invariably form. Thus, during the oxidation of isotactic polypropylene in air at 150°C, a carbonyl band was observed in the 1650 to 1850 cm^{-1} (5.4 to 6.0 μ) region (cf. Fig. 3.13). In Fig. 3.17 is shown the intensity of this band after various oxidation periods [18]. Although the typical >C==O bands due to aldehydes, ketones, and esters

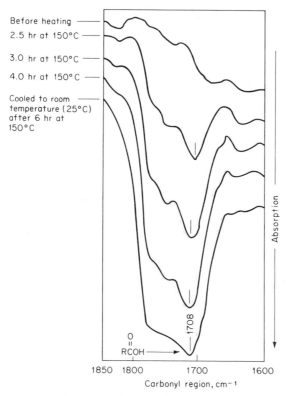

FIG. 3.17 Carbonyl region of polypropylene during and after oxidation at 150°C [18].

were merged into a single, broad, ill-defined band between 1710 and 1750 cm^{-1}, the band for the carbonyl groups due to carboxylic acids was quite distinct near 1708 cm^{-1}. This band at 1708 cm^{-1} was first to form and remained strong during the oxidation, indicating that acid groups are one of the major oxidation products of polypropylene. Jadrnicek, Stivala, and Reich [20] studied the relative amount of the carbonyl groups which formed during atactic polypropylene (APP) oxidation appearing as carboxylic acids [20]. To dilute solutions of undecan-6-one in a mixture of n-butanol and carbon tetrachloride (1:4 by volume) was added a saturated solution of 2,4-dinitrophenylhydrazine in n-butanol (total volume maintained constant). The resulting colored solutions of the hydrazones were scanned with a Beckman DU spectrophotometer, model 2400, and showed a maximum absorption at 480 mμ. A plot of absorbance versus undecanone concentration afforded a linear relationship at 480 mμ. Also, samples of APP films (7.5 mg, 2½ mils) were prepared which contained various known amounts of undecanone-6 and undecanal-1. The absorbance areas (in cm^2)

were obtained for the various admixtures in the bulk phase. The same linear relationship was found for both the aldehyde and ketone per 7.5-mg sample:

$$(AA) \; = \; 2.7 \times 10^5 \; [>C\!=\!O]_{AK} \tag{38}$$

Undecanoic acid and its ester were used in a similar manner to afford the following linear relationship per 7.5-mg sample:

$$(AA) \; = \; 6.7 \times 10^5 \; [>C\!=\!O]_{ACE} \tag{39}$$

where $[>C\!=\!O]_{AK}$ = moles of carbonyl as aldehyde and/or ketone; and $[>C\!=\!O]_{ACE}$ = moles of carbonyl as carboxylic acid and/or ester. Subsequently, APP films (7.5 mg, 2½ mils) were oxidized in an oxidation cell (Fig. 3.12) under various conditions of temperature, time, and oxygen concentration. After the appropriate carbonyl band was recorded (Fig. 3.13), the film sample was dissolved in a standard volume of a solution of n-butanol and carbon tetrachloride containing a standard amount of 2,4-dinitrophenylhydrazine. The absorbance at 480 mμ was recorded and from a calibration curve, the amount of carbonyl as aldehyde and/or ketone was determined per 7.5-mg polymer sample. It was found that Eqs. (38) and (39) were additive since the calculated and experimental values of AA agreed closely when these equations were utilized to estimate AA for a mixture of aldehyde, ketone, carboxylic acid, and ester, i.e.,

$$(AA)_T \; = \; 2.7 \times 10^5 \; [>C\!=\!O]_{AK} + 6.7 \times 10^5 \; [>C\!=\!O]_{ACE} \tag{40}$$

where $(AA)_T$ = total absorbance area (cm^2) per 7.5-mg sample. Since $(AA)_T$ and $[>C\!=\!O]_{AK}$ were known, the value of $[>C\!=\!O]_{ACE}$ could be calculated, and consequently the weight percentage of carbonyl as aldehyde and/or ketone and as carboxylic acid and/or ester. A solution of the oxidized APP gave no absorption band in the 1175 cm^{-1} region, whereas such a band occurred when a small amount of ester was added to the polymer solution. This indicated the absence of carbonyl as ester in the oxidized APP film sample. Thus, it was found that the weight percent of carbonyl as acid under various oxidation conditions was 15.5 ± 2 percent. From this value, Eq. (40) may be rewritten as

$$(AA)_T \; = \; (6.1 \pm 0.8) \times 10^5 \; [>C\!=\!O]_T \tag{41}$$

where $[>C\!=\!O]_T$ = total moles of carbonyl as aldehyde and/or ketone and as carboxylic acid.

The application of IR to the study of the formation of functional groups during the oxidative degradation of polymers will be discussed much more fully in Chap. 5.

Unsaturation

Luongo [18] assigned a pendant-type unsaturation, $R_1 R_2 C{=}CH_2$, to a band in atactic polypropylene (APP) at 888 cm^{-1}. Thus, upon chlorination of APP to remove unsaturation, the band at 888 cm^{-1} disappeared. Unsaturation due to a terminal vinyl type, $-CH{=}CH_2$, was assigned to the bands at 910 and 990 cm^{-1}. Furthermore, unsaturation depicted by $-CH{=}CH-$ groups was represented by bands at 960 cm^{-1} for polyethylene [22].

Phase Changes

Polymorphism in polymers such as polybutene-1 (PB) may be detected by IR. This polymer can exist in three distinct crystalline forms, depending upon how the film was prepared. These forms have been designated as I, II, and III. Forms I and III are stable whereas II is unstable under ordinary conditions. In Fig. 3.18 are shown the IR spectra of the three modifications [22]. In this figure, the arrows denote major spectral differences.

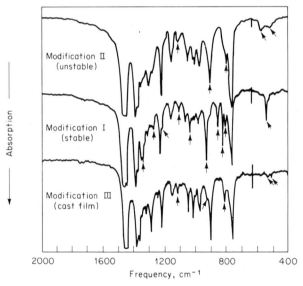

FIG. 3.18   Spectra of polybutene-1 polymorphs [22].

OTHER METHODS

Viscometric

A decrease in average molecular weight $\bar{M}$ generally accompanies thermal and/or oxidative degradation. Viscometry represents one method of measuring such changes. By comparing the efflux time t required for a given volume of polymer solution to flow through a capillary tube with the corresponding efflux time t_0 for the solvent, a measure of the solution viscosity can be made. Thus, the relative viscosity η_r is closely related to the ratio t/t_0, and the specific viscosity η_{sp} is defined as $\eta_r - 1$. The intrinsic viscosity $[\eta]$ is defined as the limit of the reduced viscosity η_{sp}/c, as the concentration of polymer solution c approaches zero. From the general Huggins equation,

$$\frac{\eta_{sp}}{c} = [\eta] + a_1 c + a_2 c^2 + \cdots \tag{42}$$

where $a_1 = k_1 [\eta]^2$, $a_2 = k_2 [\eta]^3$, etc.; k_1 and k_2 are constants (k_1 is usually referred to as the Huggins constant; and the value of $[\eta]$ may be obtained by extrapolation of linear plots of η_{sp}/c versus c to zero value of c. [Such a linear plot would obtain if powers of c of two or greater in Eq. (42) may be neglected.] The value of k_1 may be obtained from the slope of such a plot. The viscosity of polymer solutions is, among other things, dependent on the polymer-solvent system, and on temperature. The intrinsic viscosity may also be obtained from plots of the inherent viscosity, $(\ln \eta_r)/c$. Thus, inherent viscosity (I.V.) may be expressed as

$$\frac{\ln \eta_r}{c} = [\eta] - b_1 c + b_2 c^2 + \cdots \tag{43}$$

A correlation may be obtained between Eqs. (42) and (43) as follows. Upon expanding I.V. (since $\eta_r - 1 = \eta_{sp}$),

$$\text{I.V.} = \frac{\eta_{sp}}{c} - \left(\frac{\eta_{sp}}{c}\right)^2 \frac{c}{2} + \left(\frac{\eta_{sp}}{c}\right)^3 \frac{c^2}{3} + \cdots \tag{44}$$

Substituting Eq. (42) into (44) and collecting terms,

$$\text{I.V.} = [\eta] + \left(a_1 - \frac{[\eta]^2}{2}\right)c + \left(a_2 - a_1[\eta] + \frac{[\eta]^3}{3}\right)c^2$$
$$+ \left(\frac{a_1[\eta]^2}{3} + \frac{2a_1[\eta]}{3} - a_2[\eta]\right)c^3 + \cdots \tag{45}$$

Comparing Eqs. (45) and (43), it can readily be shown that

$$b_1 \equiv \frac{[\eta]^2}{2} - a_1 \tag{46a}$$

$$b_2 = a_2 + b_1[\eta] - \frac{[\eta]^3}{6} \tag{46b}$$

From Eqs. (46) and (42),

$$\frac{\eta_{sp}}{c} = [\eta] + \left(\frac{[\eta]^2}{2} - b_1\right)c + \left(b_2 - b_1[\eta] + \frac{[\eta]^3}{6}\right)c^2 + \cdots \tag{47}$$

Such an expression was obtained by Kobayashi [23]. If we let $b_1 = \beta_1 [\eta]^2$, Eq. (46a) affords,

$$k_1 + \beta_1 = \frac{1}{2} \tag{48}$$

When plots of η_{sp}/c and I.V. versus c do not yield values of k_1 and β_1 such that Eq. (48) is satisfied, it may be necessary to consider terms involving c^2 or higher in Eqs. (42) and (43). Thus, it may be necessary to construct plots of $(\eta_{sp}/c - [\eta])/c$ and $(I.V. - [\eta])/c$ versus c (utilizing values of $[\eta]$ obtained by neglecting terms in c involving powers of two and higher); and by proper adjustment of values of $[\eta]$, Eq. (48) can be satisfied [23]. This value of $[\eta]$ is considered to be more reliable than the corresponding value obtained by neglecting terms of c^2 and higher.

The relation of the intrinsic viscosity to the molecular weight M is given by the Mark-Houwink equation,

$$[\eta] = KM^{\alpha} \tag{49}$$

where K and α are constants. The exponent α varies from 1/2 to 1 for randomly coiled polymers, and approaches a value of 2 for rigid rods. Generally, the value of α lies between 0.6 and 0.8, and Eq. (49) is valid for linear polymers. The use of Eq. (49) is based on a homologous polymer series. Polymer degradation is accompanied by changes not only in molecular weight but also in molecular-weight distribution, and as in oxidative degradation, also by chemical composition. Accordingly, Eq. (49) should be viewed as an approximate relationship for highly degraded polymers. For a discussion of the various ramifications of intrinsic viscosity in regard to its relationship to molecular size, weight, and shape, the reader is referred to a review by Kurate and Stockmayer [24].

The application of viscometry to polymer degradation will be discussed more fully in later chapters of the book. In the meanwhile, we draw from the literature a specific example wherein intrinsic viscosity was employed to follow the molecular-weight decrease of atactic polybutene-1 in air at 90°C as a function of time [25] (Fig. 3.19).

Oxygen Absorption

A relatively simple method of studying the oxidation of polymers is in measuring the amount of oxygen absorbed. This method permits accurate measurements at low conversions, when the influence of oxidation products can be neglected. Generally, oxygen up-take is measured at constant pressure. In Fig. 3.20 is shown one such type of apparatus [26]. In this apparatus, a synthetic zeolite was used as an adsorbent for water and carbon dioxide. This adsorbent was held in place by a small plug of glass wool. A second plug of glass wool was placed near the open end of the reaction tube to prevent heat loss and to trap any con-densing materials which might other-wise contaminate the gas buret. After temperature equilibrium was reached, the system was adjusted to atmospheric pressure, and an initial reading was

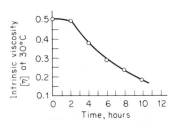

FIG. 3.19 Variation of the intrin-sic viscosity of atactic polybutene versus the oxidation time [25].

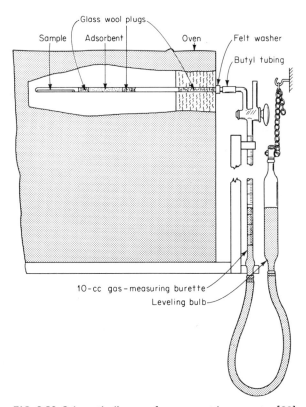

FIG. 3.20 Schematic diagram of oxygen-uptake apparatus [26].

made. Oxygen uptake was measured by subsequent readings at atmospheric pressure. Similar reaction tubes containing no sample were placed in the oven as blanks in correcting for changes due to fluctuating room and oven temperatures and atmospheric pressure. Hawkins and coworkers [26] utilized such an apparatus to study the relationship between increasing carbon black content and the oxygen uptake per gram of polyethylene as a function of reaction time at 140°C (Fig. 3.21).

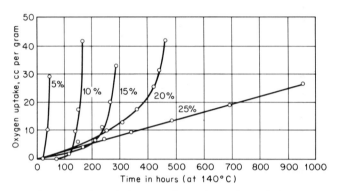

FIG. 3.21 Relationship between increasing carbon black concentration and the thermal oxidation rate of polyethylene [26].

Chemiluminescence

Chemiluminescence is the emission of light resulting from certain chemical reactions. For example, the oxidation of Grignard reagents in ether can give rise to a greenish blue glow which is visible in daylight. During the oxidation of polyolefins [27] a weak chemiluminescence has been observed. This luminescence is associated with the emission of light during electronic transitions from an excited state to the ground state. Thus, during the oxidation of simple hydrocarbons, the termination steps (50) and (51) can occur:

$$2RO_2 \cdot \longrightarrow (Products)^* + O_2 \tag{50}$$

$$(Products)^* \longrightarrow Products \tag{51}$$

In step (50), the asterisk denotes the formation of one or more products which are in an excited state, e.g., the formation of an electronically excited ketone. This ketone can undergo an electronic transition ($n \rightarrow \pi^*$ type) as indicated in step (51), which can give rise to phosphorescence (chemiluminescence). A similar situation may exist for the case of polyolefins. Thus, during the oxidation of these substances in the bulk phase, the following may occur:

$$RO_2H \longrightarrow (Products)^* \tag{52}$$

followed by reactions as step (51). Again these excited products in (52) can be ketonic in nature. Thus, De Kock and Hol [28] indicated that the luminescence occurring during polypropylene autoxidation may be interpreted as phosphorescence of carbonyl-group-containing compounds. However, it should also be noted that a much weaker chemiluminescence can result from a termination step involving recombination of radicals, R ·. Chemiluminescence will be mentioned again in this book in Chap. 5 on oxidative degradation. For a more detailed treatment of this subject, the reader is referred to Refs. 14 and 29. Suffice it at this point to say that chemiluminescence measurements may be used to determine various kinetic parameters, e.g., rate constants during hydrocarbon autoxidation. In the following is briefly described equipment used to measure chemiluminescence.

In Fig. 3.22 is depicted apparatus which was used in a study of the oxyluminescence of polymers [30]. A sample, E, was placed on the surface of a thermostatically controlled heating block, G, inside a light tight box. A ring with an optical glass cover provided a chamber, D, in which the sample

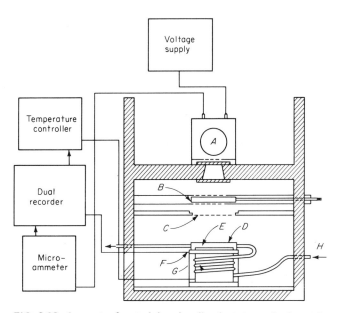

FIG. 3.22 Apparatus for studying chemiluminescence: A, phototube; B, shutter assembly; C, filter holder; D, gas chambers; E, sample; F, thermocouple; G, heating block; H, gas inlet [30].

could be heated under various gas atmospheres. The entering gas was metered into D and heated by coils around the block G. The light emitted from the polymer sample passed through glass filters C, if any were used; the shutter opening B, through a separated pair of windows which served as a thermal barrier; and finally to a photomultiplier tube cathode, A. The phototube assembly was placed in a chamber with solid CO_2 (to reduce the noise level) which was thermally insulated from the rest of the apparatus. The voltage supply to A was maintained constant, and recorders allowed the measurement of temperature and the milliamperage which came from A (proportional to the sample luminosity).

Miscellaneous Methods

There are other important methods which can be used in a study of polymer degradation but which must be classified as miscellaneous since a discussion of these methods would be outside the scope of this book. In the following are given in order the methods and related references which the interested reader may consult, plus a possible use of the methods in polymer degradation studies: polarography [31] can be employed for the estimation of hydroperoxide content; osmometry [32], light scattering [32], and ultracentrifugation [32] can be used in molecular-weight measurements; mass spectrometry [32], for the identification of degradation products; and, electron-spin resonance [33], for the qualitative and quantitative detection of free radicals associated with degradation processes.

REFERENCES

1. Cazes, J.: *J. Chem. Educ.*, **43**:A567, A625 (1966).
2. Cazes, J.: *Appl. Polymer Symp.*, **10**:7 (1969).
3. Grubisic, Z., P. Rempp, and H. Benoit: *J. Polymer Sci.*, **B-5**:753 (1967).
4. Doherty, S., and R. J. Valles: *Picatinny Arsenal Tech. Rept.* 3511, May, 1967.
5. Harmon, D. J.: *J. Polymer Sci.*, **C-8**:243 (1965).
6. Maley, L. E.: *J. Polymer Sci.*, **C-8**:253 (1965).
7. Keulemans, A. I. M.: "Gas Chromatography," Reinhold Publishing Corporation, New York, 1959.
8. Purnell, H.: "Gas Chromatography," John Wiley & Sons, Inc., New York, 1962.
9. Tsuchiya, Y., and K. Sumi: *J. Polymer Sci.*, **A-1,7**:3151 (1969).
10. Cox, B. C., and B. Ellis: *Anal. Chem.*, **36**:90 (1964).
11. Barlow, A., R. S. Lehrle, and J. C. Robb: *Polymer*, **2**:27 (1961).
12. Bombaugh, K. J., C. E. Cook, and B. H. Clampitt: *Anal. Chem.*, **35**:1834 (1963).
13. Groten, B.: *Anal. Chem.*, **36**:1206 (1964).
14. Reich, L., and S. S. Stivala: "Autoxidation of Hydrocarbons and Polyolefins," Marcel Dekker, Inc., New York, 1969.
15. Conley, R. T.: "Infrared Spectroscopy," Allyn and Bacon, Inc., Boston, 1966.
16. Stivala, S. S., L. Reich, and P. G. Kelleher: *Makromol. Chem.*, **59**:28 (1963).
17. Stivala, S. S., E. B. Kaplan, and L. Reich: *J. Appl. Polymer Sci.*, **9**:3557 (1965).
18. Luongo, J. P.: *J. Appl. Polymer Sci.*, **3**:302 (1960).
19. Balaban, L., J. Majer, and K. Vesely: *J. Polymer Sci.*, **C-22**:1059 (1969).
20. Jadrnicek, B., S. S. Stivala, and L. Reich: *J. Appl. Polymer Sci.*, **14**:2537 (1970).
21. Rugg, F. H., J. J. Smith, and R. C. Bacon: *J. Polymer Sci.*, **13**:535 (1954).
22. Luongo, J. P.: *J. Appl. Polymer Sci.*, **C-10**:121 (1969).
23. Kobayashi, H.: *J. Polymer Sci.*, **B-1**:299 (1963).
24. Kurata, M., and W. H. Stockmayer: *Advan. Polymer Sci.*, **3**:196 (1963).
25. Beati, E., F. Severini, and G. Clerici: *Makromol. Chem.*, **61**:104 (1963).
26. Hawkins, W. L., R. H. Hansen, W. Matreyek, and F. H. Winslow: *J. Appl. Polymer Sci.*, **1**:37 (1959).

27. Ashby, G. E.: *J. Polymer Sci.*, **50**:99 (1961).
28. De Kock, R. J., and P. A. H. M. Hol: *Rec. Trav. Chim.*, **85**:102 (1966).
29. Shlyapintokh, V. Y., O. N. Karpukhin, L. M. Postnikov, V. F. Tsepalov, A. A. Vichutinskii, and I. V. Zakharov: "Chemiluminescent Techniques in Chemical Reactions," Consultants Bureau, Plenum Publishing Corporation, New York, 1968.
30. Schard, M. P., and C. A. Russell: *J. Appl. Polymer Sci.*, **8**:985 (1964).
31. Kolthoff, I. M., and J. J. Lingane: "Polarography," vols. 1 and 2, Interscience Publishers, Inc., New York, 1952.
32. Schwarz, J. C. P. (ed.): "Physical Methods in Organic Chemistry," Holden-Day, Inc., Publisher, San Francisco, 1964.
33. Squires, T. L.: "An Introduction to Electron Spin Resonance," Academic Press, Inc., New York, 1964.

4

KINETICS AND
MECHANISMS OF THE THERMAL
DEGRADATION
OF POLYMERS

The degradation of polymers is of considerable importance from both practical and theoretical points of views. Knowledge of the extent of degradation of commercially available materials during use under various conditions is essential, e.g., synthetic rubber subjected to relatively high temperatures. On the other hand, a theoretical study can form the basis for the elucidation of the manner in which a material degrades. Predictions of performance of substances exposed to environmental factors often may be assessed from such studies. Thus, theory and practice are inseparable (more often than not), particularly in thermal (and oxidative) degradation of polymers. Although a vast amount of thermal degradation studies have been reported [1], including theoretical treatments [2–5], the kinetics and mechanisms of only a relatively few polymers are well understood. The reason is that a thorough analysis of the degradation mechanism would involve the determination of a number of factors not all of which can be readily obtained. These include: (1) degradation products, (2) molecular weights before and during degradation as a function of time or conversion, (3) rates of volatilization versus time or conversion, (4) the type of molecular-weight distribution before and during degradation, (5) the effects of the type of method used in polymer formation, e.g., free-radical versus anionic polymerization, etc.

The kinetics of thermal degradation of polymers is treated from an elementary viewpoint in this chapter. Accordingly, the extensive mechanism and relatively complicated mathematics which arise from treatments of Simha and coworkers [2-7] will not be considered. The reader is referred to Chap. 1 for a brief review.

INITIAL STAGES OF DEGRADATION

In the following will be presented a simplified degradation scheme which generally applies to vinyl polymers (as opposed to condensation polymers) and which is subject to the following limitations. During the initial degradation stages, it is assumed that the monomer is the only volatile species (when the monomer is the main volatile product, then the initial stage limitation need not apply), and that the polymer is monodisperse (except when indicated otherwise). We may write the following steps with the corresponding kinetic expressions for this simplified scheme.

Random Initiation

$$\text{\small{\textasciitilde\textasciitilde\textasciitilde}} \xrightarrow{k_{ir}} \text{\small{\textasciitilde\textasciitilde}} \cdot + \text{\small{\textasciitilde\textasciitilde}} \cdot \tag{1}$$

$$\frac{d[\text{R}\cdot]}{dt} = 4k_{ir}D_P[n] \tag{1a}$$

where $\text{\textasciitilde\textasciitilde\textasciitilde}$ and $\text{\textasciitilde\textasciitilde}\cdot$ denote polymer chains and radicals, respectively; $[\text{R}\cdot]$ = polymer radical concentration; k_{ir} = rate constant for random initiation; D_P = degree of polymerization; and $[n]$ = concentration of polymer chains. It is assumed in Eq. (1a) that $D_P \gg 1$, and the integer 4 is a factor arising from the formation of two polymer radical molecules per polymer molecule plus a statistical factor of 2, which arises from the fact that the two polymer radicals can form from the polymer molecule in two ways (depending upon which end of the molecule is taken as the basis).

Terminal (end) Initiation

$$\text{\small{\textasciitilde\textasciitilde\textasciitilde}} \xrightarrow{k_{ie}} \text{\small{\textasciitilde\textasciitilde}} \cdot + M\cdot\uparrow \tag{2}$$

$$\frac{d[\text{R}\cdot]}{dt} = 2k_{ie}[n] \tag{2a}$$

In Eq. (2a), the integer 2 is a statistical factor, and it is assumed that the monomer radical $M\cdot$ is rapidly volatilized out of the reaction medium, as indicated by the arrow pointing upward.

Depropagation

$$\sim\!\!\sim\!\!\sim \cdot \xrightarrow{\;k_d\;} \sim\!\!\sim \cdot + M \nearrow \tag{3}$$

$$\frac{d(M)}{dt} = k_d[R \cdot]V \tag{3a}$$

It is assumed in this step that the monomer M is also rapidly volatilized from the system. Furthermore, (M) denotes amount of monomer, e.g., moles, and V = volume of the reaction medium.

Unimolecular Termination

$$\sim\!\!\sim\!\!\sim \cdot \xrightarrow{\;k_{t1}\;} \text{Products} \tag{4}$$

$$-\frac{d[R \cdot]}{dt} = k_{t1}[R \cdot] \tag{4a}$$

Although it is often difficult to account for a unimolecular termination in the bulk phase, such a termination is often employed in order to explain kinetic features of some polymer degradations (cf. Chap. 1).

Termination by Recombination

$$2 \sim\!\!\sim\!\!\sim \cdot \xrightarrow{\;k_{tc}\;} \text{Products} \tag{5}$$

$$-\frac{d[R \cdot]}{dt} = 2k_{tc}[R \cdot]^2 \tag{5a}$$

Termination by Disproportionation

$$2 \sim\!\!\sim \cdot \xrightarrow{\;k_{td}\;} \sim\!\!\sim + \sim\!\!\sim \tag{6}$$

$$-\frac{d[R \cdot]}{dt} = k_{td}[R \cdot]^2 \tag{6a}$$

Intermolecular Transfer

$$\sim\!\!\sim\!\!\sim + \sim\!\!\sim \cdot \xrightarrow{\;k_{tr}\;} \sim\!\!\sim + \sim\!\!\sim + \sim\!\!\sim \cdot \tag{7}$$

$$\frac{d(n)}{dt} = k_{tr}D_P[n][R \cdot]V \tag{7a}$$

where (n) = number of polymer molecules.

Along with relationships shown in Eqs. (1a) to (7a), the following expressions will be employed:

$$\frac{d(M)}{dt} = \frac{1}{M_m} \left(-\frac{dW}{dt} \right) \tag{8}$$

$$[n] = \frac{W}{M_m D_P V} = \frac{\rho}{M_m D_P} \tag{9}$$

and

$$\frac{dM_n}{dt} = \frac{M_n}{W} \frac{dW}{dt} - \frac{M_n^2}{W} \frac{d(n)}{dt} \tag{10}$$

where M_m = monomeric molecular weight, W = weight of polymer at time t, ρ = polymer density, and M_n = number-average molecular weight of polymer. Equation (10) derives from the expression $M_n = W/(n)$.

Case 1: Random initiation (R.I.) followed by complete unzipping. In case 1, complete unzipping of the polymer radical will occur when the average kinetic chain length (CL) (very nearly equal to the so-called average zip length), i.e., the average number of monomer units successively released by depropagation along the length of the polymer radical, is larger than the polymer degree of polymerization, i.e., $(CL) > D_{P,0}$. The subscript zero indicates zero time, i.e., before the onset of degradation. We may write

$$\frac{d(M)}{dt} = \frac{D_P}{2} \frac{d[\text{R} \cdot]}{dt} V \tag{11}$$

Using Eqs. (1a), (8), and (9), there is obtained

$$-\frac{dW}{dt} = 2k_{ir} D_P W \tag{12}$$

We may also write [cf. Eqs. (1), (1a), and (9)]

$$-\frac{d(n)}{dt} = 2k_{ir} D_P [n] V = \frac{2k_{ir} W}{M_m} \tag{13}$$

From Eqs. (10), (12), and (13),

$$\frac{dM_n}{dt} = -2k_{ir} D_P M_n + \frac{2k_{ir} M_n^2}{M_m} = 0 \tag{14}$$

Thus, from Eqs. (13) and (14), it can be seen that for case 1 the rate of volatilization is first order in respect to W and that there is no change in M_n with time during the degradation.

A more general treatment of case 1 will now be presented in which zip length Z and molecular-weight distribution (non-monodisperse polymer) will be considered [8]. In such a treatment, it will be necessary to let Z be some multiple of D_P, i.e.,

$$Z = bD_P \tag{15}$$

(For a monodisperse polymer, $b = 1$). We may now write

$$-\frac{dW}{dt} = 2k_{ir} ZW \tag{12a}$$

where Z replaces D_P in Eq. (12). When Eqs. (12a) and (13) are substituted into Eq. (10), we obtain after slight modification

$$-\frac{dD_P}{dt} = 2k_{ir}(b - 1)D_P^2 \tag{16}$$

Upon integration, Eq. (16) yields

$$\frac{1}{D_P} - \frac{1}{D_{P,0}} = 2k_{ir}(b - 1)t \tag{17}$$

Furthermore, when Eqs. (12a), (15), and (16) are combined,

$$\frac{dD_P}{dW} = \frac{D_P(b - 1)}{bW} \tag{18}$$

Upon integration of Eq. (18), there is obtained

$$\frac{D_P}{D_{P,0}} \equiv B = \left(\frac{W}{W_0}\right)^{(b-1)/b} \tag{19}$$

or

$$\frac{W}{W_0} = 1 - C = B^{b/(b-1)} \tag{19a}$$

where C = conversion = $(W_0 - W)/W_0$. As before, the subscript zero refers to zero time: i.e., the W_0 is weight before degradation. In Fig. 4.1 are shown curves drawn in accordance with Eq. (19a) for various values of b. When Eqs. (12a) and (19) are combined,

$$-\frac{dW}{dt} = \frac{2k_{ir}bD_{P,0}W^{(2b-1)/b}}{W_0^{(b-1)/b}} \tag{20}$$

It can be seen from Eq. (20) that when $b = 1, 2$, and 3, the reaction order is 1, $3/2$, and $5/2$, respectively. For high values of b, the reaction order approaches a limiting value of 2. In the following, molecular-weight distributions will be considered in order to attach some significance to b.

A common distribution of molecular weights for vinyl-type polymers is the exponential distribution (9),

$$(n)_x^o = \frac{(n)^o}{D_{P,0}} \exp\left(\frac{-x}{D_{P,0}}\right) \tag{21}$$

where $(n)^o$ and $(n)_x^o$ denote the total number and the number of molecules of degree of polymerization x, respectively, which are initially present in the polymer and $D_{P,0}$ denotes the number average of polymerization before degradation. As previously indicated [cf. Eq. (13)], the rate of removal of molecules of degree of polymerization x is proportional to the

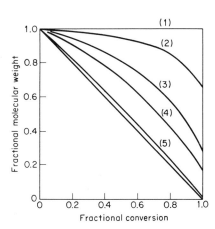

FIG. 4.1 A series of curves drawn according to Eq. (19a). The values of b for the curves are: (1) $b = 1$, (2) $b = 1.1$, (3) $b = 1.5$, (4) $b = 2$, (5) $b = 10$ [8].

total number of links associated with such molecules. Thus,

$$-\frac{d(n)_x}{dt} = 2k_{ir}(n)_x(x-1) \tag{22}$$

Upon integration, Eq. (22) becomes

$$(n)_x^t = (n)_x^o \exp[-(x - 1) \cdot 2k_{ir} t] \qquad (23)$$

where $(n)_x^t$ = number of molecules of degree of polymerization x at time t. When Eq. (21) is substituted into Eq. (23),

$$(n)_x^t \simeq \frac{(n)^o}{D_{P,0}} \exp[-(x - 1) d] \qquad (24)$$

where $d = 2k_{ir}t + 1/D_{P,0}$. The total number of molecules (n) at time t may be written as

$$\sum_2^\infty (n)_x^t \quad \text{or} \quad \int_2^\infty (n)_x^t \, dx$$

Thus,

$$(n) = \int_2^\infty \frac{(n)^o}{D_{P,0}} \exp[-(x - 1) \cdot d] \, dx = \frac{(n)^o}{D_{P,0}} \cdot d^{-1} \qquad (25)$$

Also,

$$W = M_m \sum_2^\infty x(n)_x^t = M_m \int_2^\infty \frac{(n)^o}{D_{P,0}} x \exp[-(x - 1)d] \, dx$$

$$= \frac{M_m (n)^o d^{-2}}{D_{P,0}} \qquad (26)$$

When Eqs. (9), (25), and (26) are combined,

$$\frac{1}{D_P} - \frac{1}{D_{P,0}} = 2k_{ir} t \qquad (27)$$

The differential form of Eq. (27) is

$$-\frac{dD_P}{dt} = 2k_{ir}D_P^2 \tag{28}$$

Further, when Eq. (9) is substituted into (25), there is obtained upon squaring

$$\left(\frac{W}{M_m D_P}\right)^2 = \left[\frac{(n)^o}{D_{P,0}}\right]^2 d^{-2} \tag{29}$$

Equation (29) becomes after employing Eq. (26),

$$\frac{W}{D_P^2} = \frac{W_0}{D_{P,0}^2} \tag{30}$$

or

$$1 - C = B^2 \tag{30a}$$

As indicated in Eq. (19), B is defined as the ratio of the degree of polymerization at time t to that at time zero, i.e., $D_P/D_{P,0}$. When Eq. (30) is differentiated,

$$\frac{dW}{dD_P} = \frac{2D_P W_0}{D_{P,0}^2} \tag{31}$$

Using Eq. (31), the expression $dW/dt = (dW/dD_P)(dD_P/dt)$, and Eqs. (28) and (31), there is obtained

$$-\frac{dW}{dt} = 4k_{ir}D_{P,0}\frac{W^{3/2}}{W_0^{1/2}} \tag{32}$$

When Eqs. (27), (28), (30a), and (32) are compared with Eqs. (17), (16), (19a), and (20), respectively, it can be readily seen that the value of $b = 2$. In a similar manner, it can be shown that a coupling-type distribution [9] yields kinetic expressions for which $b = 1.5$.

It is noted from the preceding that for a monodisperse polymer undergoing degradation according to case 1, $b = 1$; for a coupling-type distribution, $b = 1.5$; and for an exponential distribution, $b = 2$. Thus, values of b are a measure of the breadth of molecular-weight distribution. An application of Eq. (19a) to the thermal degradation of poly(methyl methacrylate) (PMMA) at relatively high reaction temperatures (about 500°C) follows (the main product of PMMA decomposition is monomer, about 95 percent).

In 1962, Wall [10] stated that the kinetics and mechanism of the decomposition of poly(methyl methacrylate) seemed to be reasonably well understood. This appears now to be an overstatement as numerous somewhat discordant studies have been reported since then on the kinetics and mechanism of PMMA degradation [11–15]. This lack of agreement is due to the many factors which may influence PMMA decomposition, e.g., type of tacticity [14], presence of double-bond ends [11], initial molecular weight [12, 14, 15], and temperature range used [12–15]. Nevertheless, for instructive purposes, the results and conclusions of Lehrle and coworkers [13, 15] will be used. Thus, these workers state that in the decomposition of PMMA, at the lowest temperatures used (about 300°C), the depropagating chains are predominantly end-initiated, and chain termination occurs mainly by bimolecular termination. However, at the highest temperatures used (about 500°C), initiation is predominantly by random chain scission, and the majority of chains are effectively terminated as the depropagation reaches the end of the molecule, the terminal radical distilling out the system. [Eq. (19) should apply.] In Fig. 4.2 are shown plots of $M_n/M_{n,0}$ versus conversion for unfractionated and fractionated samples of PMMA which were pyrolyzed at 500°C. From Eq. (19a) and Fig. 4.1, it can be seen that for a fractionated sample ($b = 1$), molecular weight should

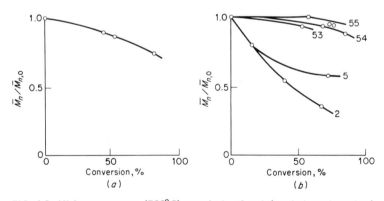

FIG. 4.2 High-temperature (500°C) pyrolysis of poly(methyl methacrylate). Fractional number-average molecular-weight changes for various percentage conversions to monomer: (a) unfractionated sample, (b) fractionated samples [15].

not change with conversion for conditions listed under case 1. This is found to hold in Fig. 4.2b for PMMA samples 53 to 55 ($M_{n,0}$ = 30,000 to 40,000), as anticipated, at the higher temperatures. Similarly, when $b > 1$, a decrease in molecular weight should occur as the conversion increases (cf. Fig. 4.1).

This trend is observed in Fig. 4.2a. The results exhibited in Fig. 4.2 generally agree with the trends expected for decompositions initiated by random chain scission followed by complete depropagation (case 1). However, the relatively high-molecular-weight fractions 2 and 5 (M_n = 150,000 and 76,000, respectively) in Fig. 4.2b show a significant decrease in M_n with conversion. Several possible explanations were offered: (1) the fractionation procedure is not perfect so that the high-molecular-weight fractions could be sufficiently heterodisperse to show some characteristics of unfractionated samples; (2) the reaction times are inevitably short, and despite experimental arrangements, the sample may spend a sufficient proportion of its reaction time at intermediate and low temperatures to display some of the characteristics of the mechanism corresponding to these temperatures; and (3) the high-temperature mechanism may be too idealistic: i.e., there is still a finite possibility of bimolecular termination which could become significant if the initial molecular weight of the fraction exceeded the mean kinetic chain length. Explanation (3) was given preference as the most likely possibility. It should also be noted here that besides the agreement between Eq. (19) and the experimental results at relatively high decomposition temperatures, Lehrle and coworkers [13] indicated that at these temperatures an expression such as Eq. (12) was also valid.

Case 2: Random initiation and unzipping without any loss of molecules. For random initiation (R.I.) with no loss of molecules [incomplete unzipping or $(CL) < D_{P,0}$], the rate of production of molecules may be expressed as [cf. Eq. (13)]

$$\frac{d(n)}{dt} = \frac{2k_{ir}W}{M_m} \tag{13a}$$

Substituting Eqs. (12a) and (13a) into (10),

$$-\frac{dD_P}{dt} = 2k_{ir}D_P(D_P + Z) \tag{33}$$

Equation (33) affords, upon integration,

$$\ln \frac{D_{P,0}(D_P + Z)}{D_P(D_{P,0} + Z)} = 2k_{ir}Zt \tag{34}$$

When Eq. (33) is divided by (12a) and the resulting expression integrated,

$$\frac{D_P(D_{P,0} + Z)}{D_{P,0}(D_P + Z)} = \frac{W}{W_0} \tag{35}$$

Upon rearrangement, Eq. (35), becomes

$$B = \frac{f(1 - C)}{f + C} \tag{35a}$$

where $f = Z/D_{P,0}$. In Fig. 4.3 are shown curves derived from a plot of Eq.

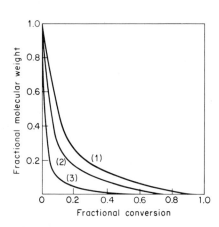

FIG. 4.3   A series of curves drawn according to Eq. (35a). The values of $Z/D_{P,0}$ for the different curves are as follows: (1) $f = 0.1$, (2) $f = 0.05$, (3) $f = 0.01$ [8].

(35a) for various values of f. It may be apropos to remark here that if the initial molecular weight exceeded the mean value of (CL) in the decomposition of PMMA (cf. explanation (3) advanced by Lehrle and coworkers [13]) and if Eq. (35a) applied, a rapid drop in molecular weight with low conversion would be expected (cf. Fig. 4.2b).

Case 3: Terminal (end) initiation (T.I.) followed by complete unzipping. Similar to case 1, we may write

$$\frac{d(M)}{dt} = D_P \frac{d[R \cdot]}{dt} V \tag{36}$$

or, more generally,

$$\frac{d(M)}{dt} = Z \frac{d[R \cdot]}{dt} V \tag{36a}$$

By employing Eqs. (2a), (8), and (36a),

$$-\frac{dW}{dt} = 2k_{ie} M_m(n)Z \tag{37}$$

where $Z = bD_P$ [cf. Eq. (15)]. For a homodisperse sample, $b = 1$. Further, when exponential [cf. Eq. (21)] and coupling molecular-weight distributions are assumed, then it may be shown, as in a previously discussed manner (cf. case 1), that b is still unity for both types of distributions. Thus, Eq. (37) can be written as follows (for fractionated and unfractionated samples, the latter possessing one of the two types of distributions mentioned previously):

$$-\frac{dW}{dt} = 2k_{ie} M_m(n)D_P \tag{37a}$$

Equation (37a) becomes, after using Eq. (9),

$$-\frac{dW}{dt} = 2k_{ie} W \tag{38}$$

We may also write [cf. Eqs. (2) and (2a)]

$$-\frac{d(n)}{dt} = 2k_{ie}(n) \tag{39}$$

Substituting Eqs. (38) and (39) into (10) yields

$$\frac{dM_n}{dt} = -2k_{ie} M_n + \frac{2k_{ie}(n)M_n^2}{W} = 0 \tag{40}$$

[since $(n) = W/M_n$]. Thus, from the preceding, the molecular weight remains constant with conversion to monomer and the rate of volatilization is first order in respect to sample weight and is independent of molecular weight [cf. Eq. (12)]. The latter characteristics may be employed to distinguish between R.I. (case 1) and T.I. (case 3) since in both these cases molecular weight should not change with conversion (cf. Ref. 13).

Case 4: Terminal initiation and unzipping without any loss of molecules. In this case, $-d(n)/dt = 0$, and Eq. (10) becomes

$$\frac{dM_n}{dt} = \frac{M_n}{W}\frac{dW}{dt} \tag{10a}$$

Substituting Eq. (37) into (10a) and using (9),

$$-\frac{dD_P}{dt} = 2k_{ie} Z \tag{41}$$

When Eq. (41) is divided by (37) and the resulting expression is integrated,

$$\frac{D_P}{D_{P,0}} \equiv B = 1 - C \tag{42}$$

When Eqs. (9), (42), and (37) are combined,

$$-\frac{dW}{dt} = \frac{2k_{ie}W_0 Z}{D_{P,0}} \tag{43}$$

Case 5: Random initiation, depropagation, and unimolecular termination.
In cases such as this, i.e., where termination reactions are specified, it is
tacitly assumed that $(CL) \ll D_{P,0}$. From Eqs. (3a) and (8),

$$-\frac{dW}{dt} = k_d M_m [R\cdot] V \tag{44}$$

Assuming steady-state conditions for $[R\cdot]$ and using Eqs. (1a) and (4a),

$$4k_{ir}D_P[n] = k_{t1}[R\cdot] \tag{45}$$

Combining Eqs. (44), (45), and (9) leads to

$$-\frac{dW}{dt} = \frac{4k_{ir}k_d W}{k_{t1}} \tag{46}$$

For the rate of change of the number of molecules, we may write

$$\frac{d(n)}{dt} = -2k_{ir}D_P(n) + k_{t1}[R\cdot]V = 2k_{ir}D_P(n) \tag{47}$$

When Eqs. (46) and (47) are substituted into Eq. (10), and Eq. (9) is used,

$$-\frac{dM_n}{dt} = \frac{4k_{ir}k_d M_n}{k_{t1}} + \frac{2k_{ir}M_n^2}{M_m} \tag{48}$$

When Eq. (48) is divided by (46), we obtain

$$\int \frac{dW}{W} = \int \frac{dM_n}{M_n(1 + AM_n)} \tag{49}$$

where $A = k_{t1}/2M_m k_d$. By the method of partial fractions,

$$\int \frac{dM_n}{M_n(1 + AM_n)} = \int \frac{dM_n}{M_n} - \int \frac{A\, dM_n}{1 + AM_n} \tag{50}$$

Substituting Eq. (50) into (49) and integrating,

$$\frac{W}{W_0} = 1 - C = \frac{M_n}{M_{n,0}}\left(\frac{1 + AM_{n,0}}{1 + AM_n}\right) \tag{51}$$

or

$$\frac{M_n}{M_{n,0}} = \frac{1 - C}{1 + AM_{n,0}C} \tag{51a}$$

where the subscript zero indicates at time zero.

An expression for the D_P as a function of time may be obtained from Eq. (48). Thus, upon integration of this equation [cf. Eq. (50)], there is obtained

$$\ln\left(1 + \frac{1}{AM_n}\right) - \ln\left(1 + \frac{1}{AM_{n,0}}\right) = Dt \tag{52}$$

where $D = 4k_{ir}k_d/k_{t1}$, and $A = k_{t1}/2M_m k_d$. If it is assumed that $AM_n \gg 1$, then Eq. (52) may be simplified to yield

$$\frac{1}{D_P} - \frac{1}{D_{P,0}} \simeq 2k_{ir}t \tag{53}$$

Equation (53) is similar to expressions which can be derived from mechanisms involving random degradation without any depropagation or termination. Furthermore, a similar expression may be obtained from a rather elaborate scheme which involves random initiation, depropagation, and termination by disproportionation (when the chain length of the largest volatile polymer chain is much less than the average D_P of the polymer during its decomposition) [16]. In connection with Eq. (53), it may be mentioned that a scheme similar to that of case 5 was employed to account for the kinetic data obtained during the photochemical degradation of poly(α-methylstyrene), cf. Chap. 1, where it was found that a relationship similar to Eq. (53) was in agreement with the experimental data obtained [17].

Case 6: Terminal initiation, depropagation, and unimolecular termination. Employing Eqs. (2a), (4a), (9), (44), and assuming steady-state conditions for [R·],

$$-\frac{dW}{dt} = \frac{2k_{ie}k_d W}{k_{t1}D_P} \tag{54}$$

Also,

$$-\frac{d(n)}{dt} = 2k_{ie}(n) - k_{t1}[R\cdot]V = 0 \tag{55}$$

Using Eqs. (10), (54), and (55),

$$-\frac{dD_P}{dt} = \frac{2k_{ie}k_d}{k_{t1}} \tag{56}$$

Upon dividing Eq. (56) by (54) and integrating,

$$B = 1 - C \tag{42}$$

As indicated, the previous expression is identical with Eq. (42). From this equation, $W/D_P = W_0/D_{P,0}$, and Eq. (54) becomes

$$-\frac{dW}{dt} = \frac{2k_{ie}k_d W_0}{k_{t1}D_{P,0}} \tag{57}$$

Equation (57) becomes identical with Eq. (43) since $Z \approx (CL) = k_d/k_{t1}$ for case 6. Thus, for cases 4 and 6, the rate of volatilization is zero order in respect to sample weight (Z = const), and the expressions for D_P as a function of conversion are identical.

Case 7: Random initiation, depropagation, and termination by disproportionation. From Eq. (44) and steady-state conditions for [R·] using Eqs. (1a) and (6a), we may write

$$-\frac{dW}{dt} = 2k_d M_m V \left(\frac{k_{ir}D_P[n]}{k_{td}}\right)^{1/2} \tag{58}$$

Substituting Eq. (9) into (58) along with the relationship for ρ ($\rho \equiv W/V$),

$$-\frac{dW}{dt} = 2k_d \left(\frac{M_m k_{ir}}{k_{td}\rho}\right)^{1/2} W \equiv k_e W \quad \text{(assuming } \rho = \text{constant)} \tag{59}$$

We may also write

$$\left(\frac{1}{V}\right)\frac{d(n)}{dt} = -2k_{ir}D_P[n] + k_{td}[R\cdot]^2$$
$$= 2k_{ir}D_P[n] \tag{60}$$

Substituting Eqs. (59) and (60) into (10) affords

$$\frac{dM_n}{dt} = -M_n\left[\frac{2k_{ir}M_n}{M_m} + 2k_d\left(\frac{k_{ir}M_m}{\rho k_{td}}\right)^{1/2}\right] \tag{61}$$

From the definition of kinetic chain length (CL), we may write

$$(CL) = \frac{k_d}{k_{td}[R\cdot]} = \frac{k_d}{2}\left(\frac{M_m}{k_{ir}k_{td}\rho}\right)^{1/2} \tag{62}$$

Combination of Eqs. (61) and (62) leads to

$$\frac{dM_n}{dt} = -2k_{ir}M_n\left[\frac{M_n}{M_m} + 2(CL)\right] \tag{63}$$

If it is assumed in Eq. (63) that $D_P \gg (CL)$ and if the resulting expression is integrated,

$$\frac{1}{D_P} - \frac{1}{D_{P,0}} = 2k_{ir}t \tag{53}$$

This expression is identical with Eq. (53) obtained previously for case 5. Further expressions may be obtained as follows. When Eq. (59) is integrated,

$$\ln\frac{W_0}{W} = 2k_d\left(\frac{M_m k_{ir}}{k_{td}\rho}\right)^{1/2}t = 4k_{ir}(CL)t \tag{64}$$

Upon combining Eqs. (53) and (64), we finally obtain

$$\frac{M_{n,0} - M_n}{M_n M_{n,0}} = \ln\frac{(W_0/W)}{2M_m(CL)} \tag{65}$$

Various workers have indicated that the expressions developed in case 7 may be applied to the thermal decomposition of poly(tetrafluorethylene) (Teflon) [18–20]. Thus, from TGA, isothermal, molecular-weight determination, etc., techniques, the following experimental observations were made: (1) the pyrolysis product is mainly monomer (the simplified treatment in case 7 should apply beyond initial degradation stages); (2) the pyrolysis rate is independent of molecular weight [cf. Eq. (59)]; (3) the pyrolysis follows first-order kinetics over the entire conversion range [cf. Eq. (59), and assume $\rho \approx$ const during degradation], and the experimental rate constant k_e could be expressed as

$$k_e = 3 \times 10^{19} \exp\left(-\frac{83,000}{RT}\right) \sec^{-1} \tag{66}$$

and (4) the molecular weight decreases during the pyrolysis (in vacuum). From Eqs. (59) and (66), it can be seen that

$$E = E_d + \frac{E_{ir} - E_{td}}{2} \tag{67}$$

where E = experimental value of overall activation energy, E_d = activation energy for the depropagation step, E_{ir} = activation energy for the random initiation step, and E_{td} = activation energy for termination by disproportionation step. If it is assumed that $E_{td} \approx 0$ and that $E_{ir} = 74$ kcal mole^{-1} (calculated from thermodynamic data for the bond energy of the C—C bond in Teflon [21]), then the value of E_d should be about 43 kcal mole^{-1} [from Eqs. (66) and (67)]. In this connection, it may be noted that the activation enthalpy for Teflon depropagation at 480°C was reported [19] to be 44 kcal mole^{-1}. The value of (CL) may be obtained from a plot of Eq. (65). Such a plot is shown in Fig. 4.4 [19]. (Molecular weights were determined by means of standard specific-gravity and/or melt-viscosity measurements.) From the value of the slope, a value of $(CL) = 720$ was calculated. For various Teflon samples studied, the value of $D_{P,0}$ varied from 3×10^3 to 4×10^5. These values support the assumption made in obtaining Eqs. (53) and (64) [that $D_P \gg (CL)$]. Furthermore, values of frequency factors A may be estimated as follows. From Eq. (59), it can be shown that at 480°C ($\rho = 1.34$ g cm^{-3}) the following applies:

$$A_{td} = A_{ir}\left(\frac{0.39 A_d}{A}\right)^2 \tag{68}$$

where the subscripts denote steps identical to those already mentioned in connection with activation energy E, and A = overall value associated with

Eq. (66). Assuming the following values (time units are in seconds): $A_{ir} = 10^{17.3}$ (similar to A for the decomposition of ethane into methyl radicals); $A_d = 10^{13}$ [similar to that reported for poly(methyl methacrylate)]; and since $A = 10^{19.5}$, the value of A_t becomes $10^{3.5}$. This latter value was used

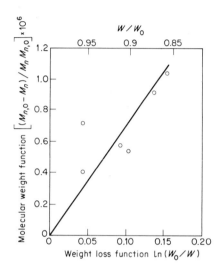

FIG. 4.4 Change of molecular weight during vacuum pyrolysis at 480°C of sample of Teflon [19].

to calculate $k_{td} = 2.5 \times 10^2$ liter mole^{-1} sec^{-1} at 360°C and 4.0×10^2 liter mole^{-1} sec^{-1} at 510°C, which are similar to values obtained for the decomposition of poly(methyl methacrylate) [22]. Finally, when various values of A and E are substituted into Eq. (62), the following relationship can be obtained:

$$\log(CL) \simeq 1.9 - \frac{306}{T} \tag{69}$$

For the calculated value of (CL) to equal the experimental value of 720 at 480°C it is necessary to assume a value of 15.94 for $\log A_{ir}$, other values remaining the same. In this case, Eq. (69) would become

$$\log(CL) \approx 3.26 - \frac{306}{T} \tag{69a}$$

Case 8: Terminal initiation, depropagation, and termination by disproportionation. From Eq. (44) and steady-state conditions for $[R\cdot]$, using Eqs. (2a) and (6a), we may write

$$-\frac{dW}{dt} = k_d M_m V \left(\frac{2k_{ie}[n]}{k_{td}}\right)^{1/2} \tag{70}$$

Substituting Eq. (9) into (58) along with the definition for ρ,

$$-\frac{dW}{dt} = k_d \left(\frac{2k_{ie} M_m}{k_{td}\rho D_P}\right)^{1/2} W \equiv k'_e W \tag{71}$$

We may also write

$$\frac{1}{V}\frac{d(n)}{dt} = -2k_{ie}[n] + k_{td}[\text{R}\cdot]^2 = 0 \tag{72}$$

Substituting Eqs. (71) and (72) into (10) leads to

$$-\frac{dM_n}{dt} = k_d M_m^{1/2} M_n^{1/2} \left(\frac{2k_{ie} M_m}{k_{td}\rho}\right)^{1/2} \tag{73}$$

When Eq. (73) is divided by (71) and the resulting expression integrated, an expression identical to Eq. (42) is obtained, i.e.,

$$\frac{D_P}{D_{P,0}} \equiv B = 1 - C \tag{42}$$

When Eq. (42) is used with Eq. (71), there is obtained

$$-\frac{dW}{dt} = k_d \left(\frac{2k_{ie} M_m W_0}{k_{td}\rho D_{P,0}}\right)^{1/2} W^{1/2} \tag{74}$$

Thus, from Eq. (74), the rate of volatilization is half order in respect to sample weight. In this connection, it may be mentioned that when the expression for Z [$\approx (CL)$] is substituted into Eq. (43), i.e.,

$$(CL) = \left(\frac{k_d}{2k_{ie} k_{td}[n]}\right)^{1/2} \tag{75}$$

and Eqs. (9) and (42) are utilized, an expression identical to Eq. (74) can be obtained.

It was previously indicated (cf. case 1) that for the relatively high-temperature decomposition of PMMA (about 500°C), the experimental data obtained were consistent with random chain scission followed by complete unzipping (or depropagation). Thus, the various kinetic expressions of case 1 were found to apply, e.g., Eq. (12). In this equation, the experimental rate constant was directly proportional to D_P. However, at relatively low temperatures (about 300°C) used in PMMA pyrolysis, another

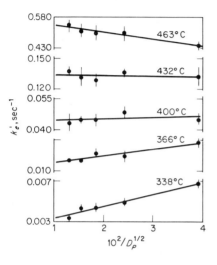

FIG. 4.5 PMMA degradation over the temperature range 338 to 463°C: dependence of specific rate on molecular weight; 75 percent confidence limits are shown [13].

mechanism was found to apply, i.e., case 8 which involves terminal (end) chain initiation, depropagation, and termination by disproportionation. Thus, Lehrle and coworkers [13] utilized an expression such as Eq. (71) to estimate values of k'_e for various values of D_P. Good correlations were obtained between k'_e and $1/D_P^{1/2}$ at the relatively low temperatures of 338 and 366°C (see Fig. 4.5) [13]. Furthermore, during the low-temperature pyrolysis of PMMA at 300°C, the value of B decreased linearly with increasing conversion, which would be anticipated from Eq. (42). At the intermediate temperatures, 400 and 432°C, k'_e becomes independent of D_P. From Eq. (59), case 7, it can be seen that k_e should be independent of D_P. Thus, a possible mechanism at these intermediate temperatures involves random chain scission, depropagation, and termination by bimolecular interaction. Also, a first-order plot of Eq. (59) at 416°C indicated no deviations from first-order behavior, even at conversions of 80 percent. On the other hand, deviations from linearity were obtained for first-order plots at 340 and 370°C [cf. Eq. (74)]. Grassie and Melville [23] also concluded that PMMA degradation is end-initiated at low temperatures. It is also interesting to note that Jellinek and Luh [14] found that in the thermal degradation of PMMA, plots of $1/D_P - 1/D_{P,0}$ versus time gave more linear relationships over a greater conversion range as the temperature rose from 300 to 350°C [cf. Eq. (53) which should apply at the intermediate reaction temperatures].

Case 9: Random initiation, depropagation, unimolecular termination, and intermolecular transfer. As in case 5, the expression for the rate of volatilization is

$$-\frac{dW}{dt} = \frac{4k_{ir}k_d W}{k_{t1}} \tag{46}$$

However, the expression for the rate of change of the number of molecules is, using Eq. (7a),

$$\frac{d(n)}{dt} = -2k_{ir}D_P(n) + k_{t1}[R\cdot]V + k_{tr}D_P(n)[R\cdot] \tag{76}$$

Using steady-state conditions for $[R\cdot]$ and Eq. (9), Eq. (76) becomes

$$\frac{d(n)}{dt} = \frac{2k_{ir}W}{M_m}\left(1 + \frac{2k_{tr}\rho}{k_{t1}M_m}\right) \tag{77}$$

When Eqs. (46) and (77) are substituted into Eq. (10), there is obtained

$$-\frac{dM_n}{dt} = \frac{4k_{ir}k_d M_n}{k_{t1}} + \frac{2k_{ir}M_n^2}{M_m} \cdot E \tag{78}$$

where $E \equiv 1 + (2k_{tr}\rho/k_{t1}M_m)$.

When Eq. (78) is divided by Eq. (46), and the resulting expression is integrated [cf. case 5, Eq. (51a)], there is finally obtained

$$\frac{M_n}{M_{n,0}} = \frac{1 - C}{1 + A'M_{n,0}C} \tag{79}$$

where $A' = k_{t1}E/2k_d M_m = AE$ [cf. Eq. (49)].

An expression for the D_P as a function of time may be obtained from Eq. (78) in a manner similar to that described in case 5. Thus,

$$\ln\left(1 + \frac{1}{A'M_n}\right) - \ln\left(1 + \frac{1}{A'M_{n,0}}\right) = Dt \tag{80}$$

where $D = 4k_{ir}k_d/k_{t1}$ [cf. Eq. (52)].

If it is assumed that $A'M_n \gg 1$, then Eq. (80) may be simplified to give

$$\frac{1}{D_P} - \frac{1}{D_{P,0}} \simeq 2k_{ir}Et \tag{81}$$

Case 10: Terminal initiation, depropagation, unimolecular termination, and intermolecular transfer. As in case 6, it can be shown that the rate of volatilization may be expressed as [cf. Eq. (54)]

$$-\frac{dW}{dt} = \frac{2k_{ie}k_dW}{k_{t1}D_P} \tag{54}$$

Further, we may write

$$\frac{d(n)}{dt} = k_{tr}D_P(n)[\text{R}\cdot] \tag{82}$$

Using steady-state conditions for $[\text{R}\cdot]$ and Eq. (9), Eq. (82) becomes

$$\frac{d(n)}{dt} = \frac{2k_{ie}k_{tr}\rho W}{k_{t1}M_mM_n} \tag{83}$$

Substituting Eqs. (54) and (83) into (10) leads to

$$-\frac{dM_n}{dt} = \frac{2k_{ie}k_dM_m}{k_{t1}} + \frac{2k_{ie}k_{tr}\rho M_n}{k_{t1}M_m} \tag{84}$$

When Eq. (84) is divided by (54),

$$\frac{dM_n}{dW} = \frac{M_n}{W} + \frac{k_{tr}\rho}{M_m^2 k_d}\frac{M_n^2}{W} \tag{85}$$

Equation (85) may now be rearranged to give an expression similar to Eq. (49). Thus, in a manner described in case 5, it can be shown that

$$\frac{M_n}{M_{n,0}} = \frac{1 - C}{1 + FM_{n,0}C} \tag{86}$$

where $F = k_{tr}\rho/M_m^2 k_d$. Further, Eq. (54) may be expressed in terms of conversion C as follows:

$$\frac{dC}{dt} = \frac{2k_{ie}k_d M_m (1 - C)}{k_{t1}M_n}$$

(54a)

When Eq. (86) is substituted into (54a),

$$\frac{dC}{dt} = \frac{2k_{ie}k_d M_m (1 + FM_{n,0}C)}{k_{t1}M_{n,0}}$$

(87)

Wall and coworkers [24] attempted to fit an expression such as Eq. (86) to the thermal degradation of monodisperse polystyrenes under high vacuum at 300°C. Values of M_n of the residue were determined by membrane osmometry. In Fig. 4.6 are shown experimental values in plots of $D_P/D_{P,0}$ (or $M_n/M_{n,0}$) versus conversion for various values of $M_{n,0}$. The curves in this figure were calculated from Eq. (86), using various values of $FM_{n,0}$ listed above the corresponding curve. A value of $FM_{n,0} = 10$ was taken for the monodisperse polystyrene of $M_{n,0} = 170,000$. Considering the relatively simple scheme used (e.g., the evaporative removal of molecules other than monomer was neglected even though only 42 weight percent of volatile products is monomer), the agreement between calculated and observed values was judged to be very good. From Fig. 4.6 it can also be seen that the value of M_n does not change with conversion for very low values of $M_{n,0}$ (5,300). This tendency follows schemes listed under cases

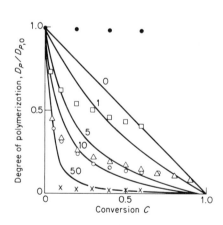

FIG. 4.6 Relative degree of polymerization as a function of conversion. The curves were calculated from Eq. (86) for various values of $FM_{n,0}$ listed above the corresponding curves [24]. The $M_{n,0}$ values for the polystyrenes are: ● 5,300; □ 37,000; △ 170,000; × 2 × 10⁶.

1 and 3 which involve terminal or random initiation, followed by complete unzipping (however, see comments under case 12). At higher values of $M_{n,0}$, intermolecular transfer becomes increasingly important, and the decrease in $D_P/D_{P,0}$ becomes greater, the greater $M_{n,0}$, as conversion

increases. From the value of $FM_{n,0} = 10$, a value of $k_{tr}/k_d = 0.63$ could be calculated. It should be mentioned here that other schemes involving intermolecular transfer can yield an expression similar or identical to Eq. (86) (cf. cases 11 and 12, respectively).

Case 11: Random initiation, depropagation, termination by dispropor- tionation, and intermolecular transfer. As in case 7, the rate of volatiliza- tion can be shown to be identical to Eq. (59),

$$-\frac{dW}{dt} = 2k_d \left(\frac{M_m k_{ir}}{k_{td}\rho}\right)^{1/2} W = k_e W \tag{59}$$

We may also write [cf. Eqs. (60) and (82)]

$$\frac{d(n)}{dt} = 2k_{ir}D_P(n) + k_{tr}D_P(n)[R\cdot] \tag{88}$$

Assuming steady-state conditions for $[R\cdot]$ (cf. case 7) and employing Eq. (9),

$$\frac{d(n)}{dt} = \frac{2k_{ir}W}{M_m}\left[1 + k_{tr}\left(\frac{\rho}{M_m k_{ir}k_{td}}\right)^{1/2}\right] \equiv PW \tag{89}$$

where P is defined in Eq. (89). Upon substituting Eqs. (59) and (89) into Eq. (10),

$$-\frac{dM_n}{dt} = k_e M_n + PM_n^2 \tag{90}$$

When Eq. (90) is divided by (59) and the resulting equation integrated [cf. Eq. (50), case 5],

$$\frac{M_n}{M_{n,0}} = \frac{1-C}{1 + PM_{n,0}C/k_e} \tag{91}$$

As in cases 5 and 9, it can be shown that if $k_e \ll PM_n$ (when inter- molecular transfer is favored, L will be large, where L is the chain length of the largest volatile chains),

$$\frac{1}{M_n} - \frac{1}{M_{n,0}} = Pt \tag{92}$$

Although Eqs. (86) and (91) may be applied equally to the data in Fig. 4.6, it will be shown that the data for the rate of volatilization of polystyrenes are more in accordance with an expression obtained from case 12. This case will also yield an expression identical to Eq. (86).

Case 12: Terminal initiation, depropagation, termination by disproportionation, and intermolecular transfer. As in case 8, the expression for rate of volatilization is

$$-\frac{dW}{dt} = k_d \left(\frac{2k_{ie} M_m}{k_{td}\rho D_P}\right)^{1/2} W \equiv k'_e W \tag{71}$$

We may also write (cf. case 10)

$$\frac{d(n)}{dt} = k_{tr} D_P (n)[R\cdot] \tag{82}$$

Using steady-state conditions for $[R\cdot]$ and Eq. (9), Eq. (82) becomes

$$\frac{d(n)}{dt} = \frac{k_{tr} W (2k_{ie}\rho/k_{td}M_n)^{1/2}}{M_m} \tag{93}$$

When Eqs. (71) and (93) are substituted into Eq. (10), there is obtained

$$-\frac{dM_n}{dt} = k_d M_m M_n^{1/2} \left(\frac{2k_{ie}}{k_{td}\rho}\right)^{1/2} + \frac{k_{tr}M_n^{3/2}(2k_{ie}\rho/k_{td})^{1/2}}{M_m} \tag{94}$$

For convenience, let us rewrite Eq. (94) as

$$-\frac{dM_n}{dt} = M_n^{1/2}(S + TM_n) \tag{94a}$$

where the definitions of the terms S and T follow from a comparison between Eqs. (94) and (94a). Thus, Eq. (71) may be rewritten as

$$-\frac{dW}{dt} = \frac{SW}{M_n^{1/2}} \tag{71a}$$

Upon dividing Eq. (94a) by (71a),

$$\frac{dM_n}{dt} = \frac{M_n}{W} + \frac{TM_n^2}{SW} \tag{95}$$

Equation (95) is of a form similar to Eq. (49), and therefore its solution becomes

$$\frac{M_n}{M_{n,0}} = \frac{1 - C}{1 + FM_{n,0}C} \tag{86}$$

since $T/S \equiv F$ [cf. Eq. (86)]. Equation (71a) may be rewritten in terms of conversion as follows:

$$\frac{dC}{dt} = \frac{S(1 - C)}{M_n^{1/2}} \tag{71b}$$

Upon substitution of Eq. (86) into Eq. (71b),

$$\frac{dC}{dt} = \frac{S(1 - C)^{1/2}(1 + FM_{n,0}C)^{1/2}}{M_{n,0}^{1/2}} \tag{96}$$

As previously mentioned in case 10, an expression of the form of Eq. (86) appeared to be in concordance with experimental data obtained for the thermal degradation of polystyrenes under vacuum [24]. Furthermore, it was found that a rate expression similar to Eq. (96) also agreed reasonably well with kinetic data obtained for polystyrene degradation [24]. This equation can be shown to possess a maximum value of dC/dt [$(dC/dt)_{max}$] when $C = (K - 1)/2K$, where $K = FM_{n,0}$. Thus, Eq. (96) becomes

$$\frac{dC/dt}{(dC/dt)_{max}} = \frac{2[K(1 + KC)(1 - C)]^{1/2}}{K + 1} \tag{97}$$

From the preceding, when $K = FM_{n,0} = 10$ (cf. case 10, Fig. 4.6, for $M_{n,0} = 170,000$) during the pyrolysis of monodisperse polystyrenes, $C_{max} \cong 0.45$. In Fig. 4.7 the dashed curves denote calculated values obtained from Eq. (97) for various values of K. The solid lines are based on an equation derived by assuming termination to be monomolecular.

It may be interesting to note here that for low values of $M_{n,0}$, molecular weight does not change with conversion. This tendency follows schemes listed under cases 1 and 3 which involve terminal or random initiation followed by complete unzipping. However, these cases also indicate that [cf. Eqs. (12) and (38)]

$$\frac{dC/dt}{(dC/dt)_{max}} = 1 - C \qquad (98)$$

and from Fig. 4.7 it can be seen that such an expression does not hold well for $M_{n,0} = 5{,}300$. On the other hand, if it is assumed that monodisperse polystyrene fractions of low $M_{n,0}$ (e.g., <4,000) vaporize molecularly without degradation and since intermolecular transfer terms for $M_{n,0} = 5{,}300$ are low in value ($K \approx 0$), then molecular weight should not change much with conversion, and from Eq. (96) it can be shown that

$$\frac{dC/dt}{(dC/dt)_{max}} \simeq (1 - C)^{1/2} \qquad (99)$$

From Fig. 4.7 it can be seen that an expression such as Eq. (99) fits the data for $M_{n,0} = 5{,}300$ much better than Eq. (98).

Richards and Salter [25] studied the thermal degradation of polystyrene (PS) in presence of poly(α-methylstyrene) (PMS). The latter degrades (95 weight percent monomer) at temperatures about 50°C lower than the

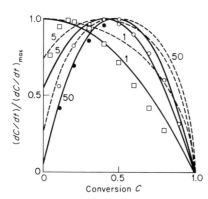

FIG. 4.7 Rates of volatilization as a function of conversion during polystyrene pyrolysis for various values of $FM_{n,0} = K$ [cf. Eq. (97)]. The $M_{n,0}$ values and the reaction temperatures used are: □ 5,300 (343°C); ● 170,100 (310°C); ○ 170,100 (331°C) [24].

former. During the degradation between 260 and 287°C, the styrene produced must have resulted from interaction of PS with degrading PMS since no styrene was obtained from pure PS under these conditions. In Fig. 4.8 are shown plots of rates of evolution of α-methylstyrene versus molecular weight of PMS in PMS-PS mixtures at various temperatures.

When log $(W_0/W)_{PMS}$ was plotted against molecular weight of PMS, a linearity was obtained up to a molecular weight of 60,000 for all temperatures studied. However, at high molecular weights, the curves became parallel to the abscissa. These results could be explained by means of

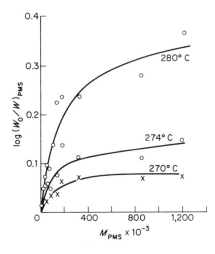

FIG. 4.8 Rates of evolution of α-methylstyrene against molecular weight of PMS in PMS-PS mixtures at various temperatures [25].

cases 1 and 7. Thus, at low values of molecular weight, PMS degradation followed case 1, cf. Eq. (12), from which a plot of log (W_0/W) versus D_P (or M_n) should be linear. At high molecular weights, the zip length of PMS becomes much smaller than the average degree of polymerization, and now case 7, Eq. (59), applies. In this case, log (W_0/W) should be independent of D_P, as observed. The kinetic results obtained are in concordance with those observed by others [26] for PMS. Thus, it is apparent that the presence of PS does not appreciably affect the degradation of PMS. However, the reverse appeared to be true. Thus, as the PMS molecular weight rose from 7,400 to 332,000, the amount of styrene which evolved over a certain time interval decreased from 0.32 to 0.00 mg. Such observations may be reconciled if only the monomer radical produced after complete unzipping of PMS can initiate PS decomposition. Then the rate of PS initiation would decrease with increasing PMS molecular weight, as observed. The α-methylstyrene monomer radicals presumably chain-transfer to PS to produce PS radicals which liberate styrene by depropagation and which undergo intermolecular transfer and termination by combination reactions, as indicated in case 12.

Case 13: Initiation via weak links, depropagation, and unimolecular (or cage) or bimolecular termination. Richards and Salter [27] synthesized polystyrene chains (PS_0) containing known numbers of weak links (X). The latter consisted of head-head links which are thermally unstable as compared with the head-tail links normally found in polystyrenes. Thus, normal polystyrene (PS) could be used as a diluent for PS_0 since degradation studies can be conducted at temperatures at which PS is stable. In this manner, problems of inhomogeneity and constant bulk viscosity may be eliminated. Richards and Salter attempted to discover whether unimolecular (or cage) or bimolecular termination would be predominant when PS_0 was degraded at relatively low temperatures (lower than those at which PS degraded). In the following will be derived expressions for both types of termination. Short zip lengths and monodisperse PS_0 will be assumed. Further, since we will be concerned only with monomer evolution, it will not be necessary to consider intermolecular transfer, and evaporative losses of molecules other than monomer will be neglected.

Cage or unimolecular termination. Cage and unimolecular termination are kinetically indistinguishable from each other; accordingly, they will be considered together. For the latter Eqs. (4) and (4*a*) apply, while for the former we may write

$$(R\cdot + R\cdot)_{cage} \xrightarrow{k_c} \text{Polymer} \tag{100}$$

and

$$-\frac{d[R\cdot]}{dt} = 2k_c[R\cdot] \tag{100a}$$

The rate of formation of radicals may be written as

$$\frac{d[R\cdot]}{dt} = 2k_{iw}[X] \tag{101}$$

where $[X]$ = concentration of weak links in the polymer chains. We may also write

$$-\frac{d[X]}{dt} = k_{iw}[X] \tag{101a}$$

assuming that once the weak links are cleaved, they are destroyed. From Eqs. (3*a*) and steady-state conditions for $[R\cdot]$,

$$\frac{d(M)}{dt} = \frac{k_{iw}k_d(X)}{k_c} \tag{102}$$

When Eq. (101*a*) is integrated $[(X) = (X)_0 e^{-k_{iw}t}]$ and the expression for the number of weak links (X) substituted into Eq. (102), there is obtained

$$\frac{d(M)}{dt} = \frac{k_{iw} k_d (X)_0 e^{-k_{iw}t}}{k_c}$$

(103)

Integration of Eq. (103) leads to

$$(M) = \frac{k_d(X)_0 \left(1 - e^{-k_{iw}t}\right)}{k_c}$$

(104)

From Eq. (104), the total monomer evolved may be expressed as

$$(M)_\infty = \frac{k_d(X)_0}{k_c}$$

(105)

and from Eqs. (104) and (105) and the definition of the reaction half-life $t_{1/2}$ [when $(M)/(M)_\infty = 1/2$],

$$k_{iw} = \frac{0.693}{t_{1/2}}$$

(106)

Finally, when Eqs. (105) and (106) are substituted into Eq. (104),

$$(M) = (M)_\infty \left(1 - e^{-0.693t/t_{1/2}}\right)$$

(107)

Bimolecular termination.   Adopting an approach similar to that in cage termination and assuming that Eq. (5a) is applicable,

$$\frac{d[M]}{dt} = \frac{k_d (k_{iw}[X]_0)^{1/2} e^{-k_{iw}t/2}}{k_{tc}}$$

(108)

Upon integration, Eq. (108) affords

$$[M] = \frac{2k_d [X]_0^{1/2} \left(1 - e^{-k_{iw}t/2}\right)}{(k_{iw}k_{tc})^{1/2}}$$

(109)

Furthermore, Eq. (109) may be converted to

$$(\overset{\bullet}{M}) = (M)_\infty \left(1 - e^{-0.693t/t_{1/2}}\right)$$

(107)

which is identical with Eq. (107).

From the expressions derived in cage and bimolecular termination it can be seen that the nature of the termination step cannot be established by studying monomer evolution as a function of reaction time. However, from Eqs. (104) and (109) it can be seen that if the weak-bond concentration $[X]_0$ is varied, then at a particular temperature and at constant degradation time, $[M] \propto [X]_0$ for cage termination and $[M] \propto [X]_0^{1/2}$ for bimolecular termination. From such expressions, the type of termination involved may be determined by measurements of monomer evolution as a function of dilution. In Fig. 4.9 are shown plots of styrene evolution versus initial weak-bond concentration over a constant time interval for various reaction temperatures. This plot should be linear when the cage reaction dominates and should curve toward the abscissa when the bimolecular termination step is significant. From Fig. 4.9 it appears that cage reactions are predominant at low weak-bond concentrations, whereas at high concentrations bimolecular termination becomes increasingly important, despite the considerable scatter of the experimental points. This trend may be explained if at low values of $[X]_0$, the average distance between radical pairs produced by weak-link scission is much greater than the average distance between the two radicals comprising each pair. An increase in $[X]_0$ reduces the average distance between radical pairs (by increasing the equilibrium radical concentration), thereby increasing the probability of bimolecular termination.

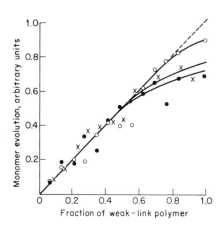

FIG. 4.9 Effect of weak-link concentration on styrene evolution [27]: ○ 276° C, × 285° C, ● 289° C.

It may be further noted here that irrespective of the termination mechanism, Eq. (107) should be valid. In Fig. 4.10 are shown curves calculated from Eq. (107). The experimental values corresponded closely to the calculated curves in all the cases studied. Finally, it should be mentioned that, as previously indicated, normal polystyrene degrades mainly by terminal scission (cf. case 12). With this type of initiation, one of the radicals formed is either monomeric or of low molecular weight. These radicals can readily diffuse away from their partners. Thus, cage effects should be must less important in this case.

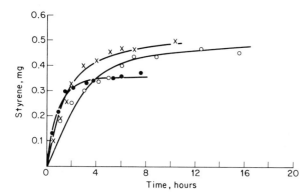

FIG. 4.10 Evolution of styrene from weak-link polymer as a function of time [27]: ○ 280°C, × 285°C, ● 292°C.

A MORE ELABORATE SCHEME FOR THERMAL DEGRADATION (cf. JELLINEK [16])

In the following will be presented a more elaborate scheme for thermal degradation of vinyl (or addition) polymers than was presented under Initial Stages of Degradation, page 165. Again, it is assumed that the polymer is monodisperse. However, the limitations of low conversions or that monomer be the only volatile species are now removed. Such a more elaborate scheme can give insight into the limitations that are to be applied in order that expressions in Initial Stages of Degradation be more applicable. The following mechanism may now be rewritten.

Random Initiation

$$P_{D_P} \xrightarrow{k_{ir}} R_{\dot{D}_{P}-n} + R_{n}^{\cdot} \tag{110}$$

where P_{D_P} denotes polymer chains of degree of polymerization D_P.

Terminal Initiation

$$P_{D_P} \xrightarrow{k_{ie}} R_{\dot{D}_{P}-1} + R_{1}^{\cdot} \tag{111}$$

Depropagation

$$R_{\dot{D}_{P}} \xrightarrow{k_d} R_{\dot{D}_{P}-1} + P_1 \text{ (monomer)} \tag{112}$$

Unimolecular Termination

$$R_{D_P} \xrightarrow{k_{t1}} P_{D_P} \tag{113}$$

Termination by Recombination

$$R_{\dot{D}_P} + R_{\dot{i}} \xrightarrow{k_{tc}} P_{D_P + i} \tag{114}$$

Termination by Disproportionation

$$R_{\dot{D}_P} + R_{\dot{i}} \xrightarrow{k_{td}} P_{D_P} + P_i \tag{115}$$

Intermolecular Transfer

$$R_{\dot{n}} + P_{D_P} \xrightarrow{k_{tr}} P_n + P_{D_P - j} + R_{\dot{j}} \tag{116}$$

From Eqs. (110) to (116), various expressions may be derived and are listed under various case numbers. These numbers do not necessarily correspond with those listed under Initial Stages of Degradation. Furthermore, owing to the complexity of the mathematical treatment, cases involving intermolecular transfer have not been included, since these would be beyond the elementary approach of this book.

Some fundamental expressions will now be given [cf. Eq. (9)]. These are:

$$\sum_{D_P = L + 1}^{D_{P,0}} [n]_{D_P} D_P = \frac{\rho}{M_m} = \frac{1}{V_m} \tag{117}$$

where V_m = monomeric-unit molar volume, and L = chain length of the largest volatile chains. Also,

$$\sum_{D_P = L + 1}^{D_{P,0}} [n]_{D_P} = \frac{1}{V_m D_P'} \tag{118}$$

where D_P' = number-average chain length of the residual polymer (not including volatiles) at time t of the degradation process. From Eqs. (117) and (118),

$$\sum_{D_P = L + 1}^{D_{P,0}} [n]_{D_P} (D_P - 1) = \frac{D_P' - 1}{V_m D_P'} \tag{119}$$

We will now consider the following cases.

Case 1': Random initiation followed by complete unzipping. In this case, as soon as a chain is initiated, it depropagates completely to monomer. Thus, the only volatile product is monomer while the residual polymer (nondepropagated) maintains its number-average chain length $D_{P,0}$ constant throughout the degradation. Accordingly, the polymer remains monodisperse. We may write [cf. Eq. (110)]

$$-\frac{d[n]_{D_{P,0}}}{dt} = 2k_{ir}[n]_{D_{P,0}}(D_{P,0} - 1) \tag{120}$$

Upon integration, Eq. (120) becomes

$$\ln\left(n_{D_{P,0;\tau=0}}/n_{D_{P,0;\tau}}\right) = (D_{P,0} - 1)\tau \tag{121}$$

where $\tau \equiv 2k_{ir}t$. Further, we may write for the total rate of monomer formation,

$$\frac{d(M)}{dt} = -\frac{d\left(n_{D_{P,0;\tau}}D_{P,0}\right)}{d\tau} = (D_{P,0} - 1)n_{D_{P,0;\tau}}D_{P,0} \tag{122}$$

Since $n_{D_{P,0;\tau}}D_{P,0} = (M)_0 - (M)$, where $(M)_0$ = total moles of monomer present in the polymer and (M) = moles of monomer which volatilized, Eq. (122) becomes [cf. Eq. (12)]

$$-\ln(1 - C) = (D_{P,0} - 1)\tau \approx D_{P,0}\tau \tag{123}$$

Case 2': Random initiation, depropagation, and unimolecular termination. From steady-state considerations for [R·], we may write

$$4k_{ir}\sum_{D_P=L+1}^{D_{P,0}}[n]_{D_P}(D_P - 1) = k_{t1}[R\cdot] \tag{124}$$

When Eq. (119) is substituted into (124), the expression for total radical concentration [R·] becomes

$$[R\cdot] = \frac{4k_{ir}(D_P' - 1)}{k_{t1}V_m D_P'} \tag{125}$$

Also, for chain lengths lower than $L + 1$,

$$\frac{d[\text{R}\cdot]_{D_P}}{dt} = k_d\left([\text{R}\cdot]_{D_P+1} - [\text{R}\cdot]_{D_P}\right) + 4k_{ir}\sum_{D_P=L+1}^{D_{P,0}}[n]_{D_P}$$

$$- k_{t1}[\text{R}\cdot]_{D_P} \tag{126}$$

Expression (126) may be transformed into [assuming that $(\text{R}\cdot)_1$ radicals are readily lost from the system],

$$\frac{d\sum_{D_P=1}^{L}[\text{R}\cdot]_{D_P}}{dt} = k_d[\text{R}\cdot]_{L+1} + \frac{4k_{ir}L}{V_m D_P'}$$

$$- k_{t1}\sum_{D_P=1}^{L}[\text{R}\cdot]_{D_P} = 0 \tag{127}$$

If the term containing k_d in Eq. (127) is neglected in comparison with other terms,

$$\sum_{D_P=1}^{L}[\text{R}\cdot]_{D_P} = \frac{4k_{ir}L}{V_m D_P' k_{t1}} \tag{128}$$

We may also write

$$\frac{d[n]_{D_P}}{dt} = -2k_{ir}(D_P - 1)[n]_{D_P} + k_{t1}[\text{R}\cdot]_{D_P} \tag{129}$$

From Eq. (129),

$$\frac{d\sum_{D_P=L+1}^{D_{P,0}}[n]_{D_P}}{dt} = -2k_{ir}\sum_{D_P=L+1}^{D_{P,0}}(D_P - 1)[n]_{D_P}$$

$$+ k_{t1}\left([\text{R}\cdot] - \sum_{D_P=1}^{L}[\text{R}\cdot]_{D_P}\right) \tag{130}$$

Substituting Eqs. (118), (119), (125), and (128) into (130) (not assuming that $D_P' \gg 1$),

$$-\frac{d(D_P')}{d\tau} = D_P'[D_P' - (1 + 2L)] \tag{131}$$

When Eq. (131) is integrated in a manner previously described [cf. Eq. (50)],

$$\tau = \frac{1}{1 + 2L}\left[\ln\left(1 - \frac{1 + 2L}{D_{P,0}}\right) - \ln\left(1 - \frac{1 + 2L}{D_P'}\right)\right] \tag{132}$$

A well-known relationship for random degradation is obtained when $L = 0$ (a closed system or when none of the polymer volatilizes in an open system), i.e.,

$$\ln\left(1 - \frac{1}{D_{P,0}}\right) - \ln\left(1 - \frac{1}{D_P'}\right) = \tau \tag{132a}$$

or, for moderate degrees of degradation,

$$\frac{1}{D_P'} - \frac{1}{D_{P,0}} \simeq \tau \tag{53}$$

The last expression is identical with Eq. (53).

The rate of formation of monomeric-unit moles of volatiles up to and including L-chain lengths may be expressed as (in terms of concentration)

$$\frac{d \sum\limits_{D_P = 1}^{L} [M]_{D_P}}{dt} = k_d[\text{R}\cdot] + k_{t1} \sum\limits_{D_P = 1}^{L} D_P[\text{R}\cdot]_{D_P} \tag{133}$$

From Eq. (126), we may write

$$\frac{d \sum\limits_{D_P = 1}^{L} D_P [\text{R} \cdot]_{D_P}}{dt} = \frac{2k_{ir} L (L + 1)}{V_m D'_P}$$

$$+ k_d \left(L [\text{R} \cdot]_{L + 1} - \sum\limits_{D_P = 1}^{L} [\text{R} \cdot]_{D_P} \right)$$

$$- k_{t1} \sum\limits_{D_P = 1}^{L} D_P [\text{R} \cdot]_{D_P} = 0 \qquad (134)$$

where steady-state conditions are assumed. Neglecting the term containing k_d in comparison with other terms [e.g., $k_d \ll k_{t1} (L + 1)/2$],

$$\sum\limits_{D_P = 1}^{L} D_P [\text{R} \cdot]_{D_P} = \frac{2k_{ir} L (L + 1)}{V_m D'_P k_{t1}} \qquad (135)$$

When Eqs. (125) and (135) are substituted into Eq. (133), there is obtained

$$\frac{d \sum\limits_{D_P = 1}^{L} [\text{M}]_{D_P}}{dt} = \frac{4k_d k_{ir} (D'_P - 1)}{k_{t1} V_m D'_P} + \frac{2k_{ir} L (L + 1)}{V_m D'_P} \qquad (136)$$

or

$$\frac{d \sum\limits_{D_P = 1}^{L} (\text{M})_{D_P}}{dt} = \left[4 (CL) k_{ir} + \frac{2k_{ir} L (L + 1)}{D'_P} \right]$$

$$\times \left[(\text{M})_{D_P, 0} - \sum\limits_{D_P = 1}^{L} (\text{M})_{D_P} \right] \qquad (136a)$$

where (CL) = kinetic chain length = k_d/k_{t1}, and $(M)_{D_P, 0}$ = initial amount of monomeric-unit moles. From Eq. (136a) it can be seen that when $(CL) \gg L(L + 1)/2D_P'$ (this assumption will be more valid at low conversions where D_P' possesses a relatively high value), then an expression may be obtained which is similar to that derived from case 5, Eq. (46), i.e.,

$$-\ln(1 - C) = 4(CL)k_{ir}t \qquad (137)$$

Case 3': Random initiation, depropagation, and termination by disproportionation. From steady-state considerations for [R·],

$$4k_{ir} \sum_{D_P = L+1}^{D_{P,0}} [n]_{D_P}(D_P - 1) = k_{td}[R·]^2 \qquad (138)$$

Upon substituting Eq. (119) into (138),

$$[R·] = 2\left[\frac{k_{ir}(D_P' - 1)}{k_{td}V_m D_P'}\right]^{1/2} \qquad (139)$$

Also, for chain lengths shorter than $L + 1$,

$$\frac{d[R·]_{D_P}}{dt} = k_d\left([R·]_{D_P + 1} - [R·]_{D_P}\right) + 4k_{ir} \sum_{D_P = L+1}^{D_{P,0}} [n]_{D_P}$$
$$- k_{td}[R·][R·]_{D_P} \qquad (140)$$

In a manner similar to that used for Eq. (126),

$$\frac{d \sum_{D_P = 1}^{L} [R·]_{D_P}}{dt} = k_d[R·]_{L+1} + \frac{4k_{ir}L}{V_m D_P'}$$
$$- k_{td}[R·] \sum_{D_P = 1}^{L} [R·]_{D_P} \qquad (141)$$

If the term containing k_d in Eq. (141) is neglected in comparison with the other terms,

$$\sum_{D_P = 1}^{L} [R·]_{D_P} = \frac{4k_{ir}L}{V_m D_P' k_{td}[R·]} \qquad (142)$$

Further, we may write [cf. Eqs. (129) and (130)]

$$\frac{d \sum\limits_{D_P = L+1}^{D_{P,0}} [n]_{D_P}}{dt} = -2k_{ir} \sum\limits_{D_P = L+1}^{D_{P,0}} (D_P - 1)[n]_{D_P}$$

$$+ k_{td}[R \cdot]\left([R \cdot] - \sum\limits_{D_P = 1}^{L} [R \cdot]_{D_P}\right) \qquad (143)$$

Upon substitution of Eqs. (118), (119), (139), and (142) into (143), there is obtained an expression identical to (131), i.e.,

$$-\frac{d(D'_P)}{d\tau} = D'_P[D'_P - (1 + 2L)] \qquad (131)$$

This equation affords Eqs. (53), (132), and (132a) upon integration and subsequent simplification.

In a manner similar to that employed in obtaining Eq. (133),

$$\frac{d \sum\limits_{D_P = 1}^{L} [M]_{D_P}}{dt} = k_d[R \cdot] + k_{td}[R \cdot] \sum\limits_{D_P = 1}^{L} D_P[R \cdot]_{D_P} \qquad (144)$$

From Eq. (140), we may write

$$\frac{d \sum\limits_{D_P = 1}^{L} D_P[R \cdot]_{D_P}}{dt} = \frac{2k_{ir}L(L + 1)}{V_m D'_P}$$

$$+ k_d\left(L[R \cdot]_{L+1} - \sum\limits_{D_P = 1}^{L} [R \cdot]_{D_P}\right)$$

$$- k_{td}[R \cdot] \sum\limits_{D_P = 1}^{L} D_P[R \cdot]_{D_P} = 0 \qquad (145)$$

where steady-state conditions are postulated. Upon substitution of Eqs. (145) (neglecting the k_d term) and (139) into (144),

$$\frac{d \sum_{D_P=1}^{L} (M)_{D_P}}{d\tau} = \left[2(CL) + \frac{L(L+1)}{D'_P} \right] \left[(M)_{D_P,0} - \sum_{D_P=1}^{L} (M)_{D_P} \right]$$

(146)

where $(CL) = k_d V_m^{1/2}/2(k_{ir}k_{td})^{1/2}$.

From Eq. (146), when $(CL) \gg L(L+1)/2D'_P$ (generally valid for low conversions), we may obtain an expression similar to Eq. (64), case 7, i.e.,

$$-\ln(1 - C) = 2(CL)\tau$$

(64')

By substituting Eq. (64') into Eq. (132), there is obtained

$$-\ln(1 - C) = \left[\frac{2(CL)}{1 + 2L} \right] \left[\ln \left(1 - \frac{1 + 2L}{D'_{P,0}} \right) \right.$$

$$\left. - \ln \left(1 - \frac{1 + 2L}{D'_P} \right) \right]$$

(147)

Equation (147) has been plotted in terms of D'_P versus C for various values of L and (CL) (Fig. 4.11) [16].

Case 4': Terminal initiation followed by complete unzipping. As in case 1', the only volatile product in case 4' is monomer with the residual polymer maintaining its number-average chain length $D_{P,0}$ constant throughout the degradation, and accordingly, the polymer remains monodisperse. We may write [cf. Eq. (111)]

$$-\frac{d[n]_{D_{P,0}}}{dt} = 2k_{ie}[n]_{D_{P,0}}$$

(148)

Upon integration, Eq. (148) becomes

$$\ln \frac{{}^n D_{P,0;\tau=0}}{{}^n D_{P,0;\tau}} = \tau$$

(149)

where $\tau \equiv 2k_{ie}t$. Also, for the total rate of monomer formation,

$$\frac{d(M)}{d\tau} = -\frac{d\left(n_{D_{P,0};\tau} D_{P,0}\right)}{d\tau}$$

$$= D_{P,0} n_{D_{P,0};\tau} = (M)_o - (M) \tag{150}$$

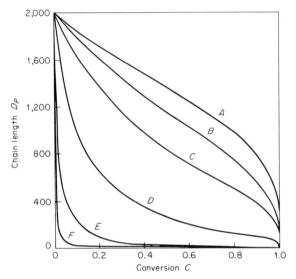

FIG. 4.11 Number-average chain length D_P as a function of fractional conversion for random initiation, depropagation, and disproportionation [16]. Evaluation of Eq. (147). Initial chain length $D_{P,0} = 2,000$; $L = 1, 5, 10,$ and 20 throughout. Number-average kinetic chain lengths (CL) are as follows: A, 1,500; B, 1,000; C, 500; D, 100; E, 10; and F, 1.

Equation (150) [cf. Eq. (38)] may be readily transformed into [cf. Eq. (123)]

$$-\ln(1 - C) = \tau \tag{151}$$

Case 5′: Terminal initiation, depropagation, and unimolecular termination. From steady-state considerations for [R·], we may write

$$2k_{ie} \sum_{D_P = L+1}^{D_{P,0}} [n]_{D_P} = k_{t1}[R\cdot] \tag{152}$$

When Eq. (118) is substituted into (152),

$$[R \cdot] = \frac{2k_{ie}}{k_{t1} V_m D_P'} \tag{153}$$

Also, for chain lengths lower than $L + 1$,

$$\frac{d[R \cdot]_{D_P}}{dt} = k_d \left([R \cdot]_{D_P+1} - [R \cdot]_{D_P}\right) + 2k_{ie} [n]_{D_P+1}$$
$$- k_{t1}[R \cdot]_{D_P} \tag{154}$$

From Eq. (154), we may write

$$\frac{d \sum_{D_P=1}^{L} D_P [R \cdot]_{D_P}}{dt} = 2k_{ie} L [n]_{L+1}$$

$$+ k_d \left(L [R \cdot]_{L+1} - \sum_{D_P=1}^{L} [R \cdot]_{D_P}\right)$$

$$- k_{t1} \sum_{D_P=1}^{L} D_P [R \cdot]_{D_P} = 0 \tag{155}$$

where steady-state conditions are assumed. Neglecting the term containing k_d in comparison with other terms (e.g., $k_d \ll k_{t1} L$),

$$\sum_{D_P=1}^{L} D_P [R \cdot]_{D_P} = \frac{2k_{ie} L [n]_{L+1}}{k_{t1}} \tag{156}$$

In Eq. (156) it should be noted that because of the presence of the $[n]_{L+1}$ term, the value of the summation should be very low. Furthermore,

$$\frac{d \sum_{D_P=1}^{L} [M]_{D_P}}{dt} = k_d [R \cdot] + 2k_{ie} \sum_{D_P=L+1}^{D_{P,0}} [n]_{D_P}$$

$$+ k_{t1} \sum_{D_P=1}^{L} D_P [R \cdot]_{D_P} \tag{157}$$

Upon substituting Eqs. (118) and (152) into Eq. (157) [neglecting the term containing k_{t1}, and letting $(CL) \gg 1$],

$$\frac{d \sum_{D_P=1}^{L} (M)_{D_P}}{dt} = \frac{2k_{ie} k_d}{k_{t1} D_P'} \left[(M)_{D_{P,0}} - \sum_{D_P=1}^{L} (M)_{D_P} \right] \qquad (158)$$

The preceding equation is similar to Eq. (54), case 6. In obtaining Eq. (156), it was assumed that $(k_d/k_{t1}) = (CL) \ll L < D_P'$. Under such conditions, the number of chains in the whole polymer system should remain nearly constant during degradation. Thus, we may write

$$\frac{(M)_{D_{P,0}} - \sum_{D_P=1}^{L} (M)_{D_P}}{D_P'} = \frac{(M)_{D_{P,0}}}{D_{P,0}'} \qquad (159)$$

or

$$1 - C = \frac{D_P'}{D_{P,0}'} \equiv B \qquad (42)$$

The last expression is identical to that obtained in case 6 [the expression following Eq. (56)]. When Eq. (159) is substituted into Eq. (158),

$$\frac{d \sum_{D_P=1}^{L} (M)_{D_P}}{dt} = \frac{2k_{ie} k_d (M)_{D_{P,0}}}{k_{t1} D_{P,0}'} \qquad (160)$$

Equation (160) is similar to Eq. (57), case 6.

Case 6': Terminal initiation, depropagation, and termination by disproportionation. As in case 5', it will be assumed here that $(CL) \ll D_P'$ and hence that Eq. (159) is applicable. Furthermore,

$$\frac{d \sum_{D_P=1}^{L} [M]_{D_P}}{dt} = k_d[R \cdot] + 2k_{ie} \sum_{D_P=L+1}^{D_{P,0}} [n]_{D_P}$$

(Continued on next page)

$$+ k_{td}[\text{R} \cdot] \sum_{D_P = 1}^{L} D_P [\text{R} \cdot]_{D_P} \tag{160'}$$

As previously indicated, the $\sum_{D_p = 1}^{L} D_P [\text{R} \cdot]_{D_p}$ term is generally of low magnitude and will be neglected in Eq. (160'). Then Eq. (160') may be converted into

$$\frac{d \sum_{D_P = 1}^{L} (\text{M})_{D_P}}{dt} = \left(\frac{2k_{ie}}{D_P'} + \frac{4k_{ie}(CL)}{D_P'} \right) \left[(\text{M})_{D_P, 0} - \sum_{D_P = 1}^{L} (\text{M})_{D_P} \right] \tag{161}$$

where $(CL) = (k_d/2)(V_m D_P'/k_{ie}k_{td})^{1/2}$. If $(CL) \gg 1/2$, Eq. (161) becomes

$$\frac{d \sum_{D_P = 1}^{L} (\text{M})_{D_P}}{dt} = 2k_d \left(\frac{k_{ie} V_m}{k_{td} D_P'} \right)^{1/2} \left[(\text{M})_{D_P, 0} - \sum_{D_P = 1}^{L} (\text{M})_{D_P} \right] \tag{162}$$

Equation (162) is similar to Eq. (71), case 8. Furthermore, Eq. (42) also applies to case 6' as well as to case 8. When Eq. (159) is substituted into Eq. (162) and the resulting expression is integrated,

$$1 - (1 - C)^{1/2} = k_d \left(\frac{k_{ie} V_m}{k_{td} D_{P,0}'} \right)^{1/2} t \tag{163}$$

By combining Eqs. (42) and (163), an expression can be obtained for D_P' as a function of t, i.e.,

$$\left(\frac{D_P'}{D_{P,0}'} \right)^{1/2} = 1 - k_d \left(\frac{k_{ie} V_m}{k_{td} D_{P,0}'} \right)^{1/2} t \tag{164}$$

It is beyond the scope of the book to treat cases in this section which involve intermolecular transfer. The resulting expressions are relatively complex. However, for didactic reasons, one such case will be treated in the following.

Case 7': Random initiation, depropagation, termination by dispropor-tionation, and intermolecular transfer. The expressions derived for $[R\cdot]$ in Eqs. (138) and (139), case 3', will be assumed to be valid here. Also, for chain lengths smaller than $L + 1$,

$$
\frac{d[R\cdot]_{D_P}}{dt} = k_d\left([R\cdot]_{D_P+1} - [R\cdot]_{D_P}\right) + 4k_{ir} \sum_{D_P=L+1}^{D_{P,0}} [n]_{D_P}
$$

$$
- k_{td}[R\cdot][R\cdot]_{D_P} - k_{tr}[R\cdot]_{D_P} \sum_{D_P=L+1}^{D_{P,0}} (D_P - 1)[n]_{D_P}
\tag{165}
$$

Further, we may write [cf. Eq. (141)]

$$
\frac{d\displaystyle\sum_{D_P=1}^{L} [R\cdot]_{D_P}}{dt} = k_d[R\cdot]_{L+1} + \frac{4k_{ir}L}{V_m D_P} - k_{td}[R\cdot]\sum_{D_P=1}^{L} [R\cdot]_{D_P}
$$

$$
- \frac{k_{tr}}{V_m} \sum_{D_P=1}^{L} [R\cdot]_{D_P} = 0
\tag{166}
$$

Neglecting the k_d term in Eq. (166),

$$
\sum_{D_P=1}^{L} [R\cdot]_{D_P} = \frac{4k_{ir}L/V_m D'_P}{k_{td}[R\cdot] + k_{tr}/V_m}
\tag{167}
$$

We may also write [cf. Eq. (143)]

$$
\frac{d\displaystyle\sum_{D_P=L+1}^{D_{P,0}} [n]_{D_P}}{dt} = -2k_{ir} \sum_{D_P=L+1}^{D_{P,0}} (D_P - 1)[n]_{D_P}
$$

(Continued on next page)

$$+ k_{td}[R\cdot]\left([R\cdot] - \sum_{D_P=1}^{L} [R\cdot]_{D_P}\right)$$

$$+ k_{tr}\left([R\cdot] - \sum_{D_P=1}^{L} [R\cdot]_{D_P}\right) \sum_{D_P=L+1}^{D_{P,0}} [(D_P - 1) - 2(L + 1)]\,[n]_{D_P}$$

$$(168)$$

In Eq. (168), the factor $[(D_P - 1) - 2(L + 1)]$ enters since the number of chain scissions necessary in order that the polymer fragments formed do not volatilize is equal to this factor [cf. Eq. (116)]. If it is assumed that $D'_P \gg 3$, then it can be shown [cf. Eq. (131)] that

$$-\frac{d(D'_P)}{dt} = \alpha D'^2_P - \beta D'_P + \gamma \qquad (169)$$

where

$$\alpha = 2k_{ir} + k_{tr}[R\cdot]$$

$$\beta = \frac{4k_{ir}k_{td}L[R\cdot]}{N} + \frac{4k_{ir}k_{tr}L}{V_m N} + 2k_{tr}L[R\cdot]$$

$$\gamma = \frac{8k_{ir}k_{tr}L^2}{V_m N}$$

where

$$N = k_{td}[R\cdot] + \frac{k_{tr}}{V_m}$$

and

$$[R\cdot] = \text{const.} = 2\left(\frac{k_{ir}}{k_{td}V_m}\right)^{1/2}$$

When $D'_P \gg 2L$, the term γ and the last term in β may be neglected so that

$$-\frac{d(D'_P)}{dt} = \alpha D'^2_P - \beta' D'_P \tag{169a}$$

where $\beta' = \beta - 2k_{tr}L[R\cdot]$.

In a similar manner used in obtaining Eq. (144),

$$\frac{d\sum\limits_{D_P=1}^{L}[M]_{D_P}}{dt} = k_d[R\cdot] + k_{td}[R\cdot]\sum\limits_{D_P=1}^{L}D_P[R\cdot]_{D_P}$$

$$+ k'_{tr}\sum\limits_{D_P=1}^{L}D_P[R\cdot]_{D_P}\sum\limits_{D_P=L+1}^{D_{P,0}}(D_P - 1)[n]_{D_P} \tag{170}$$

Utilizing Eq. (165), we may write

$$\frac{d\sum\limits_{D_P=1}^{L}D_P[R\cdot]_{D_P}}{dt} = \frac{2k_{ir}L(L+1)}{V_m D'_P}$$

$$+ k_d\left(L[R\cdot]_{L+1} - \sum\limits_{D_P=1}^{L}[R\cdot]_{D_P}\right) - k_{td}[R\cdot]\sum\limits_{D_P=1}^{L}D_P[R\cdot]_{D_P}$$

$$- \frac{k_{tr}}{V_m}\sum\limits_{D_P=1}^{L}D_P[R\cdot]_{D_P} = 0 \tag{171}$$

where steady-state conditions are employed. By using Eqs. (171) (neglecting the k_d term) and (119), Eq. (170) may be converted into

$$\frac{d\sum\limits_{D_P=1}^{L}(M)_{D_P}}{dt} = \left\{k_d[R\cdot]V_m + \frac{2k_{ir}k_{td}[R\cdot]L(L+1)}{ND'_P}\right.$$

$$\left. + \frac{2k_{ir}k'_{tr}L(L+1)}{V_m ND'_P}\right\}\left[(M)_{D_{P,0}} - \sum\limits_{D_P=1}^{L}(M)_{D_P}\right] \tag{172}$$

It may be noted that similarities exist among Eqs. (131) and (146), case 3', and Eqs. (169a) and (172), case 7', respectively.

SPECIAL CASE OF RANDOM DEGRADATION

In this section will be treated, by several methods, random degradation which involves random initiation, followed by rapid termination by disproportionation. Such a case is often applicable to the thermal degradation of condensation polymers, e.g., nylon 6.

Kinetic Methods

The mathematical treatment for the rate of change of number-average degree of polymerization D_P' as a function of time is identical with that given in case 3'. Thus, Eqs. (139), (142), and (143) are applicable, yielding Eqs. (131), (132), (132a), and (53). The mathematical treatment for the rate of volatilization is somewhat more complex. It may be carried out as follows.

Let us assume the presence of polymer chains of initial degree of polymerization $D_{P,0}$, which are undergoing random degradation to form chain fragments of length x. The general rate of formation of the concentration of such fragments may be expressed as

$$\frac{d[n]_x}{dt} = -k(x-1)[n]_x + 2k \sum_{y=x+1}^{D_{P,0}} [n]_y \tag{173}$$

At $t = 0$, $[n]_x = 0$, except when $x = D_{P,0}$. In the latter case, Eq. (173) becomes [cf. Eq. (120)]

$$\frac{d[n]_{D_{P,0}}}{dt} = -k(D_{P,0} - 1)[n]_{D_{P,0}} \tag{174}$$

Upon integration, Eq. (174) affords

$$[n]_{D_{P,0}} = \left[n\right]_{D_{P,0}}^{t=0} \exp[-k(D_{P,0} - 1)t] \tag{175}$$

When $x = D_{P,0} - 1$, Eq. (173) becomes

$$\frac{d[n]_{D_{P,0}-1}}{dt} = -k(D_{P,0} - 2)[n]_{D_{P,0}-1} + 2k[n]_{D_{P,0}} \tag{176}$$

Upon substitution of Eq. (175) into (176) and integration of the resulting expression, we obtain

$$[n]_{D_{P,0}-1} = 2\left[n\right]_{D_{P,0}}^{t=0} \exp[-k(D_{P,0} - 2)t][1 - \exp(-kt)] \tag{177}$$

When $x = D_{P,0} - 2$, Eq. (173) becomes

$$\frac{d[n]_{D_{P,0} - 2}}{dt} = -k(D_{P,0} - 3)[n]_{D_{P,0} - 2}$$
$$+ 2k[n]_{D_{P,0} - 1} + 2k[n]_{D_{P,0}} \qquad (178)$$

Upon substitution of Eqs. (175) and (177) into (178) and integration of the resulting expression,

$$[n]_{D_{P,0} - 2} = \left[n\right]_{D_{P,0}}^{t=0} \exp[-k(D_{P,0} - 3)t][1 - \exp(-kt)]$$
$$\times [3 - \exp(-kt)] \quad (179)$$

Continuing in this manner, it can be shown that

$$[n]_x = \left[n\right]_{D_{P,0}}^{t=0} \alpha(1 - \alpha)^{x-1}[2 + (D_{P,0} - 1)\alpha - \alpha x] \qquad (180)$$

where $\alpha = 1 - e^{-kt}$ = fraction of bonds broken in the original polymer chain at time t. This expression for α may be readily obtained as follows. For the random degradation described in this section, we may write

$$\frac{d\alpha}{dt} = k(1 - \alpha) \qquad (181)$$

or

$$1 - \alpha = e^{-kt} \qquad (181a)$$

Furthermore,

$$1 - C = \frac{\displaystyle\sum_{L'}^{D_{P,0}} x[n]_x}{D_{P,0}[n]_{D_{P,0}}} \qquad (182)$$

where $L' = L + 1$. When Eq. (180) is substituted into Eq. (182) and the summation carried out,

$$1 - C = (1 - \alpha)^{L'-1}\left[1 + \frac{\alpha(D_{P,0} - L')(L' - 1)}{D_{P,0}}\right] \qquad (183)$$

Also, from Eq. (183),

$$\frac{dC}{k\,dt} = (L' - 1)(1 - C) - (1 - \alpha)^{L'}\frac{(D_{P,0} - L')(L' - 1)}{D_{P,0}} \tag{184}$$

At zero time,

$$\left(\frac{dC}{k\,dt}\right)_{t \to 0} = (L' - 1) - \frac{(D_{P,0} - L')(L' - 1)}{D_{P,0}} = \frac{(L' - 1)L'}{D_{P,0}} \tag{185}$$

When $D_{P,0} \gg L'$, Eqs. (183) and (184) become respectively

$$1 - C = (1 - \alpha)^{L'-1}[1 + \alpha(L' - 1)] \tag{183a}$$

and

$$\frac{dC}{k\,dt} = (L' - 1)(1 - C) - (1 - \alpha)^{L'}(L' - 1) \tag{184a}$$

Upon substituting Eq. (183a) into (184a), differentiating, and setting the resulting expression equal to zero, we can obtain

$$\alpha_{max} = \frac{1}{L'} \tag{186}$$

and, consequently,

$$C_{max} = 1 - \left(\frac{L' - 1}{L'}\right)^{L'-1}\frac{2L' - 1}{L'} \tag{187}$$

For relatively low values of L', e.g., $L' = 4$, Eq. (187) becomes

$$C_{max} = 1 - \left(\frac{3}{4}\right)^3\left(\frac{7}{4}\right) \approx 0.26$$

For relatively high values of L', i.e., $L' \gg 1$, Eq. (187) may be written as

$$C_{max} \simeq 1 - 2\left(\frac{L' - 1}{L'}\right)^{L'}$$

or

$$C_{max} \simeq 1 - \frac{2}{e} \approx 0.26$$

Thus, from the preceding, it can be seen that during the random thermal degradation of polymers, the maximum rate of volatilization will occur at a conversion of about 26 percent. The maximum volatilization rate can be expressed as

$$\left(\frac{dC}{k\,dt}\right)_{max} = (L' - 1)\left(\frac{L' - 1}{L'}\right)^{L' - 1} \tag{188}$$

When $L' \gg 1$, Eq. (188) may be written as

$$\left(\frac{dC}{k\,dt}\right)_{max} = \frac{L'}{e} \tag{188a}$$

where $1/e \approx [(L' - 1)/L']^{L'}$.

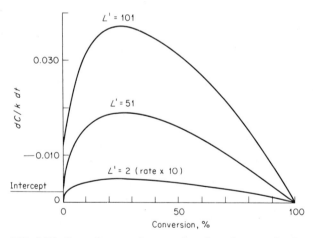

FIG. 4.12 Rate of conversion as a function of conversion for random degradation for various values of L' and $D_{P,0} = 1,000$ [3].

From Eq. (185), as L' increases, the value of $(dC/k\,dt)_{t \to 0}$ varies as L'^2 whereas the value of $(dC/k\,dt)_{max}$ varies as L' [cf. Eq. (188a)]. Furthermore, the value of C_{max} is approximately constant at 26 percent. These relationships are depicted in Fig. 4.12 [3].

Equation (183a) and (184a) may be applied to polymers (generally condensation) which undergo random thermal degradation. Thus, Ozawa [28] found from dynamic thermogravimetric analysis that the degradation of nylon 6 was random and that $L' = 2$. Under these conditions Eqs. (183a) and (184a) afford

$$\alpha = C^{1/2} \tag{189}$$

and

$$\frac{dC}{k\,dt} = 2(C^{1/2} - C) \tag{190}$$

As indicated in Chap. 2, Eq. (190) may be written as

$$\frac{1}{2} \int_0^C \frac{dC}{C^{1/2} - C} = \frac{A}{(RH)} \int_0^T e^{-E/RT}\,dT \tag{191}$$

where A = Arrhenius frequency factor, (RH) = rate of heating of sample during dynamic thermogravimetric analysis (TGA), and E = activation energy. Using an approximation given by Doyle [29] for the integral on the right-hand side of Eq. (191) (cf. Chap. 2), Eq. (191) becomes

$$\log(1 - C^{1/2})^{-1} = \frac{AE}{2.3R\,(RH)}\,p(x) \tag{192}$$

where R = gas constant, and $\log p(x) = -2.315 - 0.457E/RT$ (for values of $E/RT > 20$). Thus, we may write

$$\log\left[\log(1 - C^{1/2})^{-1}\right] = \log\frac{AE}{2.3R\,(RH)} - 2.315 - 0.457\frac{E}{RT} \tag{193}$$

From Eq. (193), it can be seen that a plot of the double-logarithm term versus $1/T$ should afford a linear relationship from whose slope the value of E can be estimated and from whose intercept the value of A can be calculated. From data provided by Ozawa, such plots have been carried out for nylon 6 (see Fig. 4.13) for various constant-heating rates. Average values from these plots were: $E = 32 \pm 2$ kcal mole^{-1} and $\log A \approx 10$ min^{-1}. Utilizing another approach, Ozawa obtained values of $E = 36 \pm 0.5$ kcal mole^{-1} and $\log A \approx 11$.

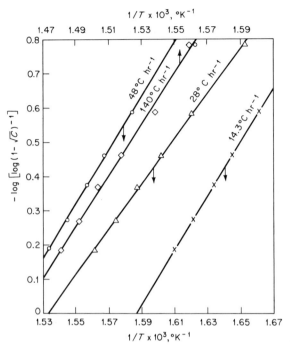

FIG. 4.13 Random degradation for nylon 6: □ 140°C hr⁻¹ ;
○ 48°C hr⁻¹ ; △ 28°C hr⁻¹ ; X 14.3°C hr⁻¹ .

For the number-average degree of polymerization of the residue remaining from a random thermal degradation, we may write

$$
D'_P = \frac{\displaystyle\sum_{x=L'}^{D_{P,0}} x[n]_x}{\displaystyle\sum_{x=L'}^{D_{P,P}} [n]_x}
\tag{194}
$$

Upon substitution of Eq. (180) into (194) and carrying out the summations,

$$
\frac{D'_P}{D_{P,0}} = \frac{1 + \alpha(D_{P,0} - L')(L' - 1)/D_{P,0}}{1 + \alpha(D_{P,0} - L')}
\tag{195}
$$

When $D_{P,0} \gg L'$, Eq. (195) becomes

$$\frac{1}{D'_P} - \frac{1}{D_{P,0}} = \frac{1 - e^{-kt}}{1 + (L' - 1)(1 - e^{-kt})} \tag{196}$$

From Eq. (196) it can be seen that for relatively low values of L' and α, Eq. (196) becomes ($kt \ll 1$),

$$\frac{1}{D'_P} - \frac{1}{D_{P,0}} = kt \tag{196a}$$

which is similar to Eq. (53).

Wall and coworkers [30] studied the thermal degradation of commercial (branched) polyethylene (PE) and of linear polymethylene (PM) prepared from diazomethane. In PM the volatilization rate exhibited a maximum as a function of conversion whereas this maximum was absent in PE. The differences between PM and PE were presumably due to the branched structure of the PE. The results obtained for PM could be fairly well described by the expressions developed for random degradation. Thus, Fig. 4.14 shows a comparison between observed and calculated values of dC/dt as a function of C for various temperatures. The agreement between the sets of observed and calculated rates is reasonably good. The two sets were fitted to each other by making the values of the heights of the respective

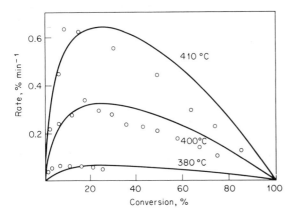

FIG. 4.14 Rates of pyrolysis of polymethylene: solid curves, calculated points; O experimental points [30].

maxima similar and letting $L' = 72$. The latter value was established from a comparison of theoretical and observed molecular-weight distribution of volatiles [31] and from an independent estimate of L' by extrapolating vapor-pressure data of C_{20} to C_{30} hydrocarbons [30]. Then from Eq.

(188a), values of k at various temperatures may be estimated: $k_{380°} = 0.41 \times 10^{-6}$ sec^{-1}; $k_{390°} = 0.98 \times 10^{-6}$ sec^{-1}; and $k_{400°} = 2.0 \times 10^{-6}$ sec^{-1}. In this manner, the following Arrhenius expression was obtained:

$$k = 1.93 \times 10^{16} e^{-68,100/RT} \text{ sec}^{-1} \tag{197}$$

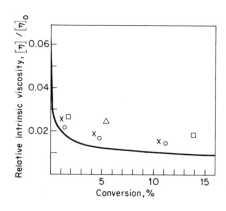

FIG. 4.15 Relative intrinsic viscosity of the residue versus conversion in pyrolysis of polymethylene: solid curves, calculated; experimental points: △ 375°; □ 380°; × 390°; ○ 400° [30].

Then values of dC/dt could be calculated for various values of C using Eqs. (183a) and (184a). Thus, for example, when $\alpha = 0.01$, the value of $C = 0.16$ [from Eq. (183a)] and the corresponding value of $dC/dt = 24.8k$. At 400°C, $dC/dt = 0.29$ percent per minute. In this manner, the curves of Fig. 4.14 were calculated.

When $D_{P,0} \gg L' \gg 1$, Eq. (195) becomes

$$\frac{D'_P}{D_{P,0}} \approx \frac{1 + \alpha L'}{1 + \alpha D_{P,0}} \tag{195a}$$

Assuming that the following is valid for PM, i.e.,

$$[\eta] = 1.35 \times 10^{-3} M_n^{0.63} \tag{198}$$

where $[\eta]$ = intrinsic viscosity, then,

$$\frac{[\eta]}{[\eta]_0} = \left(\frac{1 + \alpha L'}{1 + \alpha D_{P,0}} \right)^{0.63} \tag{199}$$

By means of Eqs. (183a) and (199), a plot of $[\eta]/[\eta]_0$ versus C may be constructed, as depicted in Fig. 4.15 [30]. Thus, for example, when $\alpha = 0.01$, $C = 0.16$ and $[\eta]/[\eta]_0 \approx 0.01$ for $D_{P,0} = 3 \times 10^5$ and $L' = 72$.

Another test of the random degradation theory for PM involves a plot of Eq. (196). As previously indicated, when L' is relatively low and $\alpha \ll 1$, a plot of $1/D_P' - 1/D_{P,0}$ versus t should be linear. Otherwise, this linearity should hold only over a limited time range. In Fig. 4.16 are shown plots of Eq. (196) (unbroken lines) using $L' = 72$ and values of k previously obtained from volatilization rates studies [cf. Eq. (197)]. As would be anticipated from Fig. 4.15, the experimental curves bend over more rapidly than the theoretical ones. The deviations are smallest at the higher temperatures and small times. The agreement between experimental and calculated values is considered to be reasonably good in view of the fact that the data used came from two independent sets of measurements and involve a conversion to number-average degree of polymerization. A better fit (dashed lines) could be obtained by varying L' empirically (there is no theoretical basis for such a result). Thus, it can be seen that the overall pattern for the thermal degradation of PM does not appreciably deviate from a random type.

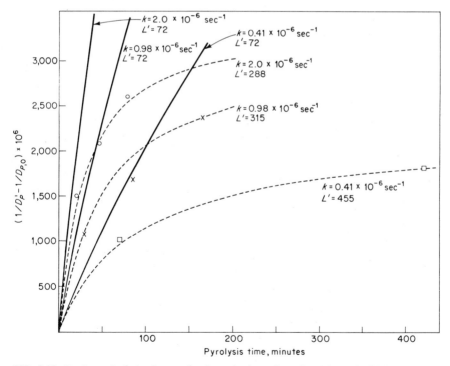

FIG. 4.16 Reciprocal of the degree of polymerization minus the reciprocal of initial degree of polymerization versus time: solid curves, calculated [Eq. (196)]; dashed curves, fitted to experimental points by adjusting L'; experimental points: ○ 400°C, × 390°C, □ 380°C [30].

Probability Methods

In addition to the use of kinetic methods, probability methods can be employed to derive the fundamental Eq. (180) which pertains to random degradation. Thus, let α denote the probability of breaking a link in an original chain [$= S/(D_{P,0} - 1) \approx S/D_{P,0}$, where S = number of links broken per original chain]. Then $1 - \alpha$ = probability of no link breakage in the original chain. Two types of chain breakage must now be considered, i.e., breakage within the molecule, and breakage from chain ends.

Breakage within the molecule. Consider the type of breakage shown in Fig. 4.17. In this case, the probability that two links are broken and that

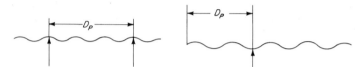

FIG. 4.17 Random breakage within a chain molecule.

FIG. 4.18 Random breakage from chain ends.

$D_P - 1$ links between the broken links are intact is

$$p' = \alpha^2(1 - \alpha)^{D_P - 1} \qquad (200)$$

The number of ways that this can occur is $(D_{P,0} - D_P) - 1$. Thus, we may write for the number of chains that may form of length D_P,

$$n'_{D_P} = n_{D_{P,0}}[(D_{P,0} - 1) - D_P]\alpha^2(1 - \alpha)^{D_P - 1} \qquad (201)$$

Breakage from chain ends. Consider now the type of breakage depicted in Fig. 4.18. The probability that $D_P - 1$ links will remain intact upon chain breakage from *one* particular chain end is

$$p''_1 = \alpha(1 - \alpha)^{D_P - 1} \qquad (202)$$

However, since there are two chain ends, the total probability may be written as

$$p'' = 2\alpha(1 - \alpha)^{D_P - 1} \qquad (203)$$

and the number of chains formed of length D_P may be written for this case as

$$n''_{D_P} = n_{D_{P,0}}2\alpha(1 - \alpha)^{D_P - 1} \qquad (204)$$

From Eqs. (201) and (204) it can be seen that the number of chains of length D_P which can form as a result of random cleavage may be expressed as [cf. Eq. (180)]

$$n_{D_P} = n'_{D_P} + n''_{D_P}$$

$$= n_{D_{P,0}} \alpha (1 - \alpha)^{D_P - 1} [2 + (D_{P,0} - 1)\alpha - \alpha D_P] \qquad (180a)$$

From Eq. (196) it can be shown that for a closed system ($L' - 1 = 0$), and for moderate amounts of conversion [cf. Eq. (53)],

$$\frac{1}{D_P} - \frac{1}{D_{P,0}} \approx kt \qquad (53')$$

Equation (53') may also be readily derived as follows. At relatively low conversions in a random scission process,

$$\alpha \approx \frac{S}{D_{P,0}} \approx kt \qquad (205)$$

where S = number of bonds broken per molecule of average chain length $D_{P,0}$ in time t. Furthermore,

$$S = \frac{D_{P,0}}{D_P} - 1 \qquad (206)$$

Upon combining Eqs. (205) and (206), Eq. (53') is obtained. Grassie and Grant [32] found that expressions such as Eq. (53') were valid in the thermal degradation of copolymers of α-chloroacrylonitrile (CAN) with styrene and with methyl methacrylate. In Fig. 4.19 are shown plots of Eq. (53') for various reaction temperatures in the thermal degradation of copolymers of CAN with styrene. From an Arrhenius plot of the slopes of the linear relationships obtained in this figure, an activation energy of 40.5 kcal mole^{-1} was estimated. It was indicated that the CAN units in the co-polymer were directly responsible for the chain-scission process. The initial step involved C—Cl bond scission, followed by chain scission and evolution of HCl (at relatively low temperatures, cf. Fig. 4.19, HCl was the main product liberated).

In CAN/methyl methacrylate copolymers thermal degradation afforded volatile products consisting mainly of HCl, methyl methacrylate, and CAN. Simultaneously, there is a decrease in molecular weight (with time). In order to test whether this decrease was the result of random scission, as for the CAN/styrene copolymers, Eq. (53') had to be modified to account for the monomer liberated. When such volatilization occurs, we may write

$$S = \frac{D_{P,0}(1 - C)}{D_P} - 1 \tag{206'}$$

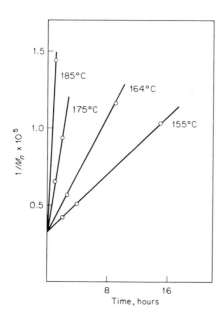

FIG. 4.19 Influence of temperature on the rate of chain scission
of a copolymer of CAN/styrene
[32].

Upon combining Eqs. (205) and (206'),

$$\frac{1 - C}{D_P} - \frac{1}{D_{P,0}} = kt \tag{53''}$$

In deriving Eq. (53''), it was assumed that chains were not eliminated from the system by complete unzipping. That the thermal degradation of CAN/methyl methacrylate copolymers involves a random chain-scission process can be seen from plots of Eq. (53'') for various reaction temperatures (Fig. 4.20). From such plots, an activation energy of 44 kcal mole^{-1} was obtained, which is similar to the value obtained for the CAN/styrene copolymers. It was also found that the amount of monomer produced was linear in respect to time. Thus, it appeared that for each chain scission a fixed amount of monomer formed, serving as strong evidence that chain scission results in radicals which immediately unzip. This represents a major difference from the thermal behavior of the CAN/styrene copolymers in which unzipping does not occur readily. The amount of monomer per chain break was found to be inversely proportional to the concentration of CAN units in the copolymer, implying that the unzipping process was terminated at CAN units. However, owing to the presence of CAN in the volatile products, apparently when a CAN unit is reached during the un-zipping process, there is direct competition between its elimination as a volatile product and a termination reaction involving it.

Bhatnagar and coworkers [33, 34] also utilized an expression similar to Eq. (53′) in studying the degradation of GR-S rubber in o-chlorotoluene (OCT) and in α-chloronaphthalene (ACN) and of butyl rubber in Decalin and p-xylene. From a plot of Eq. (53′), initial slopes should afford values of k at various temperatures. Then, an Arrhenius plot will yield values of frequency factor A and activation energy E. Thus, values of $A = 5.9 \times 10^5$ and 17.2×10^5 sec^{-1} and of $E = 24$ and 26 kcal mole^{-1} were obtained for the GR-S rubber in OCT and ACN, respectively. A similar approach for butyl rubber in solution [34] led to values of $E = 24.4$ and 18.4 kcal mole^{-1} for this rubber in Decalin (DE) and p-xylene (PX), respectively. These values were checked by use of the intrinsic-viscosity relationship [cf. Eq. (198)], $[\eta] = KM_n{}^a$. When this expression is substituted into Eq. (53′) and the resulting equation differentiated, it can readily be shown, using the Arrhenius expression, that the initial values, $(-d[\eta]/dt]_0$, are proportional to $e^{-E/RT}$. In this manner, $E = 24.6$ and 17.9 kcal mole^{-1}, respectively, for butyl rubber in DE and PX.

It may be instructive at this point to indicate how molecular-weight dis-tribution (MWD) can influence the random degradation process. Berlin

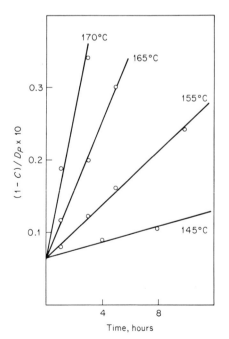

FIG. 4.20 Influence of tempera-ture on the rate of chain scission of a copolymer of CAN/methyl-methacrylate [32].

and Yenikolopyan [35] have shown theoretically that when a polymer undergoes random degradation, its MWD rapidly approaches the most probable $(D_{P,w}/D_P = 2)$, after a few ruptures per polymer chain regardless of whether its initial MWD is monodisperse, most probable, or some broad distribution $(D_{P,w} = $ weight-average degree of polymerization). Then if S chain scissions have occurred in the original polymer, we may write

$$\frac{\pi}{D_P} = \frac{\pi_0}{D_{P,0}} + S \tag{207}$$

(no chains are lost by complete unzipping) where π denotes the number of monomeric units in the polymer. Letting $X = D_{P,w}/D_P$, Eq. (207) becomes

$$\frac{X\pi}{D_{P,w}} = \frac{X_0\pi_0}{D_{P,w,0}} + S \tag{208}$$

Equation (208) may be transformed into, at low conversions ($\pi = \pi_0$),

$$\frac{D_{P,w,0}}{D_{P,w}} = \frac{X_0}{2} + \frac{D_{P,0,w}\,S}{2\pi_0} \tag{209}$$

The theoretical dependence of $D_{P,w,0}/D_{P,w}$ on $\lambda D_{P,0} (= D_{P,w,0}S/\pi_0 X_0)$ is shown in Fig. 4.21 for polymers of various initial MWD's. After an initial

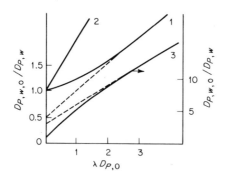

FIG. 4.21 Theoretical curves of the variation in weight-average molecular weight in breakdown of polymer molecules without depolymerization for (1) monodisperse, (2) normal, and (3) broad initial MWD [36].

period, $X = 2$, and Eq. (209) applies; i.e., a linear relationship obtains whose intercept will afford a value of $X_0/2$. Berlin and coworkers [36] have utilized such plots to determine the initial MWD during the random hydrolysis of poly-1,3-dioxolane (in acidified aqueous solution). The expression for the hydrolysis rate was found to be (in terms of scissions)

$$\frac{dS}{dt} = 2k_a[\text{HCl}][\pi_0] \tag{210}$$

where $k_a = K \cdot k_2$ = apparent hydrolysis rate constant. The following mechanism was suggested:

$$\sim\!\!\sim O - CH_2 - CH_2 - O - CH_2 \sim\!\!\sim + H_3\overset{\oplus}{O} \xrightarrow{\quad K \quad}$$

$$\sim\!\!\sim O - CH_2 - CH_2 - \underset{\underset{H}{|}}{\overset{\oplus}{O}} - CH_2 \sim\!\!\sim + H_2O \tag{211}$$

$$\sim\!\!\sim O - CH_2 - CH_2 - \underset{\underset{H}{|}}{\overset{\oplus}{O}} - CH_2 \sim\!\!\sim \xrightarrow[H_2O]{k_2}$$

$$\sim O - CH_2 - CH_2 - OH + \sim\!\!\sim CH_2 - O - CH_2 - \underset{\underset{H}{|}}{\overset{\overset{H}{|}}{O}}{}^{\oplus} \tag{212}$$

$$\sim\!\!\sim O - CH_2 - CH_2 - O - CH_2 - \underset{\underset{H}{|}}{\overset{\overset{H}{|}}{O}}{}^{\oplus} + H_2O \rightleftharpoons$$

$$\sim\!\!\sim O - CH_2 - CH_2 - O - CH_2OH + H_3O^{\oplus} \tag{213}$$

The factor 2 in Eq. (210) corresponds to the two points of attack (oxygen atoms) in the monomeric unit of polydioxolane. Upon integration, Eq. (210) yields ([HCl] and [π_0] being constant)

$$S = 2k_a[\text{HCl}]\pi_0 t \tag{214}$$

By combining Eqs. (209) and (214),

$$\frac{D_{P,w,0}}{D_{P,w}} = \frac{X_0}{2} + k_a D_{P,0,w}[\text{HCl}]t \tag{215}$$

In Fig. 4.22 are shown plots of Eq. (215) for polymer whose initial value of $X_0 = 2.14$, as determined experimentally. From the plots a value of X_0 (calc.) = 2.1. The apparent activation energy of 17.5 kcal/mole^{-1} was in

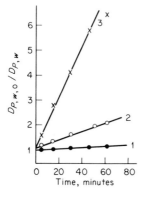

FIG. 4.22 Rate of change of molecular weight in hydrolysis of poly-1,3-dioxolane: $D_{P,w,0} = 1.79 \times 10^4$; [HCl] = 2.4 x 10^{-3} mole liter^{-1}. Curve 1, 35°C; curve 2, 60°C; curve 3, 75°C [36].

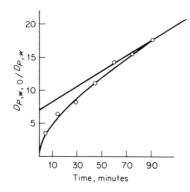

FIG. 4.23 Change of molecular weight versus time at 60°C for a mixture of polydioxolanes: $X_0 = 13.3$ [36].

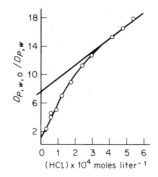

FIG. 4.24 Change of molecular weight versus [HCl] at 60°C and for a reaction time of 15 min for a mixture of polydioxolanes: $X_0 = 13.3$ [36].

satisfactory agreement with values found for the acid hydrolysis of polymeric and low-molecular-weight acetals. Further validity of Eq. (215) is to be found in plots of $D_{P,w,0}/D_{P,w}$ versus time and [HCl] (for a constant reaction time of 15 min), as depicted in Figs. 4.23 and 4.24. From the extrapolated intercepts of these plots, values of $X_0 = 12.9$ and 14 were obtained, respectively (X_0, theor., = 13.3). These values agreed well with the expected value for the polymer mixture.

REFERENCES

1. Teetsel, D. A., and D. W. Levi: *Plastec Note* 10, Plastics Technology Evaluation Center, Picatinny Arsenal, Dover, N. J., January, 1966.
2. Simha, R., L. A. Wall, and P. J. Blatz: *J. Polymer Sci.*, **5**:615 (1950).
3. Simha, R., and L. A. Wall: *J. Phys. Chem.*, **56**:707 (1952).
4. Simha, R., and L. A. Wall: In P. H. Emmett (ed.), "Catalysis," vol. 6, pp. 191ff., Reinhold Publishing Corporation, New York, 1958.
5. Simha, R., L. A. Wall, and J. Bram: *J. Chem. Phys.*, **29**:894 (1958).
6. Jellinek, H. H. G.: "Degradation of Vinyl Polymers," Academic Press, Inc., New York, 1955.
7. Burnett, G. M.: "Mechanism of Polymer Reactions," Interscience Publishers, Inc., New York, 1954.
8. MacCallum, J. R.: *European Polymer J.*, **2**:413 (1966).
9. Bamford, C. H., W. G. Barb, A. D. Jenkins, and P. F. Onyon: "The Kinetics of Vinyl Polymerization by Radical Mechanisms," pp. 266ff., Academic Press, Inc., New York, 1958.
10. Wall, L. A.: In G. M. Kline (ed.), "Analytical Chemistry of Polymers," High Polymer Series, vol. 12, p. 181, Interscience Publishers, a division of John Wiley & Sons, Inc., New York, 1962.
11. MacCallum, J. R.: *Trans. Faraday Soc.*, **59**:2099 (1963).
12. MacCallum, J. R.: *Makromol. Chem.*, **83**:137 (1965).
13. Barlow, A., R. S. Lehrle, J. C. Robb, and D. Sunderland: *Polymer*, **8**:537 (1967).
14. Jellinek, H. H. G., and M. D. Luh: "Polymer Preprints," vol. 8, p. 533, American Chemical Society Meeting, Miami Beach, April, 1967.
15. Bagby, G., R. S. Lehrle, and J. C. Robb: *Makromol. Chem.*, **119**:122 (1968).
16. Jellinek, H. H. G.: in "Encyclopedia of Polymer Science and Technology," vol. 4, pp. 740–793, Interscience Publishers, a division of John Wiley & Sons, Inc., New York, 1966.
17. Jellinek, H. H. G.: *Pure Appl. Chem.*, **4**:419 (1962).
18. Friedman, H. L.: *Aerophys. Res. Mem. 37, Tech. Inform. Ser.* R59 SD385, General Electric Co., June 19, 1959.
19. Siegle, J. C., L. T. Muus, T. -P. Lin, and H. A. Larsen: *J. Polymer Sci.*, **A-2**:391 (1964).
20. Reich, L., and D. W. Levi: In A. Peterlin, M. Goodman, S. Okamura, B. H. Zimm, and H. F. Marks (eds.), "Macromolecular Reviews," vol. 1, pp. 173ff., Interscience Publishers, a division of John Wiley & Sons, Inc., New York, 1966.
21. Patrick, C. R.: *Nature*, **181**:698 (1958).
22. Cowley, P. R. E. J., and H. W. Melville: *Proc. Roy. Soc. (London)*, **A211**:320 (1952).
23. Grassie, N., and H. W. Melville, *Proc. Roy. Soc. (London)*, **A199**:1, 14, 24 (1949).
24. Wall, L. A., S. Straus, and L. J. Fetters: "Polymer Preprints," vol. 1o, p. 1472, American Chemical Society Meeting, New York, September, 1969.
25. Richards, D. H., and D. A. Salter: *Polymer*, **8**:127 (1967).
26. Brown, D. W., and L. A. Wall: *J. Phys. Chem.*, **62**:848 (1958).
27. Richards, D. H., and D. A. Salter: *Polymer*, **8**:139 (1967).
28. Ozawa, T.: *Bull. Chem. Soc. Japan*, **38**:1881 (1965).
29. Doyle, C. D.: *J. Appl. Polymer Sci.*, **6**:639 (1962).
30. Wall, L. A., S. L. Madorsky, D. W. Brown, S. Straus, and R. Simha: *J. Am. Chem. Soc.*, **76**:3430 (1954).
31. Simha, R., and L. A. Wall: *J. Polymer Sci.*, **6**:39 (1951).
32. Grassie, N., and E. M. Grant: *European Polymer J.*, **2**:255 (1966).

33.   Gur, I. S., and H. L. Bhatnagar: *Indian J. Chem.*, **7**:495 (1969).
34.   Singh, M., and H. L. Bhatnagar: *Indian J. Chem.*, **7**:1028 (1969).
35.   Berlin, A. A., and N. S. Yenikolopyan: *Vysokomolekul. Soedin.*, **A10**:1475 (1968).
36.   Berlin, A. A., M. A. Khakimdzhanova, L. V. Karmilova, and N. S. Yenikolopyan: *Vysokomolekul. Soedin.*, **A10**:1496 (1968).

5

KINETICS
AND MECHANISM OF OXIDATIVE
DEGRADATION
OF POLYMERS

In this chapter will be emphasized the kinetics and mechanisms of oxidative degradation of polymers. In this connection, we have been necessarily selective in choice of references, considered from the vast literature available on oxidative degradation of polymers. For a more detailed background on this subject, the reader is referred to Ref. 1. Initially, the autoxidation of polyolefins will be treated in absence and presence of inhibitors and metal catalysts (additives). Then, other types of polymers will be considered. It will be assumed in the following that oxygen diffusion control is nonexistent.

POLYOLEFINS

Oxidation in Absence of Additives

The autoxidation kinetics of polyolefins in absence of additives (inhibitors and metal catalysts) has been studied from essentially two different approaches [1]. In one approach, termination in bulk-phase oxidation is postulated involving the recombination of two polymeric peroxy radicals [2–7] similar to that proposed for liquid-phase oxidations. In the other approach, such a termination step is considered unlikely and is therefore excluded, since it would depend upon two very small concentrations requiring polymer-polymer interaction in a medium of very high viscosity [8–20]. [In contrast to the free-radical mechanisms involving polymeric radical termination by recombination in the thermal degradation of polymers described in Chap. 4, the oxidative degradation of polymers in this chapter is postulated not to involve such polymeric radical termination. The difference in mechanisms can be attributed to the much lower temperatures generally used in oxidative degradation (>100°C) as opposed to the relatively high temperatures employed in thermal degradation (>300°C)].

Evidence favoring termination by recombination of two polymeric peroxy radicals consists of the following:

1. Tobolsky and coworkers [2–4], Neiman [5, 6], and Chien and Boss [7] have derived kinetic expressions using such a termination step which can explain reasonably well the rate data obtained during polyolefin autoxidation.

2. Bartlett and Gunther [21] presented strong evidence for the formation of a trioxide (from $RO_2 \cdot$ and $RO \cdot$ radicals) and for the recombination of $RO_2 \cdot$ radicals at very low temperatures (down to $-70°C$) during the oxidation of tert-butyl and cumyl hydroperoxides in methylene chloride. Since it would be expected that at such low temperatures the viscosity of the medium would be relatively high, the evidence presented by these workers tends to favor the proposal of polymeric peroxy radical recombination during oxidation in the bulk phase.

Evidence opposing this termination includes:

1. Reich, Stivala, and coworkers [1, 10–15, 17–20], Notley [8, 9], and Bawn and Chaudhri [16] excluded such a termination step in obtaining kinetic expressions to explain acquired rate data.

2. The Trommsdorf or gel effect during polymerization in the bulk phase has generally been ascribed to the much larger decrease in the termination rate than in the propagation rate as the degree of polymerization (and of viscosity) increased. In this connection, Betts and Uri [22] found that the autoxidation rate for liquid and solid hydrocarbons proceeded more rapidly for the latter than for the former. This was attributed to a marked reduction in the solid phase of the diffusion-controlled rate of termination by peroxy radical combination as compared with the propagation step.

3. Hiatt and Traylor [23] and Mayo [24] studied the effect of viscosity on the formation of decomposition products from di-tert-butylperoxy oxalate and found that changes in the medium viscosity caused changes in the amounts of the various products formed. This was attributed to the influence of viscosity upon the ability of oxy radicals within a cage to combine within or diffuse out of the cage.

4. Bresler and Kazbekov [25] reported on macroradical reactivity in the bulk phase studied by electron-spin resonance. It was found that the decay of macroradicals formed during mechanical degradation of various

polymers, e.g., polystyrene, involved an activation energy of the order of magnitude of 24 kcal mole^{-1}. This result is peculiar when compared to the value of practically zero obtained for the recombination of radicals in gaseous and liquid phases. This difference in activation energies was attributed to the greatly reduced mobility of the reacting macroradicals in the solid phase.

In the following will be presented kinetic expressions for oxidation rate which are derived from schemes which do and do not include termination by polymeric radical recombination.

Kinetics of polyolefin oxidation including polymeric peroxy radical recombination. *Approach of Tobolsky and coworkers and Chien and Boss.* Tobolsky and coworkers [2, 3] investigated the autoxidation of amorphous polypropylene with neat oxygen in the presence of benzoyl peroxide. A scheme was advanced for the solid-state oxidation which was similar to that for liquid-phase oxidations (cf. Chap. 1), i.e.,

$$I \xrightarrow{k_d} 2R' \xrightarrow{RH} 2R\cdot \tag{1}$$

$$R\cdot + O_2 \xrightarrow{k_2} RO_2\cdot \tag{2}$$

$$RO_2\cdot + RH \xrightarrow{k_3} RO_2H + R\cdot \tag{3}$$

$$2R\cdot \xrightarrow{k_4} Products \tag{4}$$

$$R\cdot + RO_2\cdot \xrightarrow{k_5} Products \tag{5}$$

$$2RO_2\cdot \xrightarrow{k_6} Products + O_2 \tag{6}$$

At relatively high oxygen pressures (generally above 200 torrs), steps (4) and (5) may be neglected in comparison with step (6). Thus, assuming steady-state conditions, we may write

$$-\frac{d[O_2]}{dt} \equiv \rho_{ox} = k_2[R\cdot][O_2] - \frac{k_6[RO_2\cdot]^2}{2} \tag{7}$$

$$R_i = 2fk_d[I] = k_6[RO_2\cdot]^2 = k_2[R\cdot][O_2] - k_3[RO_2\cdot][RH] \tag{8}$$

where f = initiator (I) efficiency, and R_i = rate of initiation. Upon combining Eqs. (7) and (8),

$$\rho_{ox} = fk_d[I] + k_3[RH]\left(\frac{2fk_d[I]}{k_6}\right)^{1/2} \tag{9}$$

or

$$\frac{\rho_{ox}}{2k_d[I]} = k_3[RH] \frac{(f/k_6)^{1/2}}{(2k_d[I])^{1/2}} + \frac{f}{2} \tag{10}$$

From Eq. (10), it can be seen that a plot of $\rho_{ox}/2k_d[I]$ versus $(2k_d[I])^{-1/2}$ should yield a linear relation whose slope is $k_3[RH](f/k_6)^{1/2}$. In Fig. 5.1 are shown such plots for amorphous polypropylene at various temperatures. The value of ρ_{ox} used was the one extrapolated to zero time so that the initial value of the initiator could be employed while at the same time complications arising from hydroperoxide decomposition were avoided. From the linear plots obtained and from an Arrhenius plot, an activation energy $E_3 - E_6/2 \approx 17$ kcal mole^{-1} was obtained. Based upon oxidation studies of simpler hydrocarbons [1], this value appears to be too high. Furthermore, the negative intercepts obtained (cf. Fig. 5.1) were attributed to the evolution of untrapped volatile products in the oxygen uptake apparatus used. A similar value for $E_3 - E_6/2$ was obtained for ethylene-propylene rubber.

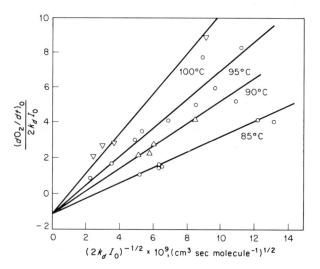

FIG. 5.1   Oxygen absorption data for polypropylene [2].

Chien and Boss [7] studied the autoxidation of polypropylene (about 50 percent crystalline) in the bulk phase, using chemical, infrared, visco-metric, and electron-spin-resonance techniques. A scheme was employed which utilized Eqs. (2), (3), and (6) plus the following steps (in the absence of initiator):

$$RO_2H \xrightarrow{k_i} RO\cdot + HO\cdot \qquad (11)$$

$$RO\cdot + RH \longrightarrow ROH + R\cdot \qquad (12)$$

$$HO\cdot + RH \longrightarrow H_2O + R\cdot \qquad (13)$$

Under the experimental conditions used, measurements were carried out during which steady-state concentrations of hydroperoxide, $[RO_2H]_s$, obtained (constant rate of oxidation and ESR signal intensity). From the scheme, we may write (steady-state conditions assumed for all radical species)

$$k_i[RO_2H]_s = k_6[RO_2\cdot]_s^2 = k_3[RO_2\cdot]_s[RH] \qquad (14)$$

and

$$k_2[R\cdot][O_2] = k_3[RO_2\cdot]_s[RH] + 2k_6[RO_2\cdot]_s^2 \qquad (15)$$

Then, using Eqs. (14), (15), and (16),

$$\rho_{ox} = k_2[R\cdot][O_2] - k_6[RO_2\cdot]_s^2 \qquad , \qquad (16)$$

the steady-state autoxidation rate becomes

$$\rho_{ox,s} = k_3[RO_2\cdot]_s[RH] + k_6[RO_2\cdot]_s^2 = 2k_i[RO_2H]_s \qquad (17)$$

Equation (17) may be easily transformed into [using Eq. (14)]

$$\rho_{ox,s} = 2k_3[RO_2\cdot]_s[RH] \qquad (18)$$

and

$$\rho_{ox,s} = 2k_6[RO_2\cdot]_s^2 \qquad (19)$$

It may also be easily shown [Eqs. (14) and (17)] that

$$\rho_{ox,s} = k_i[RO_2H]_s + k_3[RH]\left(\frac{k_i[RO_2H]_s}{k_6}\right)^{1/2} \qquad (20)$$

From values of $[RO_2H]_s$ (iodimetry), $[RO_2\cdot]_s$ (ESR), and using $[RH] = 11$ moles l^{-1}, values of k_i, k_3, and k_6 could be estimated from Eqs. (17) to (19), respectively. It was found that the values of k_i obtained at various temperatures corresponded well with values obtained in another study of the decomposition of polypropylene hydroperoxide [26]. Furthermore, the value of $k_3(f/k_6)^{1/2}$ obtained by Tobolsky and coworkers [2] for amorphous polypropylene at 110°C appeared to be in reasonable agreement with the corresponding value obtained by Chien and Boss.

In connection with Chien's work [7, 26] may be mentioned some of the comments made by Mayo [27]. These are, briefly: (1) The presence of alcohol and carbonyl groups (based on the apparent low kinetic chain length) should affect the decomposition of polypropylene hydroperoxide (cf. Ref. 26), and Mayo considered the agreement between rate constants for polypropylene hydroperoxide decomposition [7, 26] to be coincidental. (2) At the point where steady-state conditions were used during polypropylene autoxidation [7], the products present presumably consisted mainly of alcohol and carbonyl groups and their oxidation products. Secondary alcohol and ketone groups are expected to be about 30 times more reactive than tertiary C—H groups in polypropylene. Thus, at the point where rate constants were determined (about 10 percent conversion) much of the oxygen consumption should be in secondary reactions, which were ignored, and Chien and Boss were presumably measuring not the values of any single rate constants for the polypropylene oxidation but average values for an unknown mixture with its oxidation products. In rebuttal [28], Chien believed that his conversions were not too high and that his chain lengths were not too short to invalidate his conclusions.

Approach of Neiman and coworkers. Neiman [5, 6] offered the following scheme to explain his results for the oxidation of polypropylene in the absence of initiator:

$$RH + O_2 \longrightarrow R\cdot + HO_2\cdot \tag{1''}$$

$$R\cdot + O_2 \longrightarrow RO_2\cdot \tag{2}$$

$$RO_2\cdot + RH \longrightarrow RO_2H + R\cdot \tag{3}$$

$$2RO_2\cdot \longrightarrow Products \tag{6}$$

$$RO_2H + RH \longrightarrow RO\cdot + R\cdot + H_2O \tag{21}$$

The following equations may be written from this scheme:

$$R_i + k_3[RO_2\cdot][RH] + k_{21}[RO_2H][RH] = k_2[R\cdot][O_2] \tag{22}$$

$$k_2[R\cdot][O_2] = k_3[RO_2\cdot][RH] + k_6[RO_2\cdot]^2 \tag{23}$$

Combining Eqs. (22) and (23) and solving for $[RO_2\cdot]$, we obtain

$$[RO_2\cdot] = \left(\frac{R_i + k_{21}[RH][RO_2H]}{k_6} \right)^{1/2} \tag{24}$$

Upon substituting Eq. (24) into Eq. (25),

$$-\frac{d[O_2]}{dt} \equiv \rho_{ox} = k_3[RO_2\cdot][RH] + k_6[RO_2\cdot]^2 \tag{25}$$

and neglecting the R_i term in Eq. (24), there is obtained

$$-\frac{d[O_2]}{dt} = k_3 \left(\frac{k_{21}}{k_6} \right)^{1/2} [RH]^{3/2}[RO_2H]^{1/2} + k_{21}[RH][RO_2H] \tag{26}$$

Equation (26) may be converted into

$$-\frac{d[O_2]/dt}{[RH][RO_2H]} = k_{21} + k_3 \left(\frac{k_{21}[RH]}{k_6[RO_2H]} \right)^{1/2} \tag{27}$$

When the left-hand side of Eq. (27) was plotted against $([RH]/[RO_2H])^{1/2}$, a linear relation was obtained, as shown in Fig. 5.2.

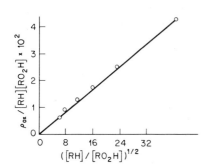

FIG. 5.2 Dependence of rate of absorption of oxygen on the concentrations [RH] and [ROOH] [5].

Kinetics of polyolefin oxidation excluding polymeric peroxy radical recombination. *Approach of Reich, Stivala, and coworkers and of Bawn and Chaudhri.* Bawn and Chaudhri [29] studied the autoxidation of atactic polypropylene (APP) in 1,2,4-trichlorobenzene (TCB) in the presence of the initiator, 2,2'-azobisisobutyronitrile (AIBN). The basic autoxidation scheme was assumed [Eqs. (1) to (6)], except that steps (4) and (5) were neglected as usual, and Eq. (1) was now modified to account for the nitrogen liberated during the AIBN decomposition; i.e.,

$$I \xrightarrow{k_d} 2R\cdot + N_2 \tag{1a}$$

When corrections are made for nitrogen evolution ($k_d[I]$), absorption of oxygen in the initiation step ($2fk_d[I]$), and liberation of oxygen in the termination step ($fk_d[I]$), ρ_{ox} becomes [cf. Eq. (9)]

$$\rho_{ox} = (3f - 1)k_d[I] + k_3[RH]\left(\frac{2fk_d[I]}{k_6}\right)^{1/2} \tag{9'}$$

or

$$\frac{\rho_{ox}}{[I]} = (3f - 1)k_d + k_3[RH]\left(\frac{2fk_d}{k_6[I]}\right)^{1/2} \tag{10'}$$

Thus, from Eq. (10'), a plot of $\rho_{ox}/[I]$ versus $[I]^{-1/2}$ should afford a linear relation ([RH] being constant). Such a plot gave a linear relation for the autoxidation of APP in TCB at 90°C for a low polymer concentration of 0.15 mole l⁻¹. However, at high APP concentrations, e.g., 1.5 moles l⁻¹, curves were obtained. Apparently, at relatively high concentrations of APP in TCB, Eq. (10') was no longer valid. Thus, the kinetic order in respect to [AIBN] was 0.85, and the order in respect to [RH] varied from 0.12 to 0.37. These results were so different from those observed with alkyl aromatic hydrocarbons that the autoxidation of a hydrocarbon structurally similar to polypropylene, 2,6,10,14-tetramethylpentadecane (pristane), was carried out in the neat liquid state. The slope of the log ρ_{ox} versus log [AIBN] plot gave a reaction order of 0.55 in respect to [AIBN], as anticipated from Eq. (9'). Furthermore, when a long kinetic chain length is assumed [1], the intercept in a plot of Eq. (10') should be $(2f - 1)k_d$, the liberation of oxygen in the termination step being neglected. From values of f and k_d appearing in the literature, the calculated value was found to be in good agreement with the experimental value of the intercept. The results with pristane and APP clearly indicated that in the autoxidation of low-molecular-weight liquid hydrocarbons there was no substantial kinetic difference in the behavior of paraffinic or of alkyl aromatic substrates and that the different kinetics with APP solutions may be dependent upon the transition from a low- to a high-molecular-weight substrate. Subsequently, Bawn and Chaudhri [16] studied the autoxidation of APP in TCB in absence of initiator, and in order to satisfactorily account for their kinetic results, they employed a general scheme first introduced by Stivala, Reich, and coworkers [1, 10] for the autoxidation of polyolefins.

Reich, Stivala, and coworkers [1] employed the following kinetic scheme to explain their results obtained during the autoxidation of polyolefins in the bulk phase by means of infrared spectroscopy. This scheme did not involve termination by recombination of polymeric peroxy radicals.

$$RH + O_2 \xrightarrow{k_i} R\cdot + HO_2\cdot \tag{1''}$$

$$R\cdot + O_2 \xrightarrow{k_2'} RO_2\cdot \tag{2}$$

$$RO_2\cdot + RH \xrightarrow{k_3} RO_2R + H\cdot \tag{3}$$
$$\text{(one R may be H)}$$

$$RO_2\cdot + RH \xrightarrow{k_4'} \text{Inactive products} \tag{3'}$$

$$RO_2R \xrightarrow{k_5'} 2RO\cdot \tag{28}$$

$$RO_2R \xrightarrow{k_6'} \text{Inactive nonvolatile products} \tag{29}$$

$$RO\cdot \xrightarrow{k_7'} R'\cdot + \text{volatile products} \tag{30}$$

$$R'\cdot + RH \xrightarrow{k_8'} R\cdot + R'H \tag{31}$$

$$R'\cdot + O_2 \xrightarrow{k_9'} \text{Less reactive products} \tag{32}$$

In the above scheme, steps (1''), (2), (3), (28), (29), (30), and (31) have been postulated in regard to hydrocarbon autoxidation [1]. Also, Fish [30] has postulated radical rearrangements in gas-phase oxidation which offer a possible explanation for step (29), namely,

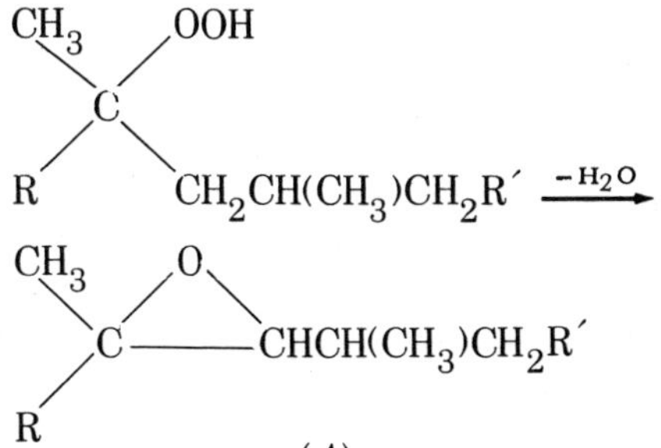

$$
\begin{array}{c}
CH_3 \quad OOH \\
\diagdown \quad \diagup \\
C \\
\diagup \quad \diagdown \\
R \qquad CH_2CH(CH_3)CH_2R' \xrightarrow{-H_2O}
\end{array}
$$

$$
\begin{array}{c}
CH_3 \quad O \\
\diagdown \quad \diagup \diagdown \\
C\text{———}CHCH(CH_3)CH_2R' \\
\diagup \\
R \\
\qquad (A)
\end{array} \tag{33}
$$

$$(A) \longrightarrow R'CH_2\overset{\cdot}{C}HCH_3 + R\overset{O}{\overset{||}{C}}\overset{\cdot}{C}HCH_3 \tag{34}$$
$$\qquad\qquad (B) \qquad\qquad (C)$$

Either radical (B) in reaction (34) can add H· to yield a saturated hydrocarbon, or H· can be removed to give an unsaturated hydrocarbon. Radical (C) can form in the same manner either a saturated or an unsaturated ketone. Step (3′) is included in the scheme since small amounts of impurities in polymers are inherently present owing to difficulties in purification. These impurities, which are proportional to [RH], may combine with RO_2· radicals, thereby inactivating them. A case in point is the inactivation of RO_2· radicals by metallic impurities (Cat), e.g., RO_2· + Cat $\longrightarrow$ RO_2·Cat. Such reactions have been reported [1]. A step similar to (30) has been advanced by various workers during hydrocarbon (e.g., cumene) autoxidation [1] (cf. Chap. 1), and for the decomposition of tertiary oxy radicals in general. Thus, for example, the tertiary cumyloxy radical may decompose as follows:

$$C_6H_5 - \underset{\underset{CH_3}{|}}{\overset{\overset{CH_3}{|}}{C}} - O \cdot \longrightarrow C_6H_5 \overset{\overset{CH_3}{|}}{C} = O + CH_3 \cdot \qquad (35)$$

In step (30), the R′· radical would be represented by the CH_3· radical of reaction (35). This radical could then react with oxygen [cf. step (32)] to afford a CH_3O_2· peroxy radical which can then rapidly terminate with itself or with an RO_2· radical, namely,

$$2CH_3O_2 \cdot \longrightarrow CH_2O + CH_3OH + O_2 \qquad (36)$$

and

$$CH_3O_2 \cdot + RO_2 \cdot \longrightarrow CH_2O + ROH + O_2 \qquad (37)$$

It is assumed that the relatively small CH_3O_2· peroxy radical can diffuse through the polymeric medium much more readily than polymeric tertiary peroxy radicals (RO_2·), and thereby undergo termination steps such as (36) [cf. step (32)]. Besides the greater ability of smaller peroxy radicals to diffuse in the polymeric medium, it has been reported that chain termination by tertiary butyl peroxy radicals is extraordinarily slow at 22°C (k = 390 l mole^{-1} sec^{-1}) as compared with chain termination by secondary or primary peroxy radicals at 30°C (k = 10^6 to 10^7 l mole^{-1} sec^{-1}) [24, 31, 32].

The following assumptions will be made before the derivation of kinetic expressions from the general scheme: steps (1″) to (3′) and (28) to (32):

1. The attack of oxygen upon the polyolefin chain is random (it will preferentially occur at the tertiary carbon).

2. The nonvolatile carbonyl- and oxygen-containing products, with the exception of peroxy compounds, arise only from the decomposition of the hydroperoxide structures [cf. step (29)].

3. Alcoholic groups react rapidly with oxygen to yield carbonyl-containing compounds, e.g., carboxylic acids. Twigg [33] found that no alcohol could be isolated in the oxidation of n-decane, and attributed this to a rapid oxidation of the intermediate alcohol that may have formed.

4. The concentration of hydrocarbon groups that can be attacked is assumed to be constant (at low conversions, there should exist a large excess of RH available at all times during oxidation).

5. It is also assumed that step (30) rapidly follows step (28).

Based upon the above assumptions and that of steady-state conditions for the radicals $R\cdot$, $R'\cdot$, and $RO_2\cdot$, the following expressions may be written for a given temperature [1]:

$$R_i = k_2'[R\cdot][O_2] - k_8'[RH][R_i'] \tag{38}$$

$$2k_5'[RO_2R] = k_8'[RH][R_i'] + k_9'[R_i'][O_2] \tag{39}$$

$$k_2'[R\cdot][O_2] = (k_3' + k_4')[RO_2\cdot][RH] \tag{40}$$

Also, the rate of hydroperoxide formation may be written as

$$\frac{d[RO_2R]}{dt} = k_3'[RO_2\cdot][RH] - (k_5' + k_6')[RO_2R] \tag{41}$$

Solving Eqs. (38) to (40) for $[RO_2\cdot]$ and substituting this expression into Eq. (41) gives

$$-\frac{d[RO_2R]}{dt} =$$

$$[RO_2R]\left\{ k_5' + k_6' - \frac{2k_3'k_5'k_8'[RH]}{(k_3' + k_4')(k_8'[RH] + k_9'[O_2])} \right\} - \frac{k_3'R_i}{k_3' + k_4'} \tag{42}$$

Upon integrating Eq. (42) and solving for $[RO_2R]$, there is obtained

$$[RO_2R] = \frac{k_3'R_i[1 - \exp(-At)]}{(k_3' + k_4')A} \tag{43}$$

where

$$A = k_5' + k_6' - \frac{2k_3'k_5'k_8'[RH]}{(k_3' + k_4')(k_8'[RH] + k_9'[O_2])}$$

The rate of formation of nonvolatile carbonyl products ρ_{CO} may be represented by $k'_6[RO_2R]$. This leads to the expression

$$\rho_{CO} = \frac{K[O_2][1 - \exp(-At)]}{1 - (K_2/K_3 + [O_2])} \tag{44}$$

where K, K_2, and K_3 are constants and

$$A = k'A' = k'\left(1 - \frac{K_2}{K_3 + [O_2]}\right) \tag{45}$$

where $k' = k'_5 + k'_6$. Equation (44) predicts that for a given oxygen concentration the value of ρ_{CO} should increase and approach a maximum value as t increases. However, as will be seen, this does not hold over the entire reaction time employed. A decrease in rate occurs as t increases to relatively large values, which may be due to various causes, e.g., the occurrence of morphological changes during the polyolefin oxidation. Therefore, an apparent maximum rate of formation ρ_m was utilized. At ρ_m the product At_m (where t_m is the time to reach ρ_m) was found to be approximately constant during polypropylene oxidation. Thus, at 140°C and for values of $[O_2]$ of 7 to 100 percent, $(At_m)_{avg} = 1.0 \pm 0.07$, while at 150°C, $(At_m)_{avg} = 1.0 \pm 0.1$. This also held for other temperatures used. That At_m should be approximately constant at any given temperature may be seen from the fact that as $[O_2]$ increases, t_m decreases while A increases [see Eq. (45)]. Therefore, we may write

$$\rho_m = \frac{K_1[O_2]}{1 - K_2/(K_3 + [O_2])} \tag{46}$$

where

$$K_1 = \frac{\text{const } k_i k'_3 k'_6[RH]}{(k'_3 + k'_4)(k'_5 + k'_6)} \tag{46a}$$

$$K_2 = \frac{2k'_3 k'_5 k'_8[RH]}{k'_9(k'_3 + k'_4)(k'_5 + k'_6)} \tag{46b}$$

$$K_3 = \frac{k'_8[RH]}{k'_9} \tag{46c}$$

and const = $1 - \exp{(-At)_m}$. Furthermore,

$$\frac{\rho_m A}{[O_2]} = \frac{\text{const } [RH] k_i k_3' k_6'}{k_3' + k_4'} \tag{47}$$

For low values of $[O_2]$, Eq. (46) reduces to

$$\rho_m = \frac{K_1 K_3 [O_2]}{K_3 - K_2} \tag{48}$$

and for high values of $[O_2]$, where $[O_2] \gg K_3 - K_2$, Eq. (46) becomes

$$\rho_m = K_1 K_3 + K_1 [O_2] \tag{49}$$

The value of A in Eq. (44) can be obtained from the derived expression,

$$-\ln \frac{\rho_m - \rho}{\rho_m} = At \tag{50}$$

where $\rho \ll \rho_m$. The value of A may also be estimated from the rate of hydroperoxide formation, ρ_{HP}. Thus, from Eq. (43)

$$\rho_{HP} = C [O_2] \exp{(-At)} \tag{51}$$

or

$$\ln \rho_{HP} = \ln{(C [O_2])} - At \tag{51a}$$

where $C = k_i k_3' [RH] /(k_3' + k_4')$. From Eq. (51a) for a given $[O_2]$ and temperature, a linear relationship should result from a plot of $\ln \rho_{HP}$ versus t, from which A may be evaluated. Furthermore, the rate of formation of volatile products ρ_{VP} may be used to estimate values of A. Thus, from the scheme and Eq. (43),

$$\rho_{VP} = S \frac{1 - \exp{(-At)}}{A} \tag{52}$$

where

$$S = \frac{k_i k_3' k_5' [RH][O_2]}{k_3' + k_4'}$$

Equation (52) may be converted into

$$\ln \rho'_{VP} = \ln S - At \tag{53}$$

where

$$\rho'_{VP} = \frac{d(\rho_{VP})}{dt}$$

The various expressions given above, including other similarly derived expressions, were applied to polypropylene and poly(butene-1) oxidations in absence of inhibitors and metal catalysts. Infrared-absorption spectroscopy techniques were employed to determine carbonyl and hydroperoxide concentrations during oxidation as a function of time [1].

Equation (44) predicts that for a given oxygen concentration, as t increases, ρ_{CO} increases and approaches a maximum value. However, as previously mentioned, this does not obtain over the entire reaction time used (Fig. 5.3). Thus, an apparent maximum rate ρ_m was used. From plots such as those depicted in Figs. 5.3 and 5.4 and from Eqs. (48) and (49), values of K_1, K_2, and K_3 could be calculated at various temperatures

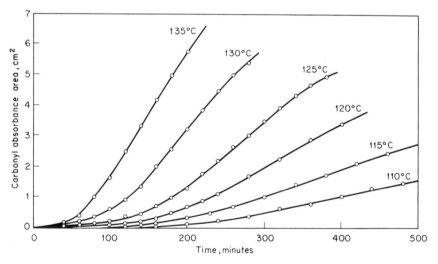

FIG. 5.3 Plots of carbonyl absorbance areas versus time for APP at an oxygen/nitrogen ratio of 10:90 by volume for various reaction temperatures [20].

from which values of ρ_m at various temperatures and oxygen concentrations may be computed. Good agreement was obtained between observed and calculated values during the bulk-phase autoxidation of APP [20], isotactic polypropylene (IPP) [10], and atactic poly(butene-1) [13].

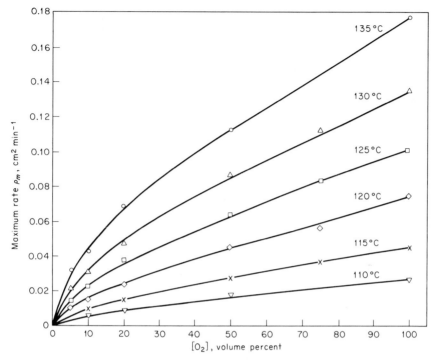

FIG. 5.4 Plots of maximum rate ρ_m versus oxygen concentrations at various temperatures for APP [20].

Furthermore, values of A [see Eq. (43)] for different experimental conditions could be estimated from plots of Eq. (50). Figure 5.5 shows such plots for APP at 115°C and various oxygen concentrations. A values having thus been obtained, estimates of k' may be made using Eq. (45). In Table 1 are listed values of A', A, and k' as a function of oxygen concentration and reaction temperature. Mean deviations from the average value of k' at a specified temperature were small, as anticipated.

From Eqs. (46a), (46b), and (46c), Arrhenius plots of log K_3 and log $(K_2/K_1 K_3)$ versus reciprocal temperature were constructed as shown in Fig. 5.6. From such plots, the following relations could be obtained: $E'_8 - E'_9 \approx 8$ kcal mole^{-1}, and $E_i + E'_6 - E'_5 \approx 17$ kcal mole^{-1}, where the subscripts correspond to the subscripts of the rate constants in the scheme. Also, from Eq. (47), an Arrhenius plot (Fig. 5.7) should afford values of $E_i + E'_6$ if k'_4/k'_3 is constant over the temperature range used (this was indicated to be approximately true [18]). Thus, a value of $E_i + E'_6 \approx 48$ kcal mole^{-1} was obtained for APP. From this and other values, $E'_5 \approx 31$ kcal mole^{-1}. This value could be checked by means of an Arrhenius plot of Eq. (54),

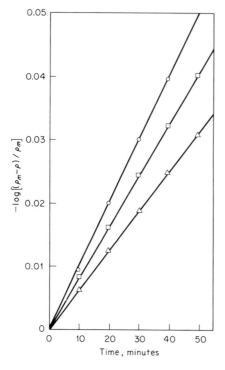

FIG. 5.5 Plots of $-\log [(\rho_m - \rho)/\rho_m]$ versus time at 115°C for APP at various oxygen concentrations: △ 20%, □ 50%, ○ 100% [20].

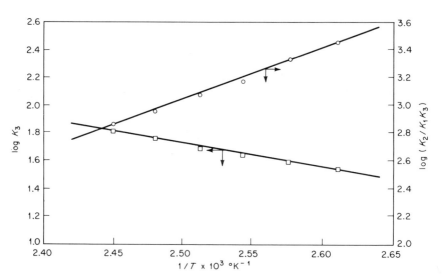

FIG. 5.6 Plots of $\log K_3$ and $\log (K_2/K_1 K_3)$ versus reciprocal temperature $1/T$ for APP [20].

TABLE 1 Values of A, A', and k' for APP at Various Temperatures and Oxygen Concentrations [20]

Temperature °C	$[O_2]$, vol. %	A'	$A \times 10^3$, min⁻¹	$k' \times 10^3$, min⁻¹	$k' \times 10^3$ (avg.), min⁻¹
	5	0.3232			
	10	0.3986	0.970	1.908	
110	20	0.5082			
	50	0.6820	1.212	1.777	1.915
	75	0.7544			
	100	0.7999	1.647	2.059	
	5	0.3121			
	10	0.3838			
115	20	0.4899	1.446	2.952	
	50	0.6638	1.874	2.823	2.874
	75	0.7381	2.065	2.798	
	100	0.7856	2.300	2.923	
	5	0.2816			
	10	0.3482			
120	20	0.4500	2.104	4.667	
	50	0.6255	2.572	4.112	4.351
	75	0.7042	3.103	4.407	
	100	0.7556	3.180	4.208	
	5	0.2569			
	10	0.3202			
125	20	0.4192	2.579	6.009	
	50	0.5959	3.739	6.274	6.297
	75	0.6776	4.312	6.364	
	100	0.7318	4.786	6.540	
	5	0.2233			
	10	0.2808			
130	20	0.3735	4.238	11.347	
	50	0.5482	5.757	10.516	10.791
	75	0.6333	6.684	10.554	
	100	0.6915	7.433	10.749	
	5	0.2048			
	10	0.2582			
135	20	0.3460	6.190	17.890	
	50	0.5178	9.540	18.438	18.183
	75	0.6039			
	100	0.6641	12.100	18.220	

$$k_5' = (k_3' + k_4')k' \frac{K_2}{2k_3'K_3} \tag{54}$$

From this plot (cf. Fig. 5.7), a value of $E_5' \approx 31$ kcal mole⁻¹ was calculated.

This value of E'_5 agrees well with that generally found for the unimolecular decomposition of hydroperoxides (30 kcal mole^{-1}) [34].

Values of various parameters obtained for APP [20] are compared with corresponding values reported for IPP [11] in Table 2. It can be seen

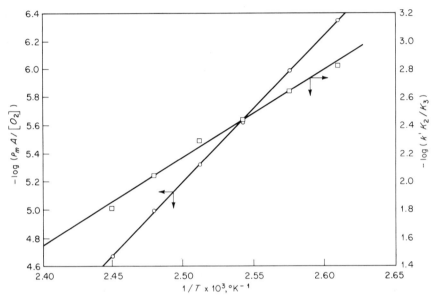

FIG. 5.7 Plots of log $(\rho_m A/[O_2])$ and log $(k'K_2/K_3)$ versus reciprocal temperature $1/T$ for APP [20].

from this table that various values of activation energies and rate constants are similar. A possible explanation follows. It is assumed that the IPP is comprised of a fringed micellar structure wherein the crystallite regions are essentially separated from the amorphous regions. The former regions presumably do not act as a contiguous barrier to oxygen penetration. Thus, diffusion control is not important at small film thicknesses both for APP and IPP. Furthermore, the isolated amorphous regions in IPP react similarly to those in APP in respect to oxidizability. Consequently, similar rate constants would be anticipated for APP and IPP. In this connection, Hawkins and coworkers [35] observed a relatively small difference in the rate of oxidation between APP and IPP. Wiles and coworkers [36] also found that the rate of photochemical oxidation of polypropylene of various tacticities did not change appreciably. Thus, IPP samples oxidized somewhat faster than APP samples.

Besides the estimation of kinetic parameters from studies of carbonyl formation by means of infrared spectroscopy (IR), such parameters have also been estimated from hydroperoxide formation, volatile products formation, oxygen absorption, oxyluminescence, and molecular-weight changes during polyolefin autoxidation. These topics will be discussed in the following paragraphs in the order listed.

Hydroperoxide formation. The scheme for polyolefin autoxidation, Eqs. (1″) to (3′) and (28) to (32), has been applied to hydroperoxide

TABLE 2   Comparison of Various Parameters Obtained during APP Oxidation with Those Reported for IPP Oxidation [20]

Kinetic parameters	APP [20]	IPP [11]
k', min^{-1} (140°C)	0.03 (extrapolated)	0.047
E'_5, kcal mole^{-1}.	31	33
$E_i + E'_6 - E'_5$, kcal mole^{-1}	17	18
$E_i + E'_6$, kcal mole^{-1}	48	51
$E'_8 - E'_9$, kcal mole^{-1}	8	6
(Ash, %)	(0.008)	(0.02)

formation [14]. Data on the concentration of hydroperoxides at a given temperature and oxygen concentration as a function of time were obtained from infrared absorption bands in the region of 2.8 to 3.0 μ. From Eq. (51a), a plot of ln ρ_{HP} versus t should afford a linear relation from which the value of A can be estimated. From such a plot (up to about 0.8 hr) the value of A was found to be 2.1 x 10^{-2} min^{-1} for IPP, which is in good agreement with the corresponding value of A estimated from carbonyl

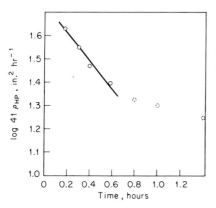

FIG. 5.8 Log 41 ρ_{HP} versus reaction time for IPP oxidation in 20:80 oxygen/nitrogen at 150°C [14].

data at 150°C and a 20:80 oxygen/nitrogen ratio (Fig. 5.8). Beyond about 0.8 hr, the plot deviated from linearity; a similar behavior has been observed from carbonyl data which may be due to morphological and other factors (cf. Chap. 6).

Volatile products formation. It can be seen from Eq. (53) that a plot of $\ln \rho_{VP}$ versus t should yield a linear relation from which values of A can be obtained. Such plots were constructed from data reported by Neiman [5] and Miller and coworkers [37] on the rate of formation of acetone, water, formaldehyde, and acetaldehyde during the autoxidation of IPP. Except for water, the values of A compared favorably with those obtained by IR techniques (cf. Table 3).

TABLE 3   Value of the Kinetic Parameter A Derived from Infrared Spectroscopy Compared to Those Values of A Obtained from Other Techniques [17]

Experimental procedure	Polymer	O_2:N_2 ratio	Temp. °C	From nonoptical methods	From infrared spectroscopy (IR)
					A, min^{-1}
IR (carbonyl).	Isotactic polypropylene (IPP)	20:80 20:80	150	- - - - - - - - -	0.021
IR (hydroperoxide) . . .	IPP	20:80	150	- - - - - - - - -	0.021
O_2 absorption	IPP	100:0	150	0.049	0.051
Oxyluminescence	IPP	100:0	150	$(0.056)_{avg}$	0.051
From rate of formation of volatile products:					
Acetone.	IPP	53:47	130	0.010	0.016
Formaldehyde	IPP	66:34	150	0.035	0.040
Intrinsic viscosity	Atactic poly(butene-1)	21:79	90	0.0037	0.0044

Oxygen absorption. An expression for the rate of oxygen absorption ρ_{ox} may be readily derived from the general scheme. Thus, it can be shown [12] that if the initiation rate is neglected,

$$\rho_{ox} = 2k_5'[RO_2R] \qquad (55)$$

(Note that this expression in comparison with that for ρ_{CO} implies that $\rho_{ox} \propto \rho_{CO}$ and that the shape of the oxygen uptake versus time curve should be similar in shape to that for the quantity of carbonyl formed versus time). When Eq. (43) is substituted into (55),

$$\rho_{ox} = \frac{K'R_i(1 - e^{-At})}{A} \qquad (56)$$

where K', R_i, and A should be approximately constant when the amount of oxygen absorbed by the polyolefin is relatively small (for a particular temperature). From Eq. (56), when $At \gg 1$, $\rho_{ox} \rightarrow \rho_{ox,\,m} \equiv K'R_i/A$. Then, Eq. (56) may be transformed into [cf. Eq. (50)]

$$\ln \frac{\rho_{ox,\,m} - \rho_{ox}}{\rho_{ox,\,m}} = A\theta - At' = At \qquad (57)$$

where θ = induction time and t' = total exposure time. By means of this expression, it was possible to obtain values of A from oxygen uptake versus time measurements. Thus, from data reported [38] for the oxidation of IPP at 150°C in oxygen (film C), a value of $A = 4.9 \times 10^{-2}$ min^{-1} was calculated ($A = 5.1 \times 10^{-2}$ min^{-1} from IR measurements, cf. Table 3) up to a reaction time of about 0.7 hr.

As mentioned earlier, Bawn and Chaudhri [16] utilized the general scheme to account for their results obtained during the autoxidation of APP in trichlorobenzene (TCB) in absence of initiator. Upon combining Eqs. (43) and (55),

$$\rho_{ox,\,m} = \frac{K_1'[O_2]_d}{1 - K_2/(K_3 + [O_2]_d)} \qquad (58)$$

where

$$K_1' = 2k_i k_3' k_5'[RH]^n[1 - \exp(-At_m)]/k'(k_3' + k_4')$$

$$n = \text{some constant}$$

$$[O_2]_d = \text{concentration of dissolved oxygen}$$

K_2 and K_3 have been previously defined.

At the relatively high oxygen pressures used by Bawn and Chaudhri, it may be assumed that $[O_2]_d \gg K_3 - K_2$; then [cf. Eq. (49)]

$$\rho_{ox,\,m} = K_1'K_3 + K_1'[O_2]_d \qquad (59)$$

Assuming that Henry's law is applicable, i.e., $[O_2]_d = Hp_{O_2}$, where H = const and p_{O_2} = partial pressure of oxygen above the solution, then Eq. (59) becomes

$$\rho_{ox,\,m} = K_1'K_3 + K_1'Hp_{O_2} \qquad (59a)$$

Furthermore, at constant p_{O_2} and temperature, Eq. (59a) may be transformed into

$$\rho_{ox,\,m} = C_1[RH]^{n+1} + C_2[RH]^n \qquad (60)$$

Equation (59a) was found to agree well with experimental results; i.e., rate was found to be proportional to oxygen pressure. Also, when $R_i \propto$ [RH][1] ($n = 1$), then from Eq. (60), the kinetic dependency of $\rho_{ox, m}$ upon [RH] should lie between 1 and 2, as was found experimentally. During the early stages of the reaction, the value of At is relatively small, and Eq. (56) becomes

$$\rho_{ox} = K'R_i t \tag{61}$$

which is in agreement with experimental results, i.e., $\rho_{ox} \propto t$. Furthermore, upon integration, Eq. (61) becomes

$$[O_2]_{abs} = \frac{K'R_i t^2}{2} \tag{62}$$

where $[O_2]_{abs}$ = amount of oxygen absorbed. When Eq. (62) is substituted into (61),

$$\rho_{ox} = C_3 R_i^{1/2} [O_2]_{abs}^{1/2} \tag{63}$$

where $C_3 = (2K')^{1/2}$. The relationship between ρ_{ox} and $[O_2]_{abs}^{1/2}$ [Eq. (63)] has been experimentally observed during the initial oxidation stages (autocatalytic region).

It was mentioned earlier that Chien and Boss [7] calculated values of rate constants from expressions derived from a relatively simple scheme [Eqs. (2), (3), (6), (11) to (13)], which appeared to agree well with reported values. Despite the comments of Mayo [27], the values of these constants appear to be of a correct order of magnitude. It will now be indicated how similar values can be obtained from expressions derived from the more general scheme [Eqs. (1'') to (3') and (28) to (32)]. Under the experimental conditions used by Chien and Boss, $d[RO_2 R]/dt = 0$, and Eq. (41) yields

$$[RO_2 \cdot] = \frac{k'[RO_2 R]_s}{k'_3[RH]} \tag{64}$$

where the subscript s denoted steady-state conditions. Also,

$$\rho_{ox, s} = k'_2[R \cdot][O_2] + k'_9[R \cdot][O_2] \tag{65}$$

Upon substituting Eqs. (39), (40), and (64) into (65), there is obtained

$$\frac{\rho_{ox, s}}{[RO_2 R]_s} = \frac{(k'_3 + k'_4)k'}{k'_3} + \frac{2k'_5[O_2]}{K_3 + [O_2]} \tag{66}$$

It can be readily shown that [cf. Eqs. (46b) and (46c)],

$$\frac{(k_3' + k_4')k'}{k_3'} = \frac{2k_5'K_3}{K_2} \tag{67}$$

Upon combining Eqs. (66) and (67),

$$\frac{\rho_{ox,s}}{[RO_2R]_s} = 2k_5' \frac{K_3/K_2 + [O_2]}{K_3 + [O_2]} \tag{68}$$

From Eq. (68), values of k_5' can be calculated since $\rho_{ox,\,s}/[RO_2R]_s$ was experimentally determined [7] and values of K_3 and K_2 were previously reported for IPP autoxidation [11]. Thus, at 120°C, $k_5' \approx 4 \times 10^{-4}$ sec^{-1} from Eq. (68), whereas Chien and Boss reported a calculated value of $(6.8 \pm 0.8) \times 10^{-4}$ sec^{-1} from their simpler scheme. Further, when Eq. (64) is substituted into (68),

$$\frac{\rho_{ox,s}}{[RO_2\cdot]_s[RH]} = k_3'\left(1 + \frac{k_4'}{k_3'}\right)\left[1 + \frac{K_2[O_2]}{K_3(K_3 + [O_2])}\right] \tag{69}$$

From Eq. (68), k_3' may be determined from experimental values of $\rho_{ox,\,s}/[RO_2\cdot]_s[RH]$ if the value of the ratio k_4'/k_3' is known. This ratio may be obtained from Eq. (67) since k_5' was calculated from Eq. (68), K_3 and K_2 were previously reported [11], and k' has been reported for IPP [11]. However, because of the limited temperature range over which k' was obtained for IPP, the values of k' for atactic poly(butene-1) were used instead [13]. In this manner, at 120°C, the value of $k_3' \cong 4\,l\,mole^{-1}$ sec^{-1}. (Chien and Boss reported a value of $k_3 = 3.3 \pm 0.1\,l\,mole^{-1}\,sec^{-1}$.) A slightly modified and more extensive treatment of the preceding has been reported [18]. In this work, slightly modified expressions and different values of rate constants were obtained when it was assumed that RO_2H formed in step (3) of the general scheme and that $RO\cdot$ and $HO\cdot$ radicals formed in Eq. (28).

The general scheme has been utilized to obtain A values for the autoxidation of APP in TCB at 135°C and [RH] = 1.63M [19]. Thus, from Eq. (57),

$$\frac{d\rho_{ox}}{dt} = A\rho_{ox,\,m}e^{-At} \tag{70}$$

or

$$\ln S = -At + \ln A\rho_{ox,\,m} \tag{70a}$$

where $S \equiv d\rho_{ox}/dt$. Thus, from a plot of Eq. (70a), values of A can be estimated. In this manner (Fig. 5.9) a value of $A = 8.5 \times 10^{-3}$ min^{-1} was obtained. From this value of A, an intercept of 5.1_6 was estimated which compared well with the experimental value of 5.1. (A more refined treatment [19] gave an intercept of 5.0.) It will be shown later (see Molecular-weight changes, page 253) that this value of A agreed well with the value

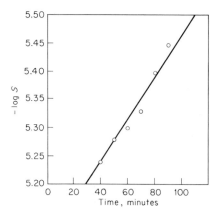

FIG. 5.9 Plot of $-\log S$ versus time for the autoxidation of atactic polypropylene in solution at 135°C: [RH] = 1.63 M [19].

obtained from data on intrinsic viscosity changes as a function of the percent of the APP oxidized [16, 19].

Oxyluminescence. Various workers [38–40] have observed the emission of weak luminescence during bulk-phase polyolefin oxidation, e.g., polypropylene. Although there still exists much speculation as to the active species responsible for this emission, it will be assumed that oxyluminescence intensity (OI) is related to the concentration of hydroperoxides [1, 12], i.e.,

$$(OI) = \alpha[RO_2R] \tag{71}$$

where α may involve one or more rate constants. Equation (71) may be combined with Eq. (55) to yield

$$\rho_{ox} = \left(\frac{2k_5'}{\alpha}\right)(OI) \tag{72}$$

Further, since $\rho_{CO} = k_6'[RO_2R]$, the expression for $\int(OI)\,dt$ becomes

$$\int (OI)\,dt = \left(\frac{\alpha}{k_6'}\right)[\text{>C=O}] \tag{73}$$

where $[>C\!\!=\!\!0]$ = carbonyl concentration. From Eq. (72), it can be seen that the ratio, $(OI)/\rho_{ox}$ should be constant at a particular temperature and oxygen concentration. For polypropylene film (film C) [38], this ratio possessed a value of $(1.5 \pm 0.1) \times 10^{-8}$ amp g min ml^{-1} for an oxygen atmosphere at 150°C and over a range of values of (OI) and ρ_{ox}. Also, Ashby [39] has indicated that an expression such as Eq. (73) was valid during the thermal oxidation of polypropylene. From Eq. (72), it may be anticipated that maximum luminosity emitted should be related to the ability of the polyolefin to absorb oxygen. Thus, the value of this luminosity should decrease in the order, polypropylene and polyethylene, as observed. Finally, it was possible to estimate values of A from luminosity data, using the general scheme which agreed well with values obtained from oxygen uptake data [12].

Molecular-weight changes. As indicated in Table 3, the kinetic parameter A may also be determined from intrinsic viscosity (molecular weight) changes during polyolefin autoxidation, e.g., atactic poly(butene-1) (APB) [17]. Thus, from the general scheme [cf. Eq. (44)] we may write

$$\rho_{CO} = C'(1 - e^{-At}) \tag{74}$$

where $C' = k'_3 k'_6 R_i/(k'_3 + k'_4)A$. It has been reported [41] that changes in intrinsic viscosity during polyolefin oxidation are related to carbonyl formation [also, cf. step (29)]. Based upon the preceding, we may assume the number of chain scissions per original chain during oxidation Δn is a function of the carbonyl concentration,

$$\Delta n = f[>C\!\!=\!\!0] = f \int_0^t \rho_{CO}\, dt \equiv F_n \tag{75}$$

Also,

$$\Delta n = \frac{M_{n,0}}{M_n} - 1 \tag{76}$$

and

$$[\eta] = K' M_v{}^a \tag{77}$$

where M_v = viscosity-average molecular weight, M_n = number-average molecular weight, $[\eta]$ = intrinsic viscosity, and K' and a are constants. If it is further assumed that the polymer samples follow a most probable distribution during oxidation, i.e., M_n/M_v = const, then Eq. (75) may be written as

$$F_n = \left(\frac{[\eta]_0}{[\eta]}\right)^{1/a} - 1 \tag{78}$$

From Eqs. (74), (75), and (78),

$$F_n = \frac{C}{A}(At + e^{-At} - 1) \tag{79}$$

where $C = \text{const} = fC'$. In order to solve Eq. (79) for the parameter A, we may utilize initially simplified expressions to obtain approximate values of C and A. Thus, from Eq. (79), we may write at high values of At,

$$F_n \approx Ct - \frac{C}{A} \tag{80}$$

Once values of C and A have been estimated from Eq. (80), they may be substituted into Eq. (79),

$$\left[F_n - \left(\frac{C}{A}\right)e^{-At}\right] \equiv \varphi = Ct - \frac{C}{A} \tag{81}$$

and values of C and A are estimated again from plots of Eq. (81). An iterative procedure such as this may be carried out until the values of C and A agree well. In Fig. 5.10 are shown plots of Eqs. (80) and (81) from data obtained for the air oxidation of APB at 90°C [17, 42] ($a = 0.70$). As anticipated, an approximately linear relationship obtained at the higher values of time in a plot of Eq. (80). From this portion of the curve values of $C = 1.7_5$ and $A = 0.25$ were obtained. When these values are substituted in Eq. (81), the following expression resulted:

$$F_n = 7.0e^{-0.25t} \equiv \varphi = Ct - \frac{C}{A} \tag{81a}$$

From several plots of φ versus t, as described previously, the first iteration gave $C = 1.45$ and $A = 0.22$ while the second iteration gave $C = 1.42$ and $A = 0.22$. Since the latter values agreed well with the former, the latter ones were employed. Thus, a value of $A = 3.7 \times 10^{-3}$ min^{-1} was calculated. This agreed well with the extrapolated value calculated from rate of carbonyl formation data obtained by IR techniques during the oxidation of APB films (cf. Table 3).

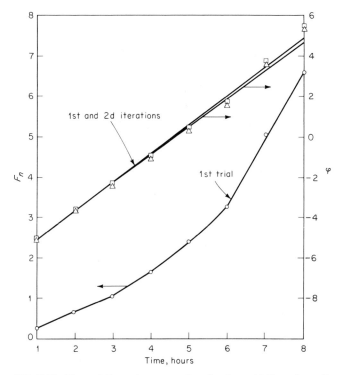

FIG. 5.10 Plots of F_n and φ versus time for the oxidation of atactic polybutene-1 in air at 90°C [17].

Equations similar to (80) and (81) have been applied to studying changes in molecular weight during autoxidation of APP in trichlorobenzene (TCB) [16, 19]. Since Bawn and Chaudhri [16] reported data for $[\eta]$ versus percent oxidation (PO), it is necessary to convert t in terms of PO before Eqs. (80) and (81) can be used $(a = 0.80)$. From Eq. (62), it can be shown that during the initial oxidation stages

$$t = \frac{(PO)^{1/2}}{K''} \tag{82}$$

where $(PO) = [O_2]_{abs} \times 100/[RH]$, and $K'' = $ const at constant temperature and oxygen pressure. By substituting Eq. (82) into (80) and (81), there are obtained respectively

$$F_n \approx \frac{C}{K''} (PO)^{1/2} - \frac{C}{A} \tag{83}$$

and

$$F_n - \left(\frac{C}{A}\right) \exp\left[-\frac{A(PO)^{1/2}}{K''}\right] \equiv \psi = \frac{C}{K''}(PO)^{1/2} - \frac{C}{A} \qquad (84)$$

As previously described, values of C/K'', C/A, and A/K'' may be initially estimated from Eq. (83). These can then be substituted into Eq. (84) to obtain new values of these parameters. When the succeeding sets of values of the corresponding parameters are in good agreement, these are adopted as the final values. In this manner, a value of $A/K'' = 0.83$ was obtained. Further, from experimental data reported [16] a value of $K'' = 1.05 \times 10^{-3}$ was estimated [19]. Thus, a value of $A \approx 8.8 \times 10^{-3}$ min^{-1} was obtained at 135°C, which is in reasonably good agreement with that previously obtained from oxygen uptake data (cf. oxygen absorption, page 248). However, the two A values obtained would not be expected to be identical since values of [RH] were different and there may have been some variation in oxygen pressure for the two sets of experimental data employed. Furthermore, Eq. (82) should be strictly applicable only at relatively low conversions.

Oxidation in Presence of Inhibitors

The general scheme in oxidation in absence of additive [Eqs. (1'') to (3'') and (28) to (32)] has been extended to take into account the presence of inhibitors (InH), e.g., phenols [43]. The mathematical treatment has been given in considerable detail [1] and, accordingly, is only briefly described here. Subsequently, the oxidation of a long-chain paraffin hydrocarbon, ceresin, in presence of the inhibitors anthracene (A) and pentacene (P) will be discussed. The activity of each of these inhibitors was found to be low, but when combined, their activity was relatively high. This high activity was attributed to the formation of a π complex between (A) and (P) [44].

In Table 4 is shown a modified general scheme for the oxidation of polyolefins, e.g., isotactic polypropylene (IPP), in presence of antioxidants (InH) [43]. This table contains the additional steps (85) to (90). However, steps (1''), (3'), (28'), (29), and (88) will be neglected in the following mathematical treatment. Furthermore, it will be assumed that $k_{85} \gg k_{86}$.

TABLE 4 Modified Kinetic Scheme for the Thermal Oxidation of Isotactic Polypropylene in Presence of Antioxidants [43]

$$RH + O_2 \xrightarrow{k_1, R_1} R\cdot + HO_2\cdot \qquad (1'')$$

$$R\cdot + O_2 \xrightarrow{k_2'} RO_2\cdot \qquad (2)$$

$$RO_2\cdot + RH \xrightarrow{k_3'} RO_2H + R\cdot \qquad (3)$$

TABLE 4 Modified Kinetic Scheme for the Thermal Oxidation of Isotactic Polypropylene in Presence of Antioxidants [43] (Continued)

$$RO_2\cdot + RH \xrightarrow{k_4'} \text{Inactive products} \tag{3'}$$

$$RO_2H \xrightarrow{k_{5a}} (*) \xrightarrow[\text{(fast)}]{} R_{\cdot}' + \text{volatile products} \tag{28'}$$

$$RO_2H + RH \xrightarrow[\text{(lnH)}]{K_5} (*)' \xrightarrow{k_{85}} \text{Products} \tag{85}$$

$$(*)' + O_2 \xrightarrow[\text{(lnH)}]{k_{86}} \delta_1 R_{\cdot}' + \text{products} \tag{86}$$

$$RO_2H \xrightarrow{k_6'} \text{Nonvolatile products} \tag{29}$$

$$R_{\cdot}' + RH \xrightarrow{k_8'} R\cdot + R'H \tag{31}$$

$$R_{\cdot}' + O_2 \xrightarrow{k_9'} \text{Less reactive products} \tag{32}$$

$$RO_2\cdot + lnH \xrightarrow{k_{87}} ln\cdot + RO_2H \tag{87}$$

$$ln\cdot + RH \xrightarrow{k_{88}} lnH + R\cdot \tag{88}$$

$$lnH + O_2 \xrightarrow[\text{(RH)}]{k_1'} \alpha R\cdot + \text{products} \tag{89}$$

$$RO_2H + R_2S \xrightarrow{k_{90}} \text{Products} \tag{90}$$

δ_1 and α define stoichiometric parameters whose values can vary with type of inhibitors: e.g., for strong inhibitors δ_1 is low whereas it is high for weak inhibitors.

Russian workers have postulated similar schemes to account for the oxidation of IPP in presence of phenols [45, 46].

Kinetic Expressions. Assuming that steady-state conditions exist for all the radicals, we may write

$$k_{87}[lnH][RO_2\cdot] + k_9'[R_{\cdot}'][O_2]$$
$$= \alpha k_4'[lnH][O_2] + \delta_4 k_5'''[lnH][RO_2H] \tag{91}$$

where $k_5''' = k_{86} K_5 [RH][O_2]$. Also,

$$\frac{d[RO_2H]}{dt} = k_3'[RH][RO_2\cdot] + k_{87}[lnH][RO_2\cdot]$$
$$- k_{90}[RO_2H][R_2S] - k_5''[RO_2H] \tag{92}$$

where $k_5'' = k_{85} K_5 [RH]$.

Depending upon whether strong or weak antioxidants are employed, the following two cases regarding hydroperoxide concentration may be written respectively.

Case 1: *Steady-state concentration of $RO_2 H$.* From Eq. (92),

$$[RO_2 H]_\infty = \frac{(k_3'[RH] + k_{87}[\ln H])[RO_2 \cdot]}{k_5'' + k_{90}[R_2 S]} \tag{92a}$$

Substituting Eq. (92a) into (91),

$$[RO_2 \cdot] = \frac{\alpha k_4'[\ln H][O_2]}{k_{87}[\ln H] - \dfrac{\delta k_5'' L (k_3'[RH] + k_{87}[\ln H])}{k_5'' + k_{90}[R_2 S]}} \tag{93}$$

where

$$\delta = \delta_4[\ln H] \qquad L = \frac{k_3'[RH]}{k_3'[RH] + k_9'[O_2]}$$

From Eq. (93), the critical inhibitor concentration ($[\ln H]_{cr}$) at which the value of $[RO_2 \cdot]$ approaches infinity, is

$$[\ln H]_{cr} = \frac{\delta k_3' k_5''' [RH]L}{k_5'''(k_5''/k_5''' - \delta L)k_{87} + k_{87}k_{90}[R_2 S]} \tag{94}$$

When $[R_2 S] = 0$,

$$[\ln H]_{cr} = \frac{\delta' k_3'[RH]}{(1 - \delta')k_{87}} \tag{95}$$

where $\delta' = \delta L(k_5''/k_5''')$.

We may also write for the rate of disappearance of antioxidant,

$$-\frac{d[\ln H]}{dt} = k_4'[\ln H][O_2] + \frac{k_{87}[\ln H]^2 \alpha k_4[O_2]}{D} \tag{96}$$

where

$$D = k_{87}[\ln H] - \frac{\delta k_5''' L (k_3'[RH] + k_{87}[\ln H])}{k_5'' + k_{90}[R_2 S]}$$

When $[R_2S]$ is very large or δ is very small, Eq. (96) becomes

$$-\frac{d[\ln H]}{dt} = K[\ln H] \tag{97}$$

Case 2: Non-steady-state concentrations of RO_2H. From Eq. (91),

$$[RO_2 \cdot] = \frac{ak_4'[O_2] + \delta_4 k_5'''[RO_2H]L}{k_{87}} \tag{91a}$$

Substituting Eq. (91a) into Eq. (92) and assuming that $[R_2S] = 0$ and $k_3'[RH] > k_{87}[\ln H]$,

$$\frac{d[RO_2H]}{dt} = \frac{k_3'[RH]}{k_{87}}(ak_4'[O_2] + \delta_4 k_5'''[RO_2H]L) - k_5''[RO_2H] \tag{98}$$

Upon integration of Eq. (98),

$$[RO_2H] = \frac{ak_4'k_3'[RH][O_2]}{k_{87}} \frac{e^{\phi t} - 1}{\phi} \tag{99}$$

where

$$\phi = \frac{\delta_4 k_3'k_5'''[RH]L - k_5''k_{87}}{k_{87}}$$

For the rate of antioxidant disappearance, we may write

$$-\frac{d[\ln H]}{dt} = k_4'[\ln H][O_2] + k_{87}[\ln H][RO_2 \cdot] \tag{96a}$$

By utilizing Eqs. (91a) and (99), Eq. (96a) becomes

$$-\frac{d[\ln H]}{dt} = k_4'[\ln H][O_2]$$

$$\left(1 + a + \frac{a\delta_4 k_3'k_5'''L[RH]}{k_{87}} \frac{e^{\phi t} - 1}{\phi}\right) \tag{100}$$

Upon integration, Eq. (100) becomes

$$\ln \frac{[\ln H]_0}{[\ln H]} = k_4'[O_2][(1 + a)t + B(e^{\phi t} - \phi t - 1)] \qquad (101)$$

where

$$B = \frac{a\delta_4 k_3' k_5''[RH]L}{k_{87}\phi^2}$$

For positive values of ϕ and for large values of t, Eq. (101) becomes

$$\ln \left(\ln \frac{[\ln H]_0}{[\ln H]} \right) = \phi t + \ln Bk_4'[O_2] \qquad (102)$$

The variation of molecular weight during oxidation in presence of a weak antioxidant may be arrived at by the following considerations [45, 46]. Let

$$\Delta n = f[RO_2H] \qquad [\text{cf. step (85)}] \qquad (103)$$

where $\Delta n \equiv$ number of scissions per chain molecule. Therefore, from Eq. (99),

$$\Delta n = f(e^{\phi t}) \qquad \text{for high values of } t \qquad (104)$$

In the case of polypropylene, Ciampa [47] has reported the following relationship between intrinsic viscosity and number-average molecular weight in Tetralin at $135°C$,

$$[\eta] = 2.5 \times 10^{-5} M_n^{1.0} \qquad (105)$$

From Eqs. (104) and (105),

$$\frac{1}{[\eta]} - \frac{1}{[\eta]_0} = f(e^{\phi t}) \qquad (106)$$

Significance of the expressions derived. Many of the expressions developed previously have been found to apply in the oxidation of IPP in presence of mono- and biphenols. Some of the applications will now be mentioned [1, 43].

Critical antioxidant concentration. It has been observed that in the oxidation of polypropylene in presence of "strong" antioxidant, e.g., certain biphenols, there is a sudden break in the curve of induction period versus biphenol concentration. Prior to this break (critical antioxidant concentration), there is a slow increase of induction period with increase of antioxidant concentration whereas, following this break, there is a rapid increase in induction time with increase of antioxidant concentration [45, 46]. However, no such critical antioxidant concentration was observed for "weak" phenolic antioxidants. But when sulfides (didecyl sulfide) were added to the weak antioxidant system, a critical antioxidant concentration obtained. Furthermore, when a weak antioxidant was added to a strong antioxidant system, the value of $[lnH]_{cr}$ increased.

The above results may be explained as follows. From Eq. (94), it can be seen that in presence of sulfide, a definite value of $[lnH]_{cr}$ will exist, even though the term $k''_s - \delta k''_s L$ may approach zero. However, even in absence of R_2S, this term will possess a finite value, as in the case of strong antioxidants, and a definite value of $[lnH]_{cr}$ will be obtained. In weak antioxidants, the same considerations, as mentioned previously, apply in presence of R_2S—a definite value of $[lnH]_{cr}$ will obtain. However, in absence of R_2S [cf. Eq. (99)], ϕ would be expected to be positive, and as reaction time increases, $[RO_2H]$ should increase considerably, indicating the existence of an autocatalytic oxidation and the absence of a strong antioxidant effect—no definite value of $[lnH]_{cr}$ would be obtained.

Rate of antioxidant disappearance. From Eq. (96) it can be seen that when R_2S is present along with inhibitor, the kinetic order in respect to inhibitor disappearance should be unity. Thus, when didecyl sulfide was added to the weak antioxidant, 2,4,6-tri-*tert*-butylphenol, the rate of consumption of the monophenol decreased, and a first-order expression could be applied to a large portion of the curve of antioxidant concentration versus time [45]. Also, for a strong antioxidant (δ has a very low value), a first-order relationship should obtain [cf. Eq. (97)]. Such a dependency has been observed for diphenols such as 2-2'-methylene-bis-(4-methyl-6-*tert*-butylphenol) [46].

In absence of sulfide, weak phenolic antioxidants tend to follow a zero-order rate of disappearance [45]. This kinetic dependency can be accounted for by means of Eq. (101). Thus, when Eq. (101) is expanded,

$$\ln \frac{[lnH]_0}{[lnH]} = (1 + \alpha)k'_4[O_2]t + Bk'_4[O_2]t^2 + Bk'_4[O_2]\phi t^2 + \cdots$$

$$(107)$$

When antioxidant is consumed at constant rate (zero order), $[lnH] = [lnH]_0 - Ct$, and

$$\ln \frac{[lnH]_0}{[lnH]} = \frac{Ct}{[lnH]_0} + \frac{C^2 t^2}{2[lnH]_0^2} + \frac{C^3 t^3}{3[lnH]_0^3} + \cdots \qquad (108)$$

From the above, it can be seen that Eqs. (107) and (108) are of the same form and that zero-order kinetics should also be applicable to the rate of disappearance of weak antioxidants.

In α-naphthol it has been indicated [45] that the value of α [cf. step (89)] is very low. From Eq. (101) it can be seen that under these conditions a first-order kinetic expression should obtain for the disappearance of antioxidant,

$$\ln \frac{[\ln H]_0}{[\ln H]} = k_4'[O_2]t = kt \tag{101a}$$

Such a dependency has been reported during polypropylene oxidation in the presence of α-naphthol [48]. Moreover, the rate constant k at 200°C was found to be directly proportional to oxygen pressure, as would be anticipated from Eq. (101a). From the temperature dependence of the rate constant, a value of activation energy for step (89) of about 20 kcal mole^{-1} could be calculated.

As previously mentioned, Berlin and coworkers [44] investigated the effectiveness of mixtures of anthracene (A) and pentacene (P) as inhibitors of the oxidation of the relatively high-molecular-weight hydrocarbon, ceresin. It was assumed that a π complex formed between (A) and (P), i.e.,

$$n(A) + m(P) \xrightarrow{K_p} [(A)_n(P)_m] \tag{109}$$

or

$$[(A)_n(P)_m] = K_p([(A)]_0 - n[(A)_n(P)_m])^n$$
$$([(P)]_0 - m[(A)_n(P)_m])^m \tag{110}$$

where $[(P)]_0$ and $[(A)]_0$ denote the initial concentrations of (P) and (A), respectively. When K_p is small,

$$[(A)_n(P)_m] = K_p[(A)]_0^n[(P)]_0^m \tag{111}$$

Assuming that only step (87) is applicable to the consumption of inhibitor,

$$-\frac{d[\ln H]}{dt} = k_{87}[\ln H][RO_2\cdot]_s \tag{112}$$

where $[RO_2\cdot]_s$ = steady-state concentration of $RO_2\cdot$ radicals (= const). Upon integration, Eq. (112) becomes

$$\ln \frac{[\ln H]_0}{[\ln H]_{cr}} = k_{87}[RO_2\cdot]_s \tau_i \tag{113}$$

where τ_i = induction time. Further, assuming that $[\text{lnH}]_0 \approx [(A)_n(P)_m]_0$, Eq. (113) becomes

$$\tau_i = \frac{2.3}{a} \log\left([(A)]_0^n [(P)]_0^m\right) - b \tag{114}$$

where $a = k_{87}[\text{RO}_2\cdot]_s$ and $b = \ln\left([\text{InH}]_{cr}/K_p\right)/k_{87}[\text{RO}_2\cdot]_s$. Equation (114) may be rewritten as

$$\tau_i = \frac{2.3m}{a} \log[(P)]_0 + \frac{2.3n}{a} \log[(A)]_0 - b \tag{114'}$$

It was found that Eq. (114') was in accord with experimental data and that $m/n = 2$.

Oxidation in Presence of Metal Accelerators

Stivala and coworkers [49] studied the autoxidation of isotactic poly-butene-1 (IPB) by IR techniques. The IPB sample used contained a relatively high ash content, 0.17 percent, which consisted mainly of Al, Ti, and Si, with lesser amounts of Mg, Pb, Fe, Ca, Mn, Sn, and Cu. Because of this, it might be anticipated that kinetic parameters and activation energies would be affected (cf. Chap. 1). Indeed, it was found that, surprisingly, induction periods for IPB autoxidation were much lower (factor of about 6 to 10) than those for APB, under similar experimental conditions. Furthermore, from the general scheme (see Oxidation in Absence of Additives, page 229), and from carbonyl formation data, it was found that the value of k' (IPB) was about twice that for k' (APB) under similar experimental conditions, and that E_5 (IPB) = 25 kcal mole^{-1} while E_5 (APB) = 29 kcal mole^{-1}. Thus, it would appear that various autoxidation processes are occurring more rapidly for IPB than for APB. Such behavior was ascribed to presence of much larger amounts of metallic impurities in IPB than in APB (0.17 versus 0.04 percent as ash). It is well known that metal salts can act as accelerators of hydrocarbon oxidation [1] and can lower the activation energy for hydroperoxide decomposition E_5. Thus, Lombard and Knopf [50] found that the cobaltic acetyl acetonate–catalyzed decomposition of cumene hydroperoxide in solution afforded a value of E_5 = 25 kcal mole^{-1}, whereas in absence of the metal salt a value of E_5 = 31 kcal mole^{-1} was obtained. Also, Osawa and coworkers [51] reported on the oxidative degradation of polypropylene (60 percent crystallinity) in presence of metallic salts of fatty acids. They found that the addition of 0.5 weight percent of fatty acid salts of transition metals, e.g., titanium (IV) stearate, to finely powdered polymer lowered the induction time of the neat polypropylene from about 125 to about 30 min, based on oxygen uptake studies at 125°C. Furthermore, some metallic salts, e.g., cupric stearate, were capable of lowering an apparent activation energy of the autoxidation from 30 for the neat polymer to about 24 kcal mole^{-1} for the metal-containing polypropylene. The order of decreasing catalytic effect of the fatty acid metallic salts was as follows, based upon oxygen absorption curves and activation energy: Co > Mn > Cu > Fe > V > Ni > Ti > Al > Mg > Ba. In general, a similar catalytic order of activity was observed in respect to the decomposition of *tert*-butyl hydroperoxide. Thus, there appears to be a relationship between metal catalyst activity during oxidation of polypropylene and metal catalyst activity during the decomposition of hydroperoxides.

Metal catalyst activity during polyolefin autoxidation has been correlated with the redox potential of the various metals studied. Thus, the following reactions have been postulated [1]:

$$RO_2H + Me^{n+} \longrightarrow RO\cdot + HO^- + Me^{(n+1)+} \qquad (115)$$

$$RO_2H + Me^{(n+1)+} \longrightarrow RO_2\cdot + H^+ + Me^{n+} \qquad (116)$$

From Eqs. (115) and (116), it can be seen that the ability of metals to decompose hydroperoxides may depend upon their redox potential. Osawa and coworkers [51] found that metal catalyst activity, as judged from oxygen uptake during polypropylene oxidation in the bulk phase and from rates of hydroperoxide decomposition, could be correlated well with redox potential. In this connection may be mentioned work by Reich, Stivala, and coworkers [52], who found that a good correlation could be obtained when log $(\rho_m - \rho_0)$ versus redox potential was plotted for various metal salts, e.g., cobaltic, manganic, ferric, and cupric, during the autoxidation of films of APP at 110°C and 100 percent oxygen, where ρ_m and ρ_0 denote maximum rate of carbonyl formation in presence and absence of metal catalysts, respectively. The order of metal activity found was similar to that observed by Osawa and coworkers [51]. However, Chaudhri [53] studied the rate of oxygen uptake during the autoxidation of APP in TCB in presence of various metal salts. Contrary to the results obtained in solid-state autoxidations of APP, Chaudhri found that the cobalt salt was the least effective, followed by nickel, iron, manganese, and copper (the activity increasing in this order). This reversal of catalyst activity was ascribed to the influence of the medium.

There have been relatively few studies on the effect of metal catalysts on various kinetic parameters during polyolefin autoxidation. Bawn and Chaudhri [54] carried out such a study on the oxidation of APP in trichlorobenzene (TCB) in presence of manganese salts, and employed a modified general scheme which had been previously used to study the kinetics of the uncatalyzed autoxidation of APP in TCB [16]. Thus, the following scheme was proposed:

$$\text{Initiation} = \phi = k_i[RH][O_2][Mn^{3+}] \qquad (1a)$$

which provides the system with free radicals.

$$R\cdot + O_2 \xrightarrow{k_2} RO_2\cdot \tag{2}$$

$$RO_2\cdot + RH \xrightarrow{k_3} RO_2R + H\cdot \ \text{ or } \ ROOH + R\cdot \tag{3}$$

$$RO_2\cdot + RH \xrightarrow{k_4} \text{Inactive products} \tag{3'}$$

$$RO_2\cdot + Cat \xrightarrow{k_{4a}} \text{Inactive products} \tag{4a}$$

$$RO_2R \ (ROOH) + Mn^{3+} \xrightarrow{k_{5a}} RO_2\cdot + Mn^{2+} + R^+ \tag{5a}$$

$$RO_2R \ (ROOH) + Mn^{2+} \xrightarrow{k_{5b}} RO\cdot + Mn^{3+} + RO^- \tag{5b}$$

$$RO_2R \ (ROOH) + Cat \xrightarrow{k_{6a}} \text{Inactive nonvolatile product}$$
$$(>C{=}O, \ -CHO, \ \text{etc.}) \tag{6a}$$

$$RO\cdot \xrightarrow{k_7} R'_. + \text{volatile products} \tag{30}$$

$$R'_. + RH \xrightarrow{k_8} R'H + R\cdot \tag{31}$$

$$R'_. + O_2 \xrightarrow{k_9} \text{Less active products} \tag{32}$$

In the initiation step, it was postulated that a loosely bound complex between the hydrocarbon molecule and the metal catalyst was formed. This complex then reacts with a molecule of oxygen to produce free radicals (cf. Reich and Stivala [1]), i.e.,

$$Mn^{3+} + RH \longrightarrow [Mn^{3+} \cdots RH] \tag{117}$$

$$[Mn^{3+} \cdots RH] + O_2 \longrightarrow Mn^{2+} + RO_2\cdot + H^+ \tag{117a}$$

Reactions (28) and (29) of the general scheme were not included. In the following we are elaborating on the above general mechanisms by discussing its various steps (cf. Kinetics of polyolefin oxidation excluding polymeric peroxy radical recombination, page 235). Steps (2), (3), (3'), (6a), (30), and (31) have been previously reported for the uncatalyzed autoxidation of various polyolefins [1]. In step (3), it has been arbitrarily maintained that RO_2R can form as well as RO_2H, although the latter is energetically favored, since in either case the derived kinetic expressions will not change much. Steps (3') and (4a) have been included to account for the deactivation of $RO_2 \cdot$ radicals by inherent impurities in the substrate RH and by the catalyst Cat, respectively. No differentiation is made as to the oxidation state of the catalyst molecule which undergoes this deactivation reaction. Other workers have also postulated a deactivation step such as (4a) to account for the metal salt–catalyzed autoxidation of Tetralin. Thus, Kamiya and Ingold [54a] studied the autoxidation of Tetralin in chlorobenzene at $65°$ as a function of cobalt (II) decanoate concentration. At relatively high metal catalyst concentrations, it was found that after a critical catalyst concentration the maximum autoxidation rate dropped catastrophically. This was attributed, in part, to a chain-termination process similar to step (4a). Steps such as (5a) and (5b) have been frequently utilized for the metal-catalyzed autoxidation of various hydrocarbons [1]. Kamiya [55] has postulated a reaction similar to step (6a) during the metal-catalyzed autoxidation of Tetralin. Furthermore, on the basis of previous work [52], the authors have found a good correlation between oxidation potential of various metals and maximum rate of non-volatile carbonyl formation during the autoxidation of APP in the bulk phase. Since metals react with RO_2H readily, it is reasonable to assume that a reaction such as step (6a) is occurring, leading to the formation of nonvolatile carbonyl-containing compounds. In Eq. (30), it is postulated that the tertiary alkoxy radical of APP decomposes into smaller radical fragments, $R' \cdot$, and volatile products. Recently, Carlsson and Wiles [55a] subjected powdered, unstabilized polypropylene to air for 1 to 5 min at $225°C$, and based upon infrared absorption spectra, they indicated that two principal ketone products were formed (presumably from the tertiary alkoxy radical of polypropylene). One of these products was apparently formed by loss of a methyl radical while the other ketone product could form by chain cleavage. Both ketones were present in about equal amounts. The formation of a methyl radical, as opposed to an alkyl radical, would not be anticipated, based upon free-radical chemistry of simple hydrocarbons in solution. In view of this and because of the much lower reaction temperatures used in bulk-phase autoxidation (100 to $130°$) [56], it was postulated that the $RO \cdot$ radical in step (30) undergoes chain scission predominantly to afford a low-molecular-weight radical (of relatively high mobility in the bulk polymer) and a relatively low-molecular-weight product which can volatilize, under the experimental conditions generally used. In respect to Eq. (32), it may be mentioned that previously [1, 10] it was indicated that $R' \cdot + O_2 \longrightarrow$ less active products. At this point, we would like to elucidate this step by showing the formation of $R'O_2 \cdot$ radicals which terminate rapidly to yield products (the $R'O_2 \cdot$ radicals are presumably of much greater mobility than polymeric $RO_2 \cdot$ radicals: $2R'O_2 \cdot \xrightarrow{\text{very fast}}$ products, and $R' \cdot + R'O_2 \cdot \xrightarrow{\text{very fast}} R'O_2R \cdot$). We [56] are also including the cross-termination step since the $R' \cdot$ is also a mobile species. However, the bimolecular reaction involving the $R'O_2 \cdot$ would be favored because of formation of peroxide radicals in an oxidizing medium. The formation of $R'O_2 \cdot$ radicals is considered to be rate controlling. The question of unimolecular versus bimolecular termination in the polymer matrix is still moot. However, bimolecular terminations are well known in free-radical chemistry and are, therefore, favored. Because of the rapid consumption of $R'O_2\cdot$ radicals in the termination step, the reaction $R'O_2\cdot +$ RH would likely be deterred and is therefore not included in the general scheme.

As the catalyst is mainly present in the Mn^{3+} state and any Mn^{2+} formed would be rapidly converted to the higher oxidation state, it was assumed that

$$k_{5a}[Mn^{3+}] = k_{5a}[Cat] = k_{5b}[Mn^{2+}] \tag{118}$$

The following equations determine the stationary-state conditions:

$$\frac{d[R\cdot]}{dt} = \phi - k_2[R\cdot][O_2] + k_8[R'][RH] = 0 \tag{119}$$

$$\frac{d[R']}{dt} = k_{5a}[RO_2R][Cat] - k_8[R'][RH] - k_9[R'][O_2] = 0 \tag{120}$$

$$\frac{d[RO_2\cdot]}{dt} = k_2[R\cdot][O_2] - (k_3 + k_4)[RO_2][RH]$$

$$- k_{4a}[RO_2][Cat] + k_{5a}[RO_2R][Cat] = 0 \tag{121}$$

and

$$\frac{d[RO_2R]}{dt} = k_3[RO_2\cdot][RH] - (2k_{5a} + k_{6a})[Cat][RO_2R] \tag{122}$$

[It should be noted that Eq. (118) is only approximate since loss of catalyst in steps (4a) and (6a) is neglected.] Upon adding Eqs. (119) and (121) and substituting the expression for [R'·] from Eq. (120), there is obtained

$$[RO_2\cdot] = \frac{\phi}{(k_3 + k_4)[RH] + k_{4a}[Cat]}$$

$$+ \frac{k_{5a}[Cat](2k_8[RH] + k_9[O_2])[RO_2R]}{\{(k_3 + k_4)[RH] + k_{4a}[Cat]\}(k_8[RH] + k_9[O_2])} \tag{123}$$

In Eq. (1a) the initiation process has been assumed to provide the system with R· radicals. If it supplied RO_2· radicals instead of R· radicals, the addition of Eq. (119) and Eq. (121) would again lead to exactly the same expression as Eq. (123).

Substituting the value of $[RO_2\cdot]$ from Eq. (123) into Eq. (122) and integrating, we get

$$
[RO_2R]
$$

$$
= \frac{\dfrac{k_3[RH]\phi}{(k_3 + k_4)[RH] + k_{4a}[Cat]}\,(1 - e^{-A't})}{\left[2k_{5a} + k_{6a} - \dfrac{k_3 k_{5a}[RH]}{(k_3 + k_4)[RH] + k_{4a}[Cat]}\,\dfrac{2k_8[RH] + k_9[O_2]}{k_8[RH] + k_9[O_2]}\right][Cat]}
$$

$$(124)$$

where A' is the value of the denominator in this expression. Since $-d[O_2]/dt \cong k_{5a}[Cat][RO_2R]$,

$$
\left(-\frac{d[O_2]}{dt}\right)_{max}
$$

$$
= \frac{\dfrac{k_3[RH]\phi}{(k_3 + k_4)[RH] + k_{4a}[Cat]}\,\dfrac{k_{5a}}{2k_{5a} + k_{6a}}\,(1 - e^{-A't_m})}{1 - \dfrac{k_3[RH]}{(k_3 + k_4)[RH] + k_{4a}[Cat]}\,\dfrac{k_{5a}}{2k_{5a} + k_{6a}}\,\dfrac{2k_8[RH] + k_9[O_2]}{k_8[RH] + k_9[O_2]}}
$$

$$(125)$$

Equation (125) is of a form similar to that obtained for the uncatalyzed autoxidation reaction [cf. Eq. (58)]. However, in this case, it is not so obvious that $A't_m$ will be approximately constant at various oxygen concentrations [cf. Eq. (45)]. From Eq. (125) various kinetic dependencies can be derived and compared with experimental evidence, as shown in the following.

Rate dependence on oxygen pressure. From Eq. (125), it can be shown that

$$
\left(-\frac{d[O_2]}{dt}\right)_{max} \equiv \rho_{ox,\,m} = \frac{\alpha\phi}{1 - \beta\,\dfrac{2K_3 + [O_2]}{K_3 + [O_2]}} \tag{125'}
$$

where

$$\alpha = \frac{k_3 k_{5a}[\text{RH}]\left(1 - e^{-A't_m}\right)}{[(k_3 + k_4)[\text{RH}] + k_{4a}[\text{Cat}]](2k_{5a} + k_{6a})}$$

$$\beta = \frac{\alpha}{1 - e^{-A't_m}} \quad \text{and} \quad K_3 = \frac{k_8[\text{RH}]}{k_9}$$

Case 1: When **[RH]**, **[Cat]**, *and temperature are constant and* **[O$_2$]** *values are high.* In this case, assume that $(1 - \beta)[\text{O}_2] \gg K_3(1 - 2\beta)$, and Eq. (125′) becomes

$$\rho_{\text{ox},m} = \alpha' K_3 + \alpha'[\text{O}_2] \tag{126}$$

where

$$\alpha' = \frac{\alpha k_i[\text{RH}][\text{Cat}]}{1 - \beta}$$

Case 2: When **[RH]**, **[Cat]**, *and temperature are constant and* **[O$_2$]** *values are low.* In this case let $(1 - \beta)[\text{O}_2] \ll K_3(1 - 2\beta)$, and Eq. (125′) becomes

$$\rho_{\text{ox},m} = \frac{\alpha'(1 - \beta)[\text{O}_2]}{1 - 2\beta} \tag{127}$$

If Henry's law applies, i.e., $[\text{O}_2] = Hp_{\text{O}_2}$, where H = const, and p_{O_2} = partial pressure of oxygen, then the rate of oxygen uptake in the catalyzed system should show a linear dependence on the partial pressure of oxygen above the solution, as observed experimentally [cf. Eqs. (126) and (127)].

Rate dependence on the catalyst concentration

Case 3: When **[O$_2$]**, **[RH]**, *and temperature are constant and* **[Cat]** *values are high.* In this case, Eq. (125) becomes

$$\rho_{\text{ox},m} = \frac{C_1\phi}{k_{4a}[\text{Cat}]} + C_2[\text{RH}] \tag{128}$$

where

$$C_1 = \frac{k_3[\text{RH}]\left(1 - e^{-A't_m}\right)k_{5a}}{2k_{5a} + k_{6a}}$$

$$C_2 = k_3 + k_4 - \frac{k_3 k_{5a}(2K_3 + [O_2])}{(2k_{5a} + k_{6a})(K_3 + [O_2])}$$

When $k_{4a}[\text{Cat}] \gg C_2[\text{RH}]$, Eq. (128) becomes

$$\rho_{\text{ox}, m} = C_3 \qquad\qquad\qquad (129)$$

where

$$C_3 = \frac{C_1 k_i [O_2][\text{RH}]}{k_{4a}}$$

Case 4: When $[O_2]$, $[\text{RH}]$, *and temperature are constant and* [Cat] *values are low.* In this case, let $k_{4a}[\text{Cat}] \ll C_2[\text{RH}]$, and Eq. (128) becomes

$$\rho_{\text{ox}, m} = C_4[\text{Cat}] \qquad\qquad\qquad (130)$$

where

$$C_4 = \frac{C_1 k_i [O_2]}{C_2}$$

Equations (129) and (130) indicate that at very low catalyst concentrations the order in respect to [Cat] should be unity and eventually tend to zero at high values of [Cat]. However, it was observed [54] that at [Cat] < 0.5×10^{-3} mole l⁻¹, the rate was almost identical with the uncatalyzed steady rate, and therefore it is improbable that Eq. (130) will apply in the system studied. However, at intermediate concentrations, the order in respect to [Cat] should lie between zero and unity [cf. Eq. (128)], as experimentally observed (0.5 to 0.6). As the value of [Cat] was increased, the rate became independent of [Cat], as predicted by Eq. (129).

Rate dependence on substrate concentration

Case 5: When $[O_2]$, [Cat], *and temperature are constant and* [Cat] *values are low.* Equation (130), case 4, should also be applicable here. Thus, for case 5,

$$\rho_{\text{ox}, m} = \frac{C_5[O_2][\text{RH}](k_8[\text{RH}] + k_9[O_2])}{k_8[\text{RH}](1 - 2\beta') + k_9[O_2](1 - \beta')} \qquad (131)$$

where

$$C_5 = \frac{k_i k_3 k_{5a}\left(1 - e^{-A't_m}\right)}{(k_3 + k_4)(2k_{5a} + k_{6a})}$$

and

$$\beta' = \frac{k_3 k_{5a}}{(k_3 + k_4)(2k_{5a} + k_{6a})}$$

At high values of $[O_2]$, i.e., $K_3(1 - 2\beta') \ll [O_2](1 - \beta')$, then Eq. (131) becomes

$$P_{ox,m} = \frac{C_5 k_8 [RH]^2}{k_9(1 - \beta')} + \frac{C_5 [O_2][RH]}{1 - \beta'} \qquad (131a)$$

From Eq. (131a), the apparent order in respect to [RH] should lie between unity and two, the exact value depending upon the relative magnitudes of the coefficients. Experimentally, it was found that at low values of [Cat] and at relatively high $[O_2]$, the rate dependence upon polypropylene concentration was of order 1.7.

Case 6: When $[O_2]$, *[Cat], and temperature are constant and [Cat] values are high.* Here Eq. (129) of case 3 should be applicable. Then we may write

$$P_{ox,m} = C_1'[RH]^2 \qquad (132)$$

where

$$C_1' = \frac{k_i [O_2] C_1}{k_{4a}[RH]} = \text{const}$$

Experimentally, it was found that at high values of [Cat] the rate dependence upon polypropylene concentration was of order 2.0₇, as anticipated. Furthermore, it can be seen from Eq. (132) that for constant [RH], [Cat], and temperature and for high values of [Cat],

$$P_{ox,m} = C_6[O_2] \qquad (132a)$$

where

$$C_6 = \frac{C_1 k_i [RH]}{k_{4a}}$$

Experimentally, it was found that at high values of [Cat] and for constant [RH], [Cat], and temperature, the rate dependence upon oxygen pressure was of order 0.9.

It may be mentioned here that if step (29) which involves the formation of nonvolatile products, e.g., ketones, in absence of metal catalysts is included, then it may be shown that [cf. Eq. (124), Kamiya [1, 55] assumed such a step in order to account for oxidation products during the autoxidation of neat Tetralin in presence of cobalt salts],

$$[RO_2R] = \frac{k_3[RH]\phi(1 - e^{-A''t})/(k_3 + k_4)[RH] + k_{4a}[Cat]}{A' + k_6'}$$

$$(124a)$$

where $A'' = A' + k_6'$. If we let $\rho_{ox, m, M}$ = maximum oxidation rate due only to presence of metal catalyst (= $k_{5a}[Cat][RO_2R]$), $\rho_{ox, m, 0}$ = maximum oxidation rate in absence of metal catalysts [= $2k_5'[RO_2R]$, cf. Eq. (55)], and $\rho_{ox, m, M, T}$ = total maximum oxidation rate observed in presence of metal catalyst (= $\rho_{ox, m, 0} + \rho_{ox, m, M}$), then Eq. (124a) can be written as

$$\frac{A_1}{k_{5a}} \rho_{ox, m, M, T} + \rho_{ox, m, 0}\left(\frac{k_6'}{2k_5'} - \frac{A_1}{k_{5a}}\right)$$

$$= \frac{k_3[RH]\phi\left(1 - e^{-A''t_m}\right)}{(k_3 + k_4)[RH]} + k_{4a}[Cat] \qquad (133)$$

where $A_1 = A'/[Cat]$. When $k_6'/2k_5' \ll A_1/k_{5a}$, Eq. (133) becomes

$$\rho_{ox, m, M, T} - \rho_{ox, m, 0}$$

$$= \frac{k_3[RH]\phi\left(1 - e^{-A''t_m}\right)/(k_3 + k_4)[RH] + k_{4a}[Cat]}{A_1/k_{5a}} \qquad (133a)$$

Equation (133a) resembles Eq. (125) except for the presence of A'' and $\rho_{ox, m, 0}$. It is interesting to note here that when $\rho_{ox, m, 0}$ is taken into account, the kinetic dependencies obtained by Bawn and Chaudhri [54] will be of higher order. Thus, employing data for the uncatalyzed oxidation of polypropylene at 135°C [16] and correcting for differences in [RH], it can be shown that instead of an order of 0.55 in respect to [Mn^{3+}], the order now is closer to unity, which would be anticipated from Eq. (130).

In a similar manner, if we let $\rho_{CO, m, M}$ = maximum rate of carbonyl formation due only to metal catalyst (= k_{6a}[Cat][RO$_2$R]), $\rho_{CO, m, 0}$ = maximum carbonyl formation rate in the absence of metal catalyst (= k'_6[RO$_2$R]), and $\rho_{CO, m, M, T}$ = total maximum carbonyl formation rate observed in presence of metal catalyst (= $\rho_{CO, m, 0} + \rho_{CO, m, M, T}$), then Eq. (124$a$) becomes, when it is assumed that $k_{6a}/A_1 \ll 1$,

$$\rho_{CO, m, M, T} - \rho_{CO, m, 0}$$

$$= \frac{k_3[\text{RH}]\phi\left(1 - e^{-A''t_m}\right)/(k_3 + k_4)[\text{RH}] + k_{4a}[\text{Cat}]}{A_1/k_{6a}} \tag{134}$$

If we denote the left-hand side of Eq. (134) by $\rho_{CO, m, net}$, then Eq. (134) should lead to various kinetic dependencies for $\rho_{CO, m, net}$, which should be similar to those found for maximum oxidation rate in presence of metal catalyst which were expressed in cases 1 to 6. In this connection may be mentioned the work of Stivala, Reich, and coworkers [56]. These workers studied the bulk-phase oxidation of APP at 110°C and [O$_2$] = 100 percent in presence of the metal catalyst, cobalt (III) acetyl acetonate. The maximum carbonyl formation rate (cm^2 of carbonyl absorbance area/min) was determined for various amounts of metal catalyst (moles of metal salt per 7.5 mg of APP film sample of about 2 1/2 mil thickness). In Fig. 5.11 is shown a plot of log ($\rho_{CO, m, M, T} - \rho_{CO, m, 0}$) versus log [Co(III)] from which the kinetic dependency in regard to [Co(III)] may be ascertained. (The value of $\rho_{CO, m, 0}$ was obtained from previous work [20].) From this plot an average kinetic order of about 1.0 was obtained with respect to [Co(III)] at low and intermediate values of [Co(III)]. As [Co(III)] was increased, an order of about −0.1 was attained after a certain value of [Co(III)]. Such kinetic dependencies would be expected, based upon Eqs. (129) and (130) of cases 3 and 4, i.e., unity at low and intermediate values of [Co(III)], and zero at relatively high values of [Co(III)]. Furthermore, $\rho_{CO, m, net}$ at high [Cat] should be directly proportional to oxygen concentration [cf. Eq. (129)]. This relationship was found to apply. Thus, when the autoxidation of APP films [56] was carried out at an O$_2$/N$_2$ ratio of 50:50 under otherwise similar experimental conditions as for the 100 percent O$_2$ autoxidation (at values of [Cat] where $\rho_{CO, m, net}$ is independent of [Cat]), $\rho_{CO, m, net}$ was reduced by a factor of 0.50, as anticipated. Also, a kinetic dependency of −0.06 was observed with respect to [Cat] for the O$_2$/N$_2$ ratio 50:50.

The kinetic dependency of $\rho_{CO, m, net}$ upon [O$_2$] was also ascertained with the aid of Eq. (134). Thus, this equation may be written as [cf. Eq. (125′)]

$$\rho_{CO, m, net} = \frac{\alpha_1\phi}{1 - \beta\{(2K_3 + [O_2])/(K_3 + [O_2])\}} \tag{135}$$

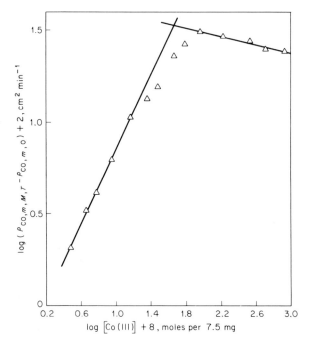

FIG. 5.11 Plot of log $\rho_{CO,\,m,\,net}$ versus log [Co (III)] for the autoxidation of atactic polypropylene films at 100°C and $[O_2]$ = 100 percent in presence of cobalt [III] acetyl acetonate [56].

where

$$\alpha_1 \equiv \frac{k_3[\text{RH}]\left(1 - e^{-A''t_m}\right)}{(k_3 + k_4)[\text{RH}] + k_{4a}[\text{Cat}]} \frac{k_{6a}}{2k_{5a} + k_{6a}}$$

and

$$\beta \equiv \frac{k_3[\text{RH}]}{(k_3 + k_4)[\text{RH}] + k_{4a}[\text{Cat}]} \frac{k_{5a}}{2k_{5a} + k_{6a}}$$

As in cases 1 and 2, Eq. (135) may be written respectively as

$$\rho_{CO,\,m,\,net} = \alpha_2 K_3 + \alpha_2[O_2] \tag{135a}$$

where

$$\alpha_2 = \frac{\alpha_1 k_i [\text{RH}][\text{Cat}]}{1 - \beta}$$

and

$$p_{\text{CO},m,\text{net}} = \frac{\alpha_2(1 - \beta)}{1 - 2\beta}[\text{O}_2] = \frac{\alpha_3[\text{O}_2]}{1 - 2\beta} \tag{135b}$$

where

$$\alpha_3 = \alpha_2(1 - \beta)$$

Furthermore, Eq. (135) may be rewritten as

$$p_{\text{CO},m,\text{net}} = \frac{\alpha_3[\text{O}_2](K_3 + [\text{O}_2])}{K_3(1 - 2\beta'') + [\text{O}_2](1 - \beta)} \tag{136}$$

Before correlating $p_{\text{CO},m,\text{net}}$ as a function of oxygen concentration, it is necessary to estimate the constants K_3, α_3, and β in Eq. (136). From Eq. (135a), it should be possible to estimate K_3 for high values of $[\text{O}_2]$ from a plot of $p_{\text{CO},m,\text{net}}$ versus $[\text{O}_2]$. These values of K_3 should be identical with corresponding K_3 values previously reported for the uncatalyzed oxidation of APP [20] under similar experimental conditions. In the following are listed K_3 values for the catalyzed and uncatalyzed oxidations and various reaction temperatures, respectively: 35 and 39 at 110°C, 42 and 44 at 120°C, and 50 and 58 at 130°C. Although the agreement is good, the K_3 values from the uncatalyzed oxidation were used since the experimental error for this oxidation is less than for the catalyzed reaction. From such values of K_3, values of α_3 and β could be estimated as follows. For the uncatalyzed APP oxidation it has been reported [cf. Eq. (46)] that maximum carbonyl formation $p_{\text{CO},m,0}$ can be expressed as

$$p_{\text{CO},m,0} = \frac{K_1[\text{O}_2]}{1 - K_2/(K_3 + [\text{O}_2])} \tag{46}$$

where Eq. (136) is divided by (46),

$$\frac{p_{\text{CO},m,\text{net}}}{p_{\text{CO},m,0}} \equiv R = \frac{\alpha_3(K_3 - K_2 + [\text{O}_2])}{K_1\{K_3(1 - 2\beta) + [\text{O}_2](1 - \beta)\}} \tag{137}$$

At high values of $[O_2]$, (137) becomes

$$R \approx \frac{\alpha_3}{K_1(1 - \beta)} \qquad (137a)$$

while at low values of $[O_2]$,

$$R \approx \frac{\alpha_3(K_3 - K_2)}{K_1 K_3(1 - 2\beta)} \qquad (137b)$$

From Eqs. (137a) and (137b), it should be possible to determine values of α_3 and β'' employing previously reported values [20] of K_1, K_2, and K_3. In this manner were obtained values of α_3 and β for various experimental conditions. From these values and employing Eq. (136), values of $\rho_{CO, m, net}$ could be calculated at various values of $[O_2]$ and reaction temperatures. Satisfactory agreement was found between calculated and observed values.

In the derivation of Eq. (134), it was assumed that the value of $[RO_2 R]_m$ was of the same order of magnitude for both catalyzed and uncatalyzed oxidations under similar experimental conditions. This was verified experimentally [56]. Furthermore, in the subsequent use of Eq. (134), it was assumed that $A''t_m \approx$ const under various experimental conditions employed. This was found to be approximately true based on calculations which employed values of β and which assumed an activation energy for step (5a) of 24 kcal mole^{-1} (also, $2k_{5a} \gg k_{6a}$).

OTHER POLYMERS

Since there are no uniform mathematical treatments of the kinetics of autoxidation of polymers, other than polyolefins, such treatments will be described in the following under the type of polymer(s) studied.

Oxidation of Polyepoxide Types

In presence of initiator only. Tobolsky and coworkers [3, 4] studied structure and reactivity during the bulk-phase autoxidation in presence of benzoyl peroxide of elastomers such as poly(ethyl acrylate) (PEA), poly-(vinyl methyl ether) (PVME), and polypropylene oxide (PPO). By using their approach, previously described [page 231 and Eq. (10)], values of $R_0 (\equiv k_3 [RH] (f/k_6)^{1/2})$ could be calculated from oxygen uptake studies at various temperatures. The overall activation energy $(E_3 - E_6/2)$ was found to be about 16 kcal mole^{-1} for these elastomers previously mentioned. The relative values of R_0 (at 80°C) for PEA, PVME, and PPO were found to be respectively 1.0, 1.9, and 5.9. In explaining these differences, it was assumed that such differences were primarily due to the effect of electronic and steric factors on the magnitude of k_3. Thus, the $RO_2\cdot$ radical would be expected to react more readily with an RH substrate which is electron-donating, and the magnitude of k_3 would be expected to be greater for PPO than for PEA based upon this factor. In the case of PVME and PPO, the steric factor should be greater for PVME than for PPO which allows for easier abstraction of the tertiary hydrogen in PPO than in PVME by the $RO_2 \cdot$ radical. This factor should also increase the relative magnitude of R_0 for PPO as compared with PVME.

Recently Reich and Stivala [57] analyzed data reported [58] for the autoxidation of poly(ethylene glycol) (PEG) and poly(propylene glycol) (PPG) in o-dichlorobenzene solution in presence of the initiator azodicyclohexanenitrile as a function of reaction time. Two schemes were employed: a relatively simple one involving termination of polymeric peroxy radicals by combination (cf. page 231) and a more general one which excluded such a termination (cf. page 235). It was generally found that the simpler kinetic scheme gave expressions which afforded values of various parameters that were in poorer agreement with those obtained from expressions derived from the more general scheme. However, despite the somewhat better agreement provided by the more general scheme between calculated and observed reaction variables, because of certain discrepancies and an apparent change in reaction mechanism, more work on PEG autoxidation is indicated before any decision can be reached as to which scheme is preferable.

Sklyarova and Lukovnikov [59] studied the oxidative degradation of powdered polyoxymethylene diacetate (POMD) in a stream of oxygen at 160°C. The major decomposition product found was formaldehyde which presumably resulted from depropagation. It was also observed that there was a correlation between the peroxide formation curve and the curve associated with the change in molecular weight. In order to determine more precisely the effect of peroxides on molecular weight during POMD oxidation, benzoyl peroxide (BP) was added to POMD, the resulting solid mixture was heated under argon at various temperatures, and the change in molecular weight was obtained as a function of time. From such changes in molecular weight, corresponding values of the number of scissions per original molecule Δn were calculated from formula [60] [cf. Eq. (78)],

$$\Delta n = \frac{([\eta]_0/[\eta])^{1/a} - 1}{p} \tag{78'}$$

where $p = [1/(a + 1)!]^{1/a}$. In this manner it was ascertained that Δn was directly proportional to the amount of BP which decomposed. (In the absence of BP no change in molecular weight was observed under the experimental conditions employed.) From this relationship it can be readily seen that $d(\Delta n)/dt = k(-d(BP)/dt) = k'(BP)$, where k and k' are constants. Thus, the rate of scissions should be directly proportional to the amount of BP present at time t. A similar relationship was employed for the oxidation of POMD, and a linear relationship was obtained when $d(\Delta n)/dt$ was plotted against the concentration of peroxide, $[RO_2 R]$, i.e. [cf. Eq. (75)],

$$\frac{d(\Delta n)}{dt} = k''[RO_2R] \tag{75a}$$

It is interesting to note here that if the general scheme (cf. page 237) is employed, Eq. (75a) may be written as

$$\frac{d(\Delta n)}{dt} = f\rho_{>C=O} \tag{75b}$$

or

$$\Delta n = f \int_0^t \rho_{>C=O}\, dt \tag{75}$$

Equation (75) has been employed to account for molecular-weight changes during polyolefin autoxidation (cf. page 253) and also for molecular-weight changes during the autoxidation of poly(alkylene glycols) in solution in presence of a free-radical initiator [57].

In absence of additives. Dulog and Storck [61] studied the autoxidation of polyepoxides such as PPO and polybutene-1-oxide (PBO), in trichlorobenzene solution at 90°C and in presence of a relatively low partial pressure of oxygen, i.e., 50 torrs. The following relatively simple scheme was postulated:

$$RO_2H \xrightarrow{k_i} \text{Radicals} \tag{11}$$

$$R\cdot + O_2 \xrightarrow{k_2} RO_2\cdot \tag{2}$$

$$RO_2\cdot + RH \xrightarrow{k_3} RO_2H + R\cdot \tag{3}$$

$$R\cdot + RO_2 \xrightarrow{k_5} \text{Products} \tag{5}$$

$$2RO_2\cdot \xrightarrow{k_6} \text{Products} \tag{6}$$

The termination step (5) was included because of the relatively low oxygen pressures employed. If steady-state concentrations are assumed for $[RO_2H]$ (maximum oxidation rate $\rho_{ox,\,m}$ obtains) along with relatively high kinetic chain lengths, we may write

$$k_i[RO_2H] = k_3[RO_2\cdot][RH] = k_5[R\cdot][RO_2\cdot] + k_6[RO_2\cdot]^2 \tag{135'}$$

$$k_2[R\cdot][O_2] = k_3[RO_2\cdot][RH] \tag{136'}$$

and

$$\rho_{ox,m} = k_3[RH][RO_2\cdot] \tag{137'}$$

Upon combining Eqs. (135') and (136') and substituting the corresponding expression for $[RO_2\cdot]$ into (137'), there is obtained

$$\rho_{ox,m} = \frac{(k_3[RH])^2 k_2[O_2]}{k_3k_5[RH] + k_2k_6[O_2]} \tag{138}$$

Equation (138) can be rearranged to yield

$$\frac{[RH]^2}{\rho_{ox,m}} = \frac{k_6}{k_3^2} + \frac{k_5[RH]}{k_2k_3[O_2]} \tag{138'}$$

The validity of Eq. (138') for PBO was ascertained from the linear plots obtained when $[RH]^2/\rho_{ox,m}$ was plotted against $[RH]$ and $1/[O_2]$. From Arrhenius plots of $\rho_{ox,m}$ it was observed that the resulting activation energies for atactic PPO and PBO varied from 15.2 to 15.6 kcal mole^{-1} (Tobolsky et al. [4] calculated an overall activation energy of 16 kcal mole^{-1} for PPO), whereas for the corresponding isotactic polymers the values varied from 22.3 to 23.0 kcal mole^{-1}.

In presence of inhibitors. Guryanova and coworkers [62] studied the oxidation of model compounds of polyformaldehyde (PFA) in presence of inhibitors and free-radical initiators, using a chemiluminescence technique [1]. Thus, dimethyl ethers of polyhydroxymethylene glycols were used, i.e., $CH_3-O-(CH_2-O)_x-CH_3$, where $x = 2$ to 6 [these ethers are liquids at room temperature ($x = 2$ to 5) whereas a solid form exists for $x = 6$ at temperatures below 26 to 28°C], along with azobisisobutyronitrile (AIBN) and inhibitors such as 4,6-di-*tert*-butyl-*o*-cresol (DTBC). Various kinetic parameters were obtained. In the following treatment, it will be assumed that termination by recombination can occur and that chemiluminescence intensity I may be expressed as [cf. Eq. (71) for the oxidation of polyolefins in the bulk phase]

$$I = \alpha'[RO_2\cdot]^2 \tag{139}$$

where $\alpha' = $ a proportionality factor. From Eq. (139), it can be seen that the assumption was made that recombination of peroxy radicals leads to the emission of a weak chemiluminescence [1].

The following relatively simple mechanism was postulated [62]:

$$Y \xrightarrow{k_i} R\cdot \tag{1}$$

$$R \cdot + O_2 \xrightarrow{\ k_2\ } RO_2 \cdot \tag{2}$$

$$RO_2 \cdot + RH \xrightarrow{\ k_3\ } RO_2 H + R \cdot \tag{3}$$

$$2RO_2 \cdot \xrightarrow{\ k_6\ } Products \tag{6}$$

$$nRO_2 \cdot + \ln H \xrightarrow{\ k_{87}\ } Products \tag{87}$$

where n = stoichiometric factor (e.g., $n = 2$ when $RO_2 \cdot + \ln H \longrightarrow RO_2 H + \ln \cdot$, and $RO_2 \cdot + \ln \cdot \longrightarrow$ Products), and Y denotes a free-radical initiator, e.g., AIBN. From this scheme we may write, assuming steady-state conditions for radicals,

$$\frac{d[RO_2 \cdot]}{dt} = 0 = R_i - k_6[RO_2 \cdot]^2 - nk_{87}[RO_2 \cdot][\ln H] \tag{140}$$

Also, from Eq. (139),

$$[RO_2 \cdot] = [RO_2 \cdot]_0 \left(\frac{I}{I_0}\right)^{1/2} = \left(\frac{R_i}{k_6}\right)^{1/2} \left(\frac{I}{I_0}\right)^{1/2} \tag{141}$$

where $[RO_2 \cdot]_0$ and I_0 denote concentration of peroxy radicals and chemiluminescence intensity, respectively, in absence of inhibitor. Upon combining Eqs. (140) and (141), it can be shown that

$$\frac{I_0 - I}{(I_0 I)^{1/2}} = \frac{nk_{87}[\ln H]}{(k_6 R_i)^{1/2}} \tag{142}$$

When the value of I is minimal, i.e., when $I = I_{min}$, $[\ln H] = [\ln H]_i$, where the subscript i denotes initial conditions, and Eq. (142) becomes

$$\frac{I_0 - I_{min}}{(I_0 I_{min})^{1/2}} = \frac{nk_{87}[\ln H]_i}{(k_6 R_i)^{1/2}} = Bn[\ln H]_i \tag{142a}$$

where $B = k_{87}/(k_6 R_i)^{1/2}$. Furthermore, upon differentiation, Eq. (142) becomes

$$2Bn\left(-\frac{d[\ln H]}{dt}\right) = \frac{1 + I/I_0}{(I/I_0)^{3/2}} \frac{d(I/I_0)}{dt} \tag{143}$$

Also

$$-\frac{d[\ln H]}{dt} = k_{87}[\ln H][RO_2\cdot] \tag{144}$$

When Eq. (141) is substituted into (144),

$$-\frac{d[\ln H]}{dt} = BR_i[\ln H]\left(\frac{I}{I_0}\right)^{1/2} \tag{145}$$

Upon substitution of Eq. (142) into (145),

$$-\frac{d[\ln H]}{dt} = \frac{R_i}{n}\left(\frac{I}{I_0}\right)^{1/2}\frac{I_0 - I}{(I_0 I)^{1/2}} \tag{146}$$

When Eq. (146) is substituted into (143),

$$\frac{d(I/I_0)}{dt} = 2BR_i\left(\frac{I}{I_0}\right)^{3/2}\frac{1 - I/I_0}{1 + I/I_0} \tag{147}$$

Upon setting the derivative of Eq. (147) equal to zero, it can be shown that $(I/I_0)_{infl} \approx 0.54$, where the subscript infl refers to the inflection point, and that consequently

$$\left[\frac{d(I/I_0)}{dt}\right]_{max} \approx 0.24BR_i \tag{148}$$

where the subscript, max denotes maximum value. When Eqs. (142a) and (148) are combined,

$$n = \frac{0.24R_i}{[\ln H]_i}\frac{(I_0 - I_{min})/(I_0 I_{min})^{1/2}}{[d(I/I_0)/dt]_{max}} \tag{149}$$

From kinetic data shown in Fig. 5.12, the value of n may be estimated for various inhibitors, i.e., 4,6-di-*tert*-butyl-*o*-cresol, S-1 [dimethyl-di(phenyl-aminophenoxy)silane], 22-46 [2,2-methylene-bis(4-methyl-6-*tert*-butyl-phenol], bucremet [4,4-methylene-bis(2-methyl-6-*tert*-butylphenol)], and diphenylamine, which were employed during the autoxidation of a PFA model compound whose value of $x = 3$, by the utilization of Eq. (149). Thus, for instance, for the first inhibitor listed an average value of $n = 1.8$ was obtained. Also, from Eq. (148) a value of $k_{87}/k_6^{1/2}$ can be calculated. For the first inhibitor listed, $k_{87}/k_6^{1/2} \approx 2.5$ (l mole^{-1} sec^{-1})$^{1/2}$. In a

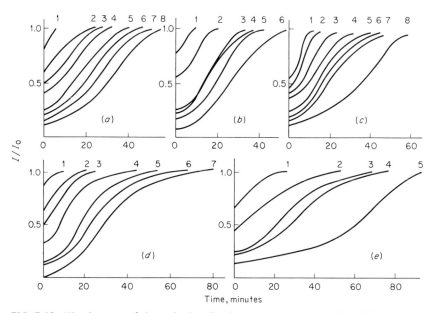

FIG. 5.12   Kinetic curves of change in chemiluminescence intensity for $CH_3O(CH_2O)_3CH_3$ in presence of antioxidants: $R_i = 5 \times 10^{-7}$ mole l^{-1} sec^{-1} at 60°C; initiator, AIBN; concentration, $c \times 10^4$ moles l^{-1}. (a) 4,6-di-tert-butyl-o-cresol: (1) 0.3; (2) 0.6; (3) 1.25; (4) 1.8; (5) 2.4; (6) 2.85; (7) 3.3; (8) 4.2. (b) S-1: (1) 0.2; (2) 0.3; (3) 0.5; (4) 0.6; (5) 1.5; (6) 2.85. (c) 22-46: (1) 0.54; (2) 1.0; (3) 1.4; (4) 1.75; (5) 2.0; (6) 2.3; (7) 2.8; (8) 4.0. (d) Bucremet: (1) 0.33; (2) 0.54; (3) 0.67; (4) 1.0; (5) 1.75; (6) 2.0; (7) 2.3. (e) Diphenylamine: (1) 0.25; (2) 0.5; (3) 0.98; (4) 1.8; (5) 6.87 [62].

similar manner, values of about $n = 3\ 1/2$ were obtained for the difunctional inhibitor bucremet. This inhibitor should possess a value of $n = 4$. Guryanov et al. [62] reported a value of $n = 2\ 1/2$ for bucremet and ascribed this low value to experimental error. However, low values for n were obtained for the difunctional inhibitor 22-46 which are in agreement with the low value of $n = 2$ reported [62] for this inhibitor.

Values of n and $k_{87}/k_6^{1/2}$ may be checked by a method used by Bamford and coworkers for polymerization studies [63]. Thus, a scheme identical with Eqs. (1) to (3), (6), and (87) may be employed. Further, if it is assumed that Eq. (140) is only approximately true ($d[RO_2\cdot]/dt$ is very small but not zero), then

$$R_i \approx k_6[RO_2\cdot]^2 + nk_{87}[RO_2\cdot][\ln H] \qquad (140a)$$

Upon differentiation, Eq. (140a) becomes

$$-\frac{d[\ln H]}{dt} = \frac{2k_6[RO_2\cdot] + nk_{87}[\ln H]}{nk_{87}[\ln H]}\frac{d[RO_2\cdot]}{dt} \tag{150}$$

Upon substituting Eqs. (140a) and (144) into Eq. (150),

$$\frac{d[RO_2\cdot]}{dt}\frac{R_i + k_6[RO_2\cdot]^2}{k_{87}[RO_2\cdot]^2} = R_i - k_6[RO_2\cdot]^2 \tag{151}$$

Upon integration, Eq. (151) yields

$$-\frac{1}{[RO_2\cdot]} + \left(\frac{k_6}{R_i}\right)^{1/2} \ln\frac{R_i + (k_6 R_i)^{1/2}[RO_2\cdot]}{R_i - (k_6 R_i)^{1/2}[RO_2\cdot]} = \int k_{87}\, dt \tag{152}$$

Since $R_i = k_6[RO_2\cdot]_0^2$, where the subscript 0 refers to conditions in absence of inhibitor, Eq. (152) becomes

$$-\frac{[RO_2\cdot]_0}{[RO_2\cdot]} + \ln\frac{[RO_2\cdot]_0 + [RO_2\cdot]}{[RO_2\cdot]_0 - [RO_2\cdot]} = k_{87}[RO_2\cdot]_0 t + C \tag{153}$$

where C = integration constant. When $t = 0$, $[RO_2\cdot] = [RO_2\cdot]_i$, and Eq. (153) affords

$$\frac{1}{\phi_0} - \frac{1}{\phi} + \ln\left(\frac{1 + \phi}{1 - \phi}\frac{1 - \phi_0}{1 + \phi_0}\right) = k_{87}[RO_2\cdot]_0 t \tag{154}$$

where ϕ_0 and ϕ denote $[RO_2\cdot]_i/[RO_2\cdot]_0$ and $[RO_2\cdot]/[RO_2\cdot]_0$, respectively. When $\phi_0 \ll 1$ (often the case), Eq. (154) becomes

$$\frac{1}{\phi_0} - \frac{1}{\phi} + \ln\frac{1 + \phi}{1 - \phi} = k_{87}[RO_2\cdot]_0 t \tag{154a}$$

If we select a value of ϕ such that $-1/\phi + \ln[(1 + \phi)/(1 - \phi)] = 0$ ($\phi_\tau = 0.648$), then

$$\frac{1}{\phi_0} = k_{87}[RO_2\cdot]_0\tau = k_{87}\left(\frac{R_i}{k_6}\right)^{1/2}\tau \tag{155}$$

where τ refers to the time at which $\phi = 0.648$.

If at $t = 0$ it is assumed that $I_{min} = I_i$, then Eq. (142a) becomes, employing the relationship between I and $[RO_2 \cdot]$, i.e., Eq. (139),

$$\frac{1}{\phi_0} - \phi_0 = Bn[\ln H]_i \tag{156}$$

Thus, from Eqs. (155) and (156) it can be seen that once values of ϕ_0 and τ are known, then values of $k_{87}/k_6{}^{1/2}$ and n may be determined. In the following is shown such a method of calculation.

Equation (153) may be rewritten as

$$-\frac{1}{\phi} + \ln \frac{1 + \phi}{1 - \phi} = k_{87}[RO_2 \cdot]_0 t + C \tag{153a}$$

which enables φ to be plotted against $k_{87}[RO_2 \cdot]_0 t + C$, as depicted in Fig. 5.13 [63]. From data in Figs. 5.12 and 5.13, Table 5 was constructed for the autoxidation of a PFA model compound ($x = 3$) in presence of 4,6-di-tert-butyl-o-cresol (see Fig. 5.12a, curve 8). From a plot of $k_{87}[RO_2 \cdot]_0 t + C$ versus t (Fig. 5.14) a value of $k_{87}[RO_2 \cdot]_0$ may be obtained from the slope of the linear relationship obtained. In this manner $k_{87}[RO_2 \cdot]_0 = 1.94 \times 10^{-3}$. Also, since τ is known (cf. Table 5), a value of $1/\phi_0 = 3.25$ was estimated using Eq. (155). Consequently, the following values were calculated: $k_{87}/k_6{}^{1/2} = 2.7$ (reported, 2.5); $n = 1.8$ (reported, 1.8).

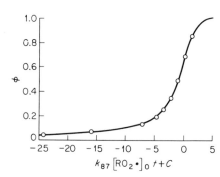

FIG. 5.13 Dependence of reduced rate ϕ on reaction time [63].

In the preceding, it was assumed that ϕ_0 was much less than unity. When this is not valid, Eq. (154) may be better approximated as

$$\frac{1}{\phi_0} - 2\phi_0 - \frac{1}{\phi} + \ln \frac{1 + \phi}{1 - \phi} = k_{87}[RO_2 \cdot]_0 t \tag{154b}$$

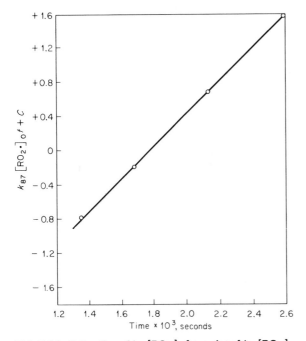

FIG. 5.14 Estimation of $k_{87}[RO_2\cdot]_0$ from plot of $k_{87}[RO_2\cdot]_0$ $t + C$ versus time t.

Then, we may write

$$\frac{1}{\phi_0} - 2\phi_0 = k_{87}[RO_2\cdot]_0\tau = k_{87}\left(\frac{R_i}{k_6}\right)^{1/2}\tau \qquad (155a)$$

From Eq. (155a) and Fig. 5.14, $\phi_0 = 0.27$. Then from Eq. (156), the value of n increases to $n = 2.0$. In general, the induction period (time to reach maximum oxidation rate) increased as the value of $k_{87}/k_6^{1/2}$ increased. In this sense, 4,6-di-*tert*-butyl-*o*-cresol was least effective whereas dimethyl-di(phenylaminophenoxy)silane (S-1) was the most effective inhibitor.

Initiation by irradiation. Neiman and coworkers [64] investigated the decay of radicals formed during the gamma irradiation (Co^{60} source) of powdered polyformaldehyde (PFA), in the presence of air, by means of electron-spin resonance (ESR). When reciprocal radical concentration ($[R\cdot]$) was plotted against time for different oxygen pressures at 55°C, linear relationships obtained. This implied that radical decay involved a

TABLE 5   Data Derived from Figs. 5.12 and
5.13 for the Autoxidation of a PFA Model
Compound ($x = 3$) at $60°C$; $R_i = 5 \times 10^{-7}$ mole
l^{-1} sec^{-1}; $[InH]_i = 4.2 \times 10^{-4}$ mole l^{-1}
(4,6-di-*tert*-butyl-*o*-cresol)

$I/I_0 = \phi^2$	ϕ	$k_{87}[RO_2\cdot]_0 t + C$	t, seconds
0.32_0	0.56_5	-0.78_0	1,344
0.42_2	0.65_0	-0.19_6	1,680 ($\equiv \tau$)
0.62_0	0.78_5	$+0.68_5$	2,130
0.80_0	0.89_0	$+1.57$	2,581

second-order termination, i.e.,

$$-\frac{d[R\cdot]}{dt} = k_e[R\cdot]^2 \tag{157}$$

Furthermore, it was found that the value of k_e at relatively high oxygen pressures was linearly dependent upon oxygen pressures; i.e.,

$$k_e = K[O_2]\exp\left(-\frac{17,000}{RT}\right) \tag{158}$$

where k_e is in milliliters per second, K denotes a constant, and $[O_2]$ = concentration of solute oxygen.

In order to account for the previous results, the following scheme was postulated,

$$R\cdot + O_2 \xrightarrow{k_2} RO_2\cdot \tag{2}$$

$$RO_2\cdot \xrightarrow{k_{-2}} R\cdot + O_2 \tag{-2}$$

$$2R\cdot \xrightarrow{k_4} \text{Products} \tag{4}$$

$$R\cdot + RO_2\cdot \xrightarrow{k_5} \text{Products} \tag{5}$$

Assuming steady-state conditions for $[RO_2\cdot]$, it can be shown that

$$[RO_2\cdot] = \frac{k_2[R\cdot][O_2]}{k_{-2}} + k_5[R\cdot] \tag{159}$$

Also,

$$-\frac{d[R\cdot]}{dt} = k_2[R\cdot][O_2] - k_{-2}[RO_2\cdot] + k_4[R\cdot]^2 + k_5[R\cdot][RO_2\cdot]$$

$$(160)$$

Upon substituting Eq. (159) into (160),

$$-\frac{d[R\cdot]}{dt} = \left(k_4 + \frac{2k_2k_5[O_2]}{k_{-2}} + k_5[R\cdot] \right)[R\cdot]^2 \qquad (161)$$

If it is assumed that $k_{-2} \gg k_5 [R\cdot]$, then

$$-\frac{d[R\cdot]}{dt} = k_e[R\cdot]^2 \qquad (157)$$

where

$$k_e = k_4 + \frac{2k_2k_5[O_2]}{k_{-2}} \qquad (158')$$

At very high oxygen pressures, k_4 can be neglected, and Eq. (158') becomes [cf. Eq. (158)]

$$k_e = \frac{2k_2k_5[O_2]}{k_{-2}} \qquad (158'')$$

It should be noted here that if step (−2) is replaced by the more generally accepted step (3), i.e.,

$$RO_2\cdot + RH \xrightarrow{\ k_3\ } RO_2H + R\cdot \qquad (3)$$

then an expression similar to Eq. (161) is obtained except that k_{-2} is replaced by $k_3 [RH]$. Then if $k_3 [RH] \gg k_5 [R\cdot]$, an expression similar to Eq. (157) obtains, where k_e is still defined by Eqs. (158') and (158'') except that the term k_{-2} is now replaced by $k_3 [RH]$. Moreover, the termination activation energy of 17 kcal mole[-1] [cf. Eq. (158)] appears to be too high, and this may be due to viscosity effects [25].

Oxidation of Rubbers

Styrene-butadiene-styrene rubber. Rode and coworkers [65] studied the thermal oxidative degradation of a graft copolymer synthesized by grafting styrene to a butadiene-styrene rubber in the presence of benzoyl peroxide. The overall graft copolymer (PBS) comprised 85 percent of styrene homopolymer (PS) and 15 percent of graft copolymer (GCP) in the form of a gel fraction. The PS could be separated from the GCP by means of centrifuging. Thus, oxidation studies could be carried out on films of PBS and PS and on powdered samples of GCP. In all cases it was established that there was no diffusion control and the oxygen pressure was 200 torrs. Also, various types of oxidation products were accounted for. Thus, peroxides, ketones, aldehydes, carboxylic acids, water, and oxides of carbon were identified from oxidized samples of PBS, PS, and GCP.

It was found that during the oxidative degradation of PBS, PS, and GCP there were two characteristic periods. The first was one of rate autoacceleration whereas during the second the oxidation rate changed relatively little. As compared with the oxidation of PBS and PS, the oxidation of GCP was associated with a relatively small initial interval during which oxidation rates were small. Subsequently, the rates of GCP oxidation during the autoacceleration period became much greater than the oxidation rates of PBS and PS. This behavior was ascribed to the presence of easily oxidizable groups in the GCP rubber. Thus, in the butadiene segments of GCP exist hydrogen atoms on tertiary carbon atoms at branching positions. Much less activation energy would be expected to remove these atoms than to remove a tertiary hydrogen atom in PS. Therefore, PBS at low temperatures should oxidize more readily than PS but less readily than GCP (which contains a greater percentage of butadiene segments). Further, when Arrhenius plots were made (assuming a first-order oxidation process), it was found that at relatively low temperatures (up to about 220°C) PBS and GCP possessed lower values of activation energy E than PS, whereas at temperatures above 220°C the values of E for all three substances became approximately equal. This was further evidence that the oxidation of PBS occurs at low temperatures by means of the oxidation of butadiene fragments, but at high temperatures by the oxidation of tertiary hydrogen atoms in the polystyrene chain.

In order to explain the kinetic data obtained, a scheme similar to that previously mentioned was employed. [Cf. Eqs. (1''), (2), (3), (6), and (21) under Approach of Neiman and coworkers, page 234.] From this scheme, the following expression could be derived [cf. Eq. (27)]:

$$\frac{\rho_{ox}}{[RH][RO_2H]} = k_{21} + k_3 \left(\frac{k_{21}[RH]}{k_6[RO_2H]} \right)^{1/2} \tag{27}$$

Thus, from Eq. (27), a plot of $\rho_{ox}/[RO_2H]$ versus $1/[RO_2H]^{1/2}$ should yield a linear relationship. Such a relationship was observed for PS, PBS, and GCP at all the temperatures used. However, in the case of GCP and PBS, relatively large and small negative intercepts obtained respectively at 225°C, which would not be anticipated from Eq. (27). Furthermore, the increase in network formation during GCP oxidation was ascribed to increased crosslinking of the existing network, whereas the increase in the gel fraction during PBS oxidation was associated with the formation of cross-linked structures from ruptured chains.

It may also be mentioned here that May and Bsharah [66] studied the oxidation of styrene-butadiene rubber in presence of metals by means of differential thermal analysis (DTA) (cf. Chap. 2) under 300 psi of oxygen, and in the bulk phase. It was found that the degree of catalytic activity of the metal naphthenates was generally in accordance with their oxidation potential (cf. Ref. 52). Thus, the activity decreased in the following order: Co > Mn, Cu, Fe > Pb, Zn.

Oxidation of Resins

Phenolformaldehyde resin (PFR). Conley and coworkers [67, 68] have studied the oxidative degradation of PFR by means of infrared spectroscopy in the relatively low temperature range of 100 to 200°C. When the PFR was cured by heating nitrogen at 102°C for 3 hr, little or no volatilization was detected. Thus, there should presumably be little thermal effect during the oxidation studies. The starting material was considered to be predominantly phenol rings connected by methylene groups and substituted at the 2,4- or 6-positions, or any combination of these, as depicted in Fig. 5.15.

When the cured resin was further heated in vacuum or nitrogen, no changes could be detected in the infrared spectrum. However, when the PFR was heated in air at 200°C, the infrared spectra (IR) could detect the formation of carbonyl groups and decreases in band intensities of —CH$_2$—, —OH, and 1,2,4-trisubstituted aromatic ring groups. It was assumed that initially oxidation occurs on the methylene linkage and a dihydroxy-substituted benzophenone structure forms (this was confirmed by independent syntheses of ketone-containing polymers). Although peroxide intermediates could not be detected by IR, their existence was shown by

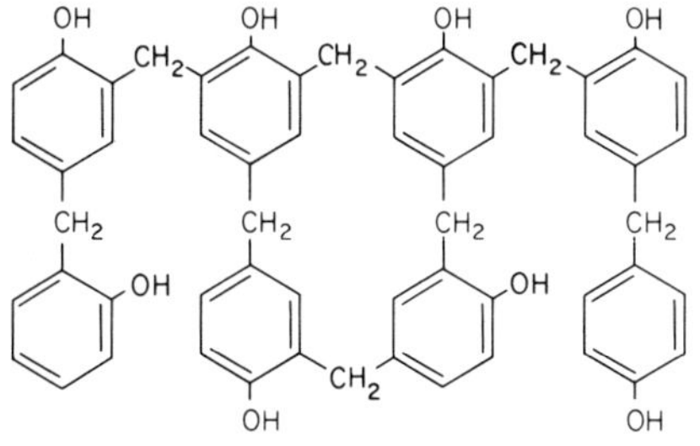

FIG. 5.15 Structure of typical phenolic resin indicating possible substitution types and major functional groups present in the cured resin [67].

chemical methods. Thus, the first step in the oxidation of PFR was depicted as shown in Fig. 5.16. Upon further oxidation, a shoulder appeared on the initial carbonyl band. This was attributed to the formation of quinone-type structures. Other IR bands also formed which indicated that the phenolic hydroxy band was decreasing and an acid hydroxyl appearing. These subsequent steps are depicted in Fig. 5.17. After the appearance of acid fragments, no further chemical changes could be detected by IR. Subsequent work [68] has indicated that the hydroperoxide formed initially

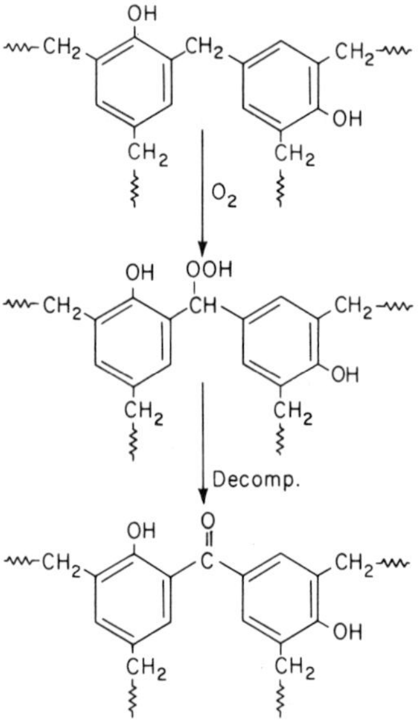

FIG. 5.16 Primary oxidation route of phenolic resins in which the diphenylmethane unit is converted to benzophenone linkages [67].

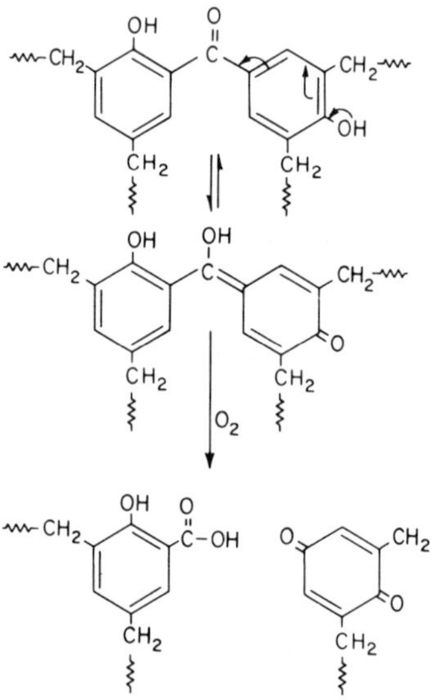

FIG. 5.17 Secondary oxidation involving chain scission through quinoid linkages to give quinone and acid fragments [67].

may undergo decomposition (cf. Fig. 5.16) to also afford dihydroxy-benzhydrol species as well as benzophenone linkages. Furthermore, methylol groups present (partially cured PFR) may oxidize to form carboxylic groups.

When absorbance data obtained for carbonyl formation as a function of time were analyzed, it was found that at 140 to 160°C, the oxidation was first order in respect to polymer over the first 50 min of reaction time. However, at higher temperatures a break resulted for the first-order plot at about 10 min at 180°C. A most likely explanation for this break was that the formation of quinoid structures (cf. Fig. 5.17) was interfering with initial oxidation reactions. Presumably, these structures are not readily formed at temperatures below about 170°C.

Berlin and coworkers [69] studied the oxidative degradation of a PFR at the relatively high temperatures of 300 to 400°C. In order to estimate the effects of thermal degradation on oxidation, two PFR samples were studied. The first PFR sample was not pretreated before oxidation (PFRU), whereas the second sample was preheated in vacuum at temperatures of 350 to 400°C (PFRT). When the IR spectra were taken of PFRU samples subjected to thermal oxidation for short and long reaction intervals, it was found that the sample exposed for the short time possessed an IR spectrum which was very similar to that obtained for a PFRU sample. This indicated that thermal reactions play an important role at the beginning of the oxidation. However, at fairly long oxidation times, this role apparently becomes insignificant and the thermal oxidation reaction becomes the predominant factor which determines the PFRU structure. Thus, it may be anticipated that the oxidative process at 350 to 400°C might involve two stages: an initial nonstationary stage followed by a stationary stage. During the initial stage, weight losses observed by thermogravimetric methods should be independent of oxygen pressure, as found experimentally up to 0.6 atm. This has been attributed to the inhibition of oxidation by structural groups in PFR at relatively low oxygen pressures. During the second or stationary stage, oxidative processes predominate, and the oxidation rate of PFRU could be expressed in the form

$$\rho_{ox} = k(PFRU)P_{O_2} \qquad (162)$$

where (PFRU) = amount of resin, k = a constant, and P_{O_2} = partial pressure of oxygen. In order to account for these results, the following scheme was postulated:

$$PFRU + O_2 \xrightarrow{k_i} R\cdot \qquad (1a)$$

$$R\cdot + O_2 \xrightarrow{k_2} RO_2\cdot \qquad (2)$$

$$RO_2\cdot + PFRU \xrightarrow{k_{3a}} RO_2H + R\cdot \qquad (3a)$$

$$RO_2\cdot + (PFRU)' \xrightarrow{k_{3b}} Products \qquad (3b)$$

where (PFRU)' denotes structural groups present in PFRU which acts as oxidation inhibitors. From this scheme, we may write, assuming relatively long kinetic chain lengths and steady-state conditions,

$$\rho_{ox} = k_2[R\cdot][O_2] = k_{3a}(PFRU)[RO_2\cdot] \qquad (163)$$

Also,

$$R_i = k_i(PFRU)[O_2] = k_{3b}(PFRU)'[RO_2\cdot] \qquad (164)$$

Upon combining Eqs. (163) and (164),

$$\rho_{ox} = \frac{k_i k_{3a}(PFRU)^2[O_2]}{k_{3b}(PFRU)'} \qquad (165)$$

Assuming that $(PFRU) \approx (PFRU)'$, Eq. (165) becomes [cf. Eq. (162)]

$$\rho_{ox} = k(PFRU)P_{O_2} \tag{162}$$

where $k = k_i k_{3a} k_s / k_{3b}$ and $k_s = [O_2]/P_{O_2}$. From an Arrhenius plot of k a value of $E = 29$ kcal mole^{-1} was obtained. This value was much higher than corresponding values previously reported from initial oxidation-rate data (cf. [67]). This lower value of E was not considered to be reliable since it was obtained during initial stages where thermal reactions may occur [69].

REFERENCES

1. Reich, L., and S. S. Stivala: "Autoxidation of Hydrocarbons and Polyolefins," Marcel Dekker, Inc., New York, 1969.
2. Tobolsky, A. V., P. M. Norling, N. H. Frick, and H. Yu: *J. Am. Chem. Soc.*, **86**:3925 (1964).
3. Norling, P. M., T. C. P. Lee, and A. V. Tobolsky: *Rubber Chem. Technol.*, **38**:1198 (1965).
4. Norling, P. M., and A. V. Tobolsky: *Rubber Chem. Technol.*, **39**:278 (1966).
5. Neiman, M. B.: *Russ. Chem. Rev.*, **33**:13 (1964).
6. Neiman, M. B. (ed.): "Aging and Stabilization of Polymers," Consultants Bureau, Plenum Publishing Corporation, New York, 1965.
7. Chien, J. C. W., and C. R. Boss: *J. Polymer Sci.*, **5**:3091 (1967).
8. Notley, N. T.: *Trans. Faraday Soc.*, **58**:66 (1962).
9. Notley, N. T.: *Trans. Faraday Soc.*, **60**:88 (1964).
10. Stivala, S. S., L. Reich, and P. G. Kelleher: *Makromol. Chem.*, **59**:28 (1963).
11. Stivala, S. S., and L. Reich: *Polymer Eng. Sci.*, **5**:179 (1965).
12. Reich, L., and S. S. Stivala: *J. Polymer Sci.*, A-**3**:4299 (1965).
13. Stivala, S. S., E. B. Kaplan, and L. Reich: *J. Appl. Polymer Sci.*, **9**:3557 (1965).
14. Reich, L., and S. S. Stivala: *J. Polymer Sci.*, B-**3**:227 (1965).
15. Reich, L., and S. S. Stivala: *Rev. Macromol. Chem.*, **1**:249 (1966).
16. Bawn, C. E. H., and S. A. Chaudhri: *Polymer*, **9**:123 (1968).
17. Reich, L., and S. S. Stivala: *J. Appl. Polymer Sci.*, **12**:2033 (1968).
18. Reich, L., and S. S. Stivala: *J. Appl. Polymer Sci.*, **13**:17 (1969).
19. Reich, L., and S. S. Stivala: *J. Appl. Polymer Sci.*, **13**:23 (1969).
20. Jadrnicek, B. R., S. S. Stivala, and L. Reich: *J. Applied Polymer Sci.*, **14**:2537 (1970).
21. Bartlett, P. D., and P. Gunther: *J. Am. Chem. Soc.*, **88**:3288 (1966).
22. Betts, A. T., and N. Uri: *Nature*, **199**:568 (1963).
23. Hiatt, R., and T. G. Traylor: *J. Am. Chem. Soc.*, **87**:3766 (1965).
24. Mayo, F. R.: "Polymer Preprints," vol. 8, no. 1, pp. 11ff., American Chemical Society Meeting, Miami Beach, April, 1967.
25. Bresler, S. E., and E. N. Kazbekov: *Fortschr. Hochpolymer. Forsch*, **3**:688 (1964).
26. Chien, J. C. W., and H. Jabloner: *J. Polymer Sci.*, A-1, **6**:393 (1968).
27. Mayo, F. R.: *Stanford Res. Inst. Proj.* PRC-8012, *Progr. Rept.* 2, September, 1969.
28. Mayo, F. R.: Private communication, Aug. 18, 1969.
29. Bawn, C. E. H., and S. A. Chaudhri: *Polymer*, **9**:113 (1968).
30. Fish, A.: *Quart. Rev. (London)*, **18**:243 (1964).
31. Howard, J. A., and K. U. Ingold: *Can. J. Chem.*, **44**:1119 (1966).
32. Howard, J. A., and K. U. Ingold: *Can. J. Chem.*, **45**:793 (1967).
33. Twigg, G. H.: *Chem. Ind. (London)*, **1962**:4.

34. Uri, N.: In W. O. Lundberg (ed.), "Autoxidation and Antioxidants," vol. 1, p. 90, Interscience Publishers, a division of John Wiley & Sons, Inc., New York, 1961.
35. Hawkins, W. L., W. Matreyek, and F. H. Winslow: *J. Polymer Sci.*, **41**:1 (1959).
36. Kato, Y., D. J. Carlsson, and D. M. Wiles: *J. Applied Polymer Sci.*, **13**:1447 (1969).
37. Miller, V. B., M. B. Neiman, V. S. Pudov, and L. I. Lafer: *Vysokomolekul. Soedin.*, **1**:1696 (1959).
38. Schard, M. P., and C. A. Russell: *J. Appl. Polymer Sci.*, **8**:985, 997 (1964).
39. Ashby, G. E.: *J. Polymer Sci.*, **50**:99 (1961).
40. Barker, R. E., J. H. Deane, and P. M. Rentzepis: *J. Polymer Sci.*, **A-3**:2033 (1965).
41. Quackenbos, H. M.: *Polymer Eng. Sci.*, **6**:117 (1966).
42. Beati, E., F. Severini, and G. Clerici: *Makromol. Chem.*, **61**:104 (1963).
43. Stivala, S. S., and L. Reich: *Polymer Eng. Sci.*, **7**:253 (1967).
44. Berlin, A. A., A. A. Ivanov, and A. P. Firsov: *Vysokomolekul. Soedin.*, **A10**:2321 (1968).
45. Gromov, B. A., V. B. Miller, M. B. Neiman, Y. S. Torsuyeva, and Y. A. Shlyapnikov: *Vysokomolekul. Soedin.*, **6**:1895 (1964).
46. Shlyapnikov, Y. A., V. B. Miller, M. B. Neiman, and Y. S. Torsuyeva: *Vysokomolekul. Soedin.*, **5**:1507 (1963).
47. Ciampa, G.: *Chim. Ind. (Milan)*, **38**:298 (1956).
48. Gromov, B. A., V. B. Miller, and Y. A. Shlyapnikov: *Vysokomolekul. Soedin.*, **6**:470 (1964).
49. Stivala, S. S., G. Yo, and L. Reich: *J. Appl. Polymer Sci.*, **13**:1289 (1969).
50. Lombard, R., and J. Knopf: *Bull. Chim. Soc. France*, **1966**:3926.
51. Osawa, Z., T. Shibamiya, and K. Matsuzaki: *Kogyo Kagaku Zasshi*, **71**:552 (1968).
52. Reich, L., B. R. Jadrnicek, and S. S. Stivala: *J. Polymer Sci.*, **A-1**, **9**:231 (1971).
53. Chaudhri, S. A.: *Polymer*, **9**:604 (1968).
54. Bawn, C. E. H., and S. A. Chaudhri: *Polymer*, **9**:81 (1968).
54a. Kamiya, Y., and K. U. Ingold: *Can. J. Chem.*, **42**:2424 (1964).
55. Kamiya, Y.: *Bull. Chem. Soc., Japan*, **38**:2156 (1965).
55a. Carlsson, D. J., and D. M. Wiles: *Macromolecules*, **2**:587 (1969).
56. Stivala, S. S., B. R. Jadrnicek, and L. Reich: *Macromolecules*, **4**:61 (1971).
57. Reich, L., and S. S. Stivala: *J. Appl. Polymer Sci.*, **13**:977 (1969).
58. Grosborne, P., I. S. de Roch, and L. Sajus: *Bull. Chim. Soc. France*, **1968**:2020.
59. Sklyarova, E. G., and A. F. Lukovnikov: *Dokl. Akad. Nauk SSSR*, **175**:1297 (1967).
60. Mizutani, Y., K. Yamamoto, S. Matsuoka, and H. Ihara: *Chem. High Polymers (Tokyo)*, **22**:97 (1965).
61. Dulog, L., and G. Storck: *Makromol. Chem.*, **91**:50 (1966).
62. Guryanova, V. V., M. B. Neiman, B. M. Kovarskaya, V. B. Miller, and G. V. Maksimova: *Vysokomolekul. Soedin.*, **A9**:2165 (1967).
63. Bamford, C. H., A. D. Jenkins, and R. Johnston: *Proc. Roy. Soc. (London)*, **A239**:214 (1957).
64. Neiman, M. B., T. S. Fedoseyeva, G. V. Chubarova, A. L. Buchachenko, and Y. S. Lebedev: *Vysokomolekul. Soedin.*, **5**:1339 (1963).
65. Rode, V. V., Y. P. Novichenko, and S. R. Rafikov: *Vysokomolekul. Soedin.*, **A10**:2471 (1968).
66. May, W. R., and L. Bsharah: *Ind. Eng. Chem., Prod. Res. Develop.*, **9**:73 (1970).
67. Conley, R. T., and J. F. Bieron: *J. Appl. Polymer Sci.*, **7**:103, 177 (1963).
68. Conley, R. T.: *J. Appl. Polymer Sci.*, **9**:1117 (1965).
69. Berlin, A. A., V. V. Yarkina, and A. P. Firsov: *Vysokomolekul. Soedin.*, **A10**:2157 (1968).

6

FACTORS AFFECTING
POLYMER STABILITY

There are a vast number of factors which can affect the resistance (or susceptibility) of polymers to thermal, oxidative, chemical, mechanical, and radiative attack. In order to limit the number discussed, this chapter will not include the effects of additives, e.g., biocides, free-radical inhibitors, metal salt accelerators of oxidation, metal deactivators, curing agents, and fillers (a discussion of the effects of some of these additives may be found in Ref. 1).

POLYMER MORPHOLOGY: CRYSTALLINITY
AND AMORPHICITY

Polymers undergoing thermal degradation generally soften or melt before degradation. Thus, one might expect that a crystalline polymer with a higher melting point than its amorphous counterpart would require a larger energy input to degrade it (all other conditions being similar). The ability of polymers to crystallize depends, among other things, upon structural regularity and chemical nature of the chains. Generally, relatively low-melting crystalline polymers possess well-packed and flexible chains (the former may be readily seen from the relatively high densities). Although strong intermolecular forces may exist, these by themselves may not be sufficient to cause very high melting points. This may be better understood from the following simplified thermodynamic treatment of the melting process.

For melting processes involving equilibrium phase changes, we may write

$$
T_m = \frac{\Delta H_f}{\Delta S_f} = \frac{(\Delta U/\Delta V)_T + P}{(\Delta S_f/\Delta V)_T} \tag{1}
$$

where T_m = melting temperature; ΔH_f and ΔS_f denote enthalpy and entropy of fusion, respectively; P = external pressure (assumed constant); and U and V denote respectively internal energy and volume. Since $(\Delta U/\Delta V)_T = P_i$ (internal pressure), and $(\Delta S/\Delta V)_T = (\partial P/\partial T)_V$,

$$
P_i + P = T\left(\frac{\Delta S_f}{\Delta V}\right)_T = T\left(\frac{\partial P}{\partial T}\right)_V \tag{2}
$$

The existence of a larger internal pressure, such as will exist in the presence of stronger intermolecular forces (due, for example, to polar, ionic, or hydrogen-bonding molecules), may well increase the value of ΔH_f. However, as may be seen from the above expression, an increase in ΔS_f may also occur, so that the effects of ΔH_f and ΔS_f on T_m may tend to cancel. Thus, in order to achieve high melting points, an increase in ΔH_f should be accompanied by a relatively smaller increase in ΔS_f (or the degrees of vibrational and rotational freedom). Besides strong intermolecular forces, such factors as structurally ordered melts (or chain stiffness), which would reduce ΔS_f without affecting structural regularity too much, would tend to raise T_m a great deal. (The effect of chain stiffness or backbone rigidity will be discussed under Polymer Chain Structure, page 296, in connection with thermal stability.) It may be stated from the preceding that, in general, crystalline polymers of relatively very high melting points would be more thermally stable than their amorphous counterparts. This is due to the higher energy input required in overcoming both strong intermolecular (van der Waals, for instance) and intramolecular forces, leading, for example, to chain stiffening. These forces undoubtedly play an important role in the behavior of crystalline (versus amorphous) polymers toward thermal, chemical, and oxidative attack. Relative to thermal stability, we may mention an example of the effect of these forces on melting and decomposition temperatures. Thus, Preston and coworkers [2] studied the thermal stability of a new class of aromatic polyamides (crystalline in the fiber form). As the order of chain stiffness increased (from m-substituted phenyl groups to 2,6-naphthalene groups), the melting temperature increased [as observed by differential thermal analysis (DTA) in nitrogen], along with a corresponding increase in the decomposition temperature [as observed from dynamic thermogravimetric (TGA) weight losses in nitrogen].

Besides affecting thermal stability, crystallinity can affect chemical and oxidative stability. Thus, it has been amply demonstrated that the attack of a chemical reagent on an amorphous structure generally occurs more readily than on a crystalline structure. Several examples follow. When cellulose was heated with mineral acids, the polymer was ultimately converted to glucose. However, morphology was found to influence the reaction rate during the early reaction stages [3]. Ravens [4] studied the acid hydrolysis of poly(ethylene terephthalate) and found that the crystalline forms were attacked more slowly than the amorphous forms. Others [5] found similar results when aqueous solutions of methylamine were employed. It was also observed that a crystalline polyethylene underwent chlorination much less readily than its less crystalline counterpart [6].

An increase in polymer crystallinity also generally implies an increase in oxidative stability. Hawkins and coworkers [7] found that oxidation rate in bulk polyethylene was inversely proportional to the percentage of crystallinity. Only the amorphous regions of semicrystalline polyethylene appeared to be sensitive to attack by oxygen. Other investigators [8] have carried out further work on polypropylene and found that similar results obtained. However, exceptions have been noted, as in the case of poly(4-methyl-1-pentene) (PMP) [9]. Thus, the crystalline regions of PMP were readily oxidized, along with the amorphous regions. This has been attributed to chain folds present in crystalline PMP [10]. Oxidation of crystalline PMP may occur preferentially at these folds owing primarily to two factors: distorted bond configurations in the fold and changes in the environment of side groups in the fold.

Oxidation of amorphous regions may lead to an apparent increase in crystallinity in these regions (chemicrystallization) [8–10]. It was found that during the oxidation of polyethylene there was an increase in polymer density corresponding to an increase in oxygen uptake. Luongo [10a] also observed an apparent increase in crystallinity during polyethylene oxidation from infrared-absorption measurements. He attributed this increase to the formation of polar groups (carbonyl, hydroxyl, etc.), which could increase intermolecular forces of attraction, thereby resulting in a higher polymer density.

POLYMER CHAIN STRUCTURE

Backbone Rigidity

As indicated in the preceding section, an increase in backbone rigidity (or chain stiffness) in a series of aromatic polyamides led to an increase in thermal stability. In general, polymers with more cyclic structure in their chains tend to be stiffer and more resistant to deformation. Hence, melting points increase and are often accompanied by an increase in the glass-transition temperature T_g. A few examples of ring systems which have pronounced chain-stiffening properties are those to be found in the high-melting polybenzimidazoles (I), aromatic polyimides (II), and polyphenylenes (III), namely

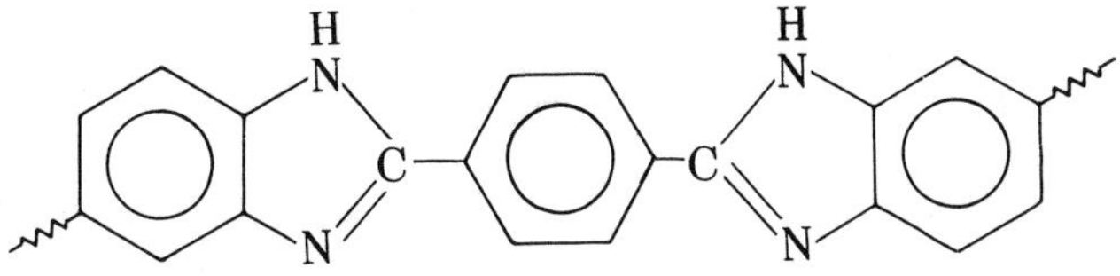

(I)

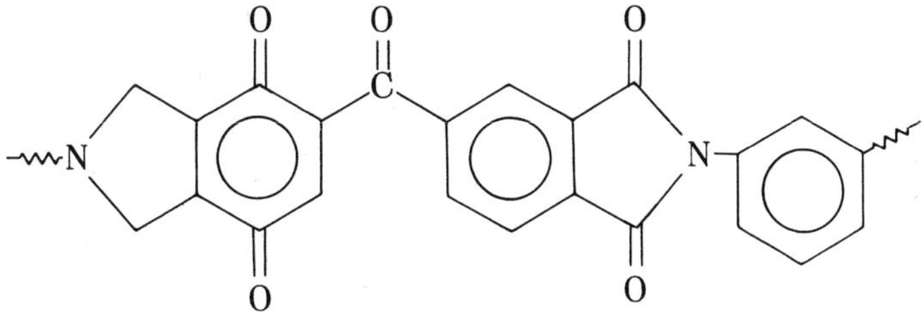

(II)

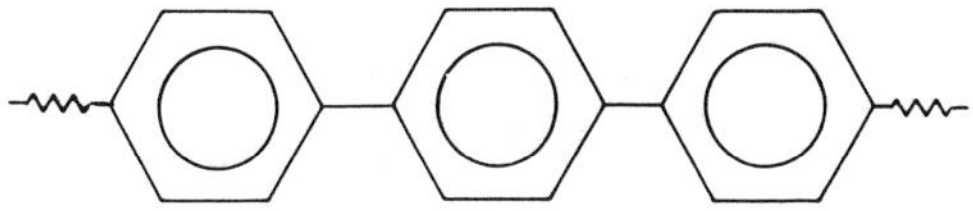

(III)

When side groups are introduced into (III), e.g., CH$_3$ groups as in polytoluylene, the chain stiffness decreases resulting in a noticeable decrease in crystallinity and softening range. However, solubility increases are often a desirable property when filaments and films are desired.

Although the presence of ring compounds enhances the thermal stability of polymers, the presence of such groups often means poor stability to ultraviolet irradiation. As mentioned previously (cf. Chap. 1), initiation of photochemical degradative processes generally involves the presence of (ultraviolet) light-absorbing groups: carbonyl, phenyl, etc. These are present in the type of compounds (I) to (III) mentioned previously. Thus, we may expect that such polymers would be vulnerable to photochemical degradation in presence or absence of oxygen. In this connection, Lilyquist and Holsten [11] found that the inherent stability of poly(phenylenetriazole) fibers and film to the effects of ultraviolet radiation was poor. Another example involves aromatic copolyamide fibers which possess a thermal stability much greater than that of nylon 66 [12]. However, the stability of these fibers on exposure to ultraviolet light was similar to that of unstabilized nylon 66.

Besides relatively high thermal stability, polymers having large intermolecular forces and chain stiffness (which does not impede structural
regularity much, i.e., they can be converted into crystalline material upon
mechanical orientation) generally demonstrate relatively excellent resistance
to oxidative degradation as well as to chemical and hydrolytic degradation.

Branched versus Linear Structures

The presence of branching in a macromolecule (generally a side group
greater than two carbon atoms in length) generally leads to diminished
thermal stabilities. If we assume that relative thermal stability of a polymer
is related to the temperature of half-decomposition T_h (the temperature
at which 50 percent weight loss occurs on heating the polymer for about
45 min), then it can be stated that as the length of the branch increases,
the thermal stability of the polymer decreases [13]. In the following are
listed polymer and T_h (°C), respectively: polymethylene (415), polyethylene (406), polypropylene (387), and polyisobutylene (348). A similar
variation occurs in the perfluorocarbons: i.e., polytetrafluorethylene (509),
polyperfluoropropylene (297), and polyperfluoroheptene (248).

Besides affecting the value of T_h, branching may also cause decomposition to proceed with a decreasing conversion rate as the percent conversion
increases [14]. Thus, linear polypropylene (prepared from Ziegler-Natta
catalysts) exhibited a maximum volatilization rate (Fig. 6.1) when pyrolyzed at 375°C. However, a branched polypropylene (prepared with an
$AlBr_3$-HBr catalyst system) no longer exhibited such a maximum when
pyrolyzed at 365°C. Among factors which may contribute to this effect of
branching in polyolefins may be: (1) an increase in intermolecular transfer
would bring about more immediate fragmentation of the polymer molecules
into small volatile products (thus, an increase in monomer production was

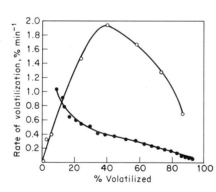

**FIG. 6.1 Pyrolysis of linear and
branched polypropylene:** ○ **linear
at 375°C,** ● **branched at 365°C [14].**

observed for the branched polypropylene) and (2) the branches themselves increase the number of potential short fragment chain ends that
would volatilize upon scission. (Intermolecular transfer would also be
facilitated by the presence of polymer radicals possessing tertiary carbon
atoms in their chains.)

Branched polymers tend to oxidize more readily than linear ones. Thus, Willbourn [15] obtained a quantitative correlation between rate of oxidation of branched polymethylenes and their degree of branching. As the latter increased, the oxidation rate increased. Similar trends have been also observed by others [9, 16].

The effects of branching upon polymer (e.g., a polyolefin) oxidation have been attributed to steric factors, which tend to weaken intra- and inter-molecular forces, and to the presence of tertiary carbons as a result from branching. In the latter case, hydrogens bonded to tertiary carbons would be expected to be more susceptible to chemical attack than hydrogens bonded to primary or secondary carbons. Steric factors may be illustrated by reference to polyisobutylene. Although polypropylene oxidizes more readily than polyethylene, polyisobutylene is more stable to oxidation than polyethylene. This may be due to the shielding (steric) effect of the inert methyl groups present.

Crosslinking Effects

Flexibility in chains may be diminished through crosslinking them by means of reactive agents that are capable of forming covalent bonds between the chains. Chain stiffening that accompanies crosslinking endows the resulting crosslinked polymer with higher softening temperature, coupled with increased thermal and chemical stability. The extent of stiffening depends on such factors as crosslinking density (number of crosslinks per unit weight or volume) and the nature of the crosslinks, e.g., length, chemical structure, flexibility versus rigidity.

Rubber is a well-known example of a crosslinked polymer, where the long flexible chains are crosslinked chemically, as with sulfur. The resulting rubber is more rigid and of higher softening temperature than the uncrosslinked chains. Short crosslinks between the carbon atoms of two chains endow greater stiffness than longer links. As for the chemical nature of the crosslinks, carbon-carbon crosslinks tend to be more stable to thermal and chemical attack than sulfur-sulfur linkages. The effect of the crosslinking agent divinylbenzene on polystyrene may also be cited. For example, as the amount of divinylbenzene in this copolymer is increased, from 0 to 56 percent, the value of T_h, half-decomposition temperature, increases from about 364 to 401°C [17].

In addition to the type of crosslinking obtained as the result of a reactive additive, another type of crosslinking may be said to exist in polymeric structures. The latter arises from the chemical nature of the starting materials used in the polymer synthesis. Thus, we may consider "ladder" (IV) and "parquet" (V) polymers to be special cases of crosslinked polymers:

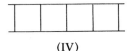

(IV)

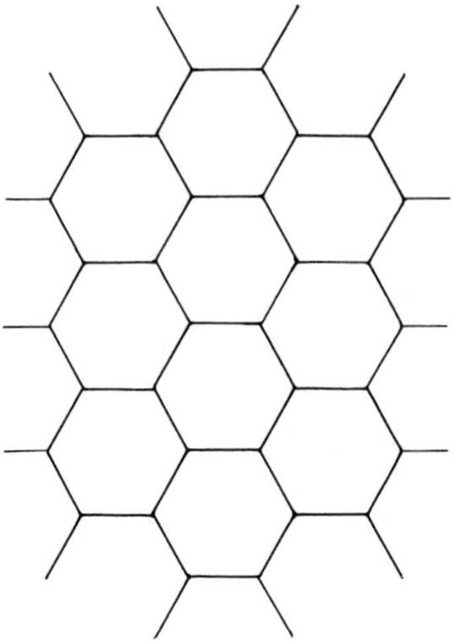

(V)

Van Deusen [18] synthesized a ladder-type polymer (pyrrone) (VI) as shown below:

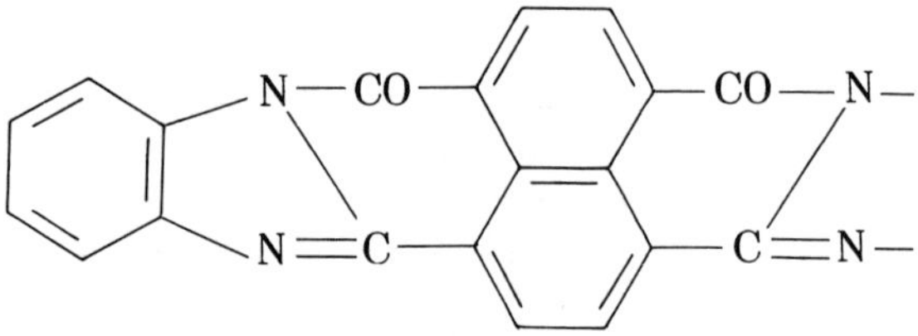

(VI)

This type of polymer has excellent thermal (little decomposition up to 600°C) and oxidative (little weight loss up to 400°C) stability. Many other such polymers have been described [19, 20]. Besides the effects of "crosslinks" in such polymers, these materials exhibit high thermal and oxidative stability by virtue of the need of breaking both the two bonds opposite each other and between the same crosslinks [cf. (IV)] in order for the molecular weight to decrease.

Whereas ladder polymers may be described as essentially crosslinked two-strand polymers, parquet polymers may be similarly described as crosslinked multiple-strand polymers [cf. (V)]. A typical representative of parquet polymers is graphite. It consists of fused six-membered carbon rings. The planar networks of carbon atoms form parallel layers. Such polymers are extremely refractory and are used in the manufacture of electrodes. A considerable number of parquet polymers are found among the silicates, e.g., talc. A parquet-type polymer has also been obtained [21] by the reaction of copper acetyl acetonate with tetracyanoethylene under vacuum [see structure (VII)]. This polymer is an infusible black product, of low solubility, chemically stable, and capable of withstanding prolonged heating at 500°C without degradation.

(VII)

Although crosslinking generally results in polymers having thermal, oxidative, and chemical stability, the oxidative stability of the polyolefin, polyethylene, was bound to decrease as the degree of crosslinking increased. Thus, Winslow and coworkers [8, 22] reported that oxidative crystallization (cleaved oxidized chains relax into more orderly arrangements at a much faster rate than the cleavage reaction) was hindered by radiation-induced crosslinkage in high-molecular-weight linear polyethylene. As a consequence, crosslinking increased the cumulative oxygen consumption and decreased the ultimate degree of crystallinity in polyethylene (see Fig. 6.2).

Copolymerization

The melting point of a polymer can be affected by the reduction of regularity with which monomer groups are spaced along the backbone chain, as in random copolymerization. It can be seen from Fig. 6.3 that when the melting points of random copolymers of nylon 610 and nylon 66 were plotted against composition material, a minimum melting point was obtained [23]. By disturbing the regularity of amide group spacings, intermolecular forces are diminished, and the melting temperature is decreased [cf. Eq. (1)]. Block copolymers can often be synthesized which are of higher melting point than the corresponding random copolymer, e.g., copolymers prepared from ethylene glycol terephthalate and ethylene glycol sebacate [23]. Besides affecting intermolecular forces and backbone rigidity, the presence of units differing in structure in copolymers often interferes with the cleavage of molecules during thermal degradation.

Thermal stability is thereby enhanced. A case in point is represented in Eq. (3):

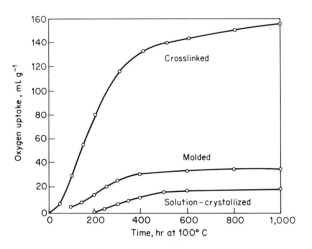

$$(3)$$

When the silaphenylene-siloxane copolymer undergoes thermal degradation, six-membered rings are split off. This process proceeds along the chain containing consecutive dimethylsiloxane units. However, when a silaphenylene moiety is reached, such ring formation is no longer possible, and

FIG. 6.2 Effect of morphology on the oxidation behavior of linear polyethylene at 100°C [8].

degradation is deterred. Thus, the thermal stability of the copolymer was found to be much higher than that of the homopolymer, polydimethylsiloxane. Generally speaking, the higher melting point and thermal stability of the copolymers indicates an enhancement in intra- and intermolecular forces and/or backbone rigidity, with a corresponding increase in chemical and oxidative stability.

Structural Order, Tacticity, and Mechanical Orientation

It was previously indicated (see Backbone Rigidity, page 296) that highly aromatic groups present in a polymer tend to increase chain stiffness. The

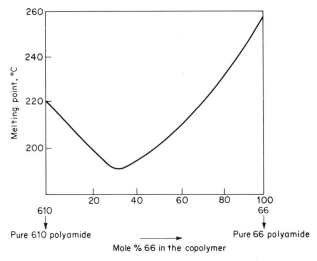

FIG. 6.3 **Melting points of polyamides as a function of the distance of the CO—NH— groups along the length of the back-bone chain [23] .**

melting point should increase (owing to a decrease in ΔS_f), provided that the groups do not greatly impede the intermolecular forces ΔH_f because of a decrease in structural regularity (or orientation). In short, the presence of the relatively bulky aromatic moieties should not inhibit the chain segments from fitting into a crystal lattice or prevent laterally bonding groups from approaching each other to a distance of optimum interaction (when no polar bonding groups are present, intermolecular attraction is due to dispersion forces, e.g., as in polyethylene). Obviously, the highest-melting-point polymers (crystals) would possess not only chain stiffness but also concomitant high chain regularity and orientation. The latter may often be obtained upon mechanical stretching of polymers possessing highly regular chains. The occurrence of crystallization upon stretching sufficiently regular elastomers has been demonstrated from x-ray and electron diffraction and from infrared absorption. It is thus apparent that mechanical orientation of regular chain structures should lead to polymers of (e.g., fibers) high termal, oxidative, and chemical stability.

Structural order of polymeric chains may also affect polymer stability. It was reported [24] that the initial thermal breakdown of polyvinyl chloride having head-to-head structures occurred more readily than that of polyvinyl chloride containing head-to-tail structures. Subsequently, however, the head-to-tail structures degraded more readily. This was explained as follows. Initially, the head-to-head structure degrades faster owing to lower thermal stability. However, after the initial degradation (release of HCl), a chloroprene-like structure forms containing halogen attached to a double-bonded carbon. This interferes with further chain dehydrochlorination, namely,

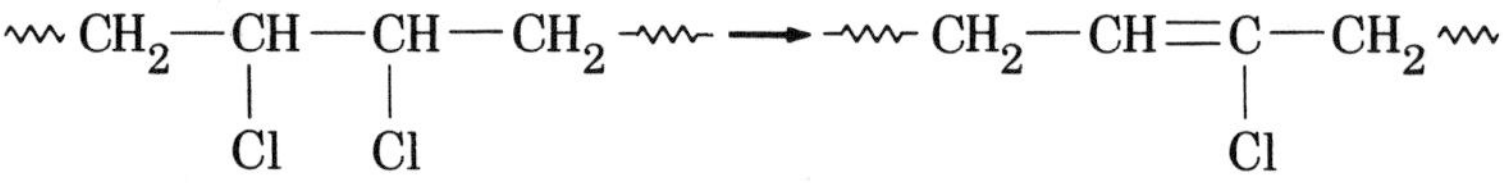

$$\text{\Large $\wedge\!\wedge$ } CH_2\!-\!\underset{\underset{\displaystyle Cl}{|}}{CH}\!-\!\underset{\underset{\displaystyle Cl}{|}}{CH}\!-\!CH_2\text{ $\wedge\!\wedge\!\wedge$ } \longrightarrow \text{ $\wedge\!\wedge\!\wedge$ } CH_2\!-\!CH\!=\!\underset{\underset{\displaystyle Cl}{|}}{C}\!-\!CH_2 \text{ $\wedge\!\wedge$}$$

$$(4)$$

Besides head-to-head and head-to-tail considerations, structural order may involve *cis* and *trans* structures (microstructure). Thus, when a butadiene molecule becomes the unit of a long chain, 1,4-*cis* (VIII); 1,4-*trans* (IX); and, 1,2- (X) structures may form, namely,

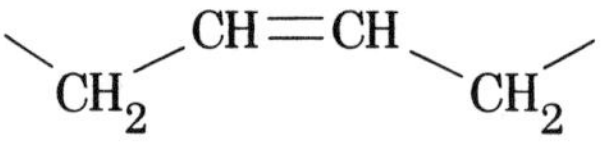

(VIII) 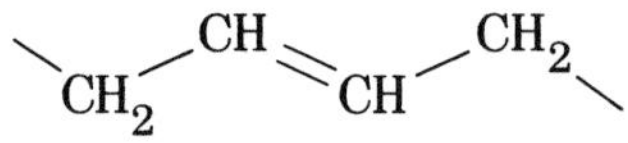   (IX)

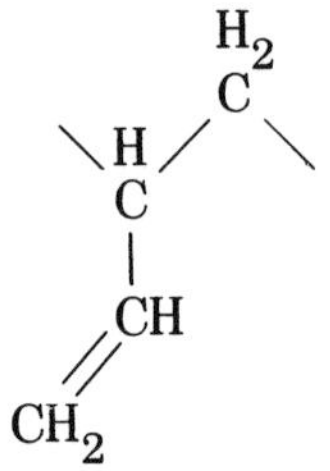

(X)

The three different structural arrangements (orders) shown above give noticeably different behavior. Thus, (VIII) is soft and crystallizes on pronounced mechanical stretching, and the crystalline phase has a low melting point. Structure (IX) is hard, crystallizes readily with stretching, and possesses a relatively high melting point.

Structure (X) may exist in an isotactic, a syndiotactic, or an atactic form. Generally, the stereoregular forms are rigid and crystalline while the atactic form is soft and noncrystalline.

Relative to tacticity, Kern and coworkers [25], for example, reported that the autoxidation of isotactic polypropylene and poly(butene-1) in solution showed a kinetic behavior different from that of the atactic fractions. It would not be expected that crystallinity would play an important role since the reaction occurred in solution. The different kinetic behavior was attributed to differences in spatial configuration between atactic and isotactic polypropylenes in solution (cf. Oxidation Products, page 334). However, Bawn and Chaudhri [26] have recently indicated that a kinetic interpretation, which does not include spatial configuration, could lead to the results obtained by Kern et al. for the atactic polyolefins (cf. Ref. 1).

It may be instructive here to discuss some of the kinetic aspects of the results obtained by Kern et al. The basic autoxidation scheme (BAS), for relatively high oxygen pressures, may be written as

Production of R· rads. (5)

$$R\cdot \ + \ O_2 \ \xrightarrow{k_2} \ RO_2\cdot \tag{6}$$

$$RO_2\cdot \ + \ RH \ \xrightarrow{k_3} \ RO_2H \ + \ R\cdot \tag{7}$$

$$2RO_2\cdot \ \xrightarrow{k_6} \ Products \tag{8}$$

When steady-state conditions obtain and when essentially all the oxygen absorbed is present as hydroperoxide (high kinetic chain lengths), we may write

$$-\frac{d[O_2]}{dt} \ \equiv \ \rho \ = \ k_3[RH][RO_2\cdot] \tag{9}$$

and

$$R_i \ = \ k_6[RO_2\cdot]^2 \tag{10}$$

From Eqs. (9) and (10),

$$\rho \ = \ \frac{k_3[RH]R_i^{1/2}}{k_6^{1/2}} \tag{11}$$

After sufficient RO_2H has accumulated, the decomposition of RO_2H will be the predominant initiation step. Thus, when the initiation step (5) involves a unimolecular decomposition, $\rho \propto [RO_2H]^{1/2} \propto [O_2]_a^{1/2}$, and when a bimolecular reaction is involved, $\rho \propto [RO_2H] \propto [O_2]_a$; where $[O_2]_a$ is the concentration of oxygen absorbed. For atactic polypropylene (random configuration), RO_2H groups will not be consistently vicinal to one another, but this will be the case for isotactic polypropylene [cf.

structure (XI)] ,

$$
\begin{array}{ccc}
\mathrm{CH_3} & \mathrm{CH_3} & \mathrm{CH_3} \\
| \quad \mathrm{H} & | \quad \mathrm{H} & | \quad \mathrm{H} \\
\text{-\!\!\!\wedge\!\!\!\wedge-\!C-C-C-C-C-C-\!\!\!\wedge\!\!\!\wedge-} \\
\mathrm{H} \quad \mathrm{H} \quad | \quad \mathrm{H} \quad | \quad \mathrm{H} \\
\mathrm{O_2H} \quad \mathrm{O_2H}
\end{array}
$$

(XI)

Therefore, it may be expected that the spatial configuration (or tacticity) of polypropylene would affect the oxidation kinetics in solution; i.e., $\rho \propto [O_2]_a^{1/2}$ for the atactic form and $\rho \propto [O_2]_a$ for the isotactic form.

Ring Structures versus Aliphatic Moieties

When a chain backbone is comprised of essentially large aromatic groups, chain stiffening can result because of chain inflexibility, resonance effects (which are enhanced by coplanarity), steric factors, etc. When the aromatic groups do not cause a large decrease in intermolecular chain forces, then a higher-melting polymer should obtain which will probably be more resistant to thermal, oxidative, and chemical degradation. However, this type of polymer may be more susceptible to radiative (UV) degradation (cf. Backbone Rigidity, page 296). Aliphatic moieties will therefore be less stable than aromatic groups. Furthermore, aliphatic moieties provide hydrogen atoms which can undergo thermal, oxidative, and chemical reactions more readily. A case in point is polyethylene versus polybenzyl. Thus, polyethylene has a half-life decomposition temperature T_h of 406°C whereas T_h for polybenzyl is about 430°C [17].

As may also be expected, the nature of the bridging groups joining arylene units in polymers can have a substantial effect on polymer stability. It has been found that thermal stability of polymers containing aromatic units in the chain decreases according to the nature of the bridging group in the following order:

$$
-\!\!\!\left\langle\!\!\bigcirc\!\!\right\rangle\!\!->-\mathrm{CH_2}->-\mathrm{O}->-\mathrm{SO_2}-
$$

Besides replacing aliphatic by aromatic groups in the chain unit to increase thermal stability, an increase in thermal stability can also be achieved by introducing heterocyclic groups of aromatic character into the chain unit to yield such polymers as polybenzimidazoles and polypyrrones.

Isomeric and Steric Effects

The type of aromatic substitution, e.g., meta or para isomers, can affect polymer stability. Thus, if T_2 is denoted as the temperature at which a polymer sample loses half its weight in 2 hr, then T_2 = 650°C for poly-p-phenylene whereas it is 540°C for poly-m-phenylene [13]. Also, Krasnov and coworkers [27] studied the thermal stability of isomeric aromatic polyamides, i.e., poly-m-phenyleneisophthalamide (PMPI), poly-m-phenyleneterephthalamide (PMPT), poly-p-phenyleneisophthalamide (PPPI), and poly-p-phenyleneterephthalamide (PPPT). Based on dynamic thermogravimetric analysis (TGA), they concluded that the polyamides derived from terephthalic acid were more thermally stable and that a similar effect was also observed for the polyamides containing p- and m-phenylenediamines. In the latter case, the para isomer was more stable than the meta isomer. The polyamides could be arranged in the following order of decreasing heat stability:

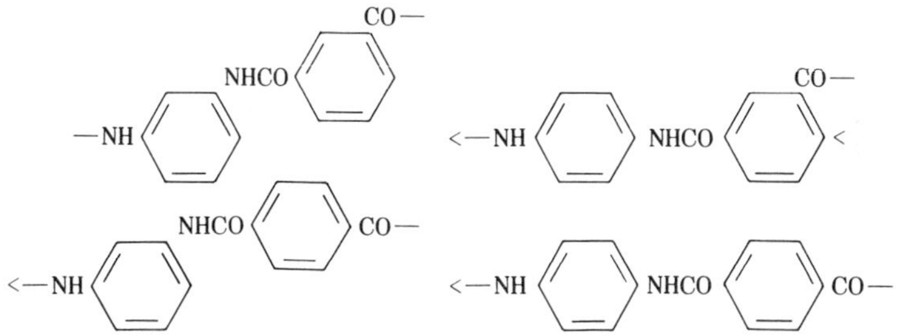

Isomeric effects may be due to a combination of various factors, e.g., resonance, steric factors. If an isomeric structure results in a lowering of the resonance effect (meta versus para isomer) and/or an increase in steric effects, then greater chain mobility would be anticipated owing to a decrease in intermolecular forces and chain rigidity. Thus, the polyamide of lowest thermal stability, PMPI, showed intense molecular mobility based on nuclear magnetic resonance data in constrast to the other isomeric polyamides. Other workers [28, 29] have observed similar results. For example, when poly-Schiff bases containing azulene [28] were examined for thermal stability, structure (XIII) was more thermally stable than (XII):

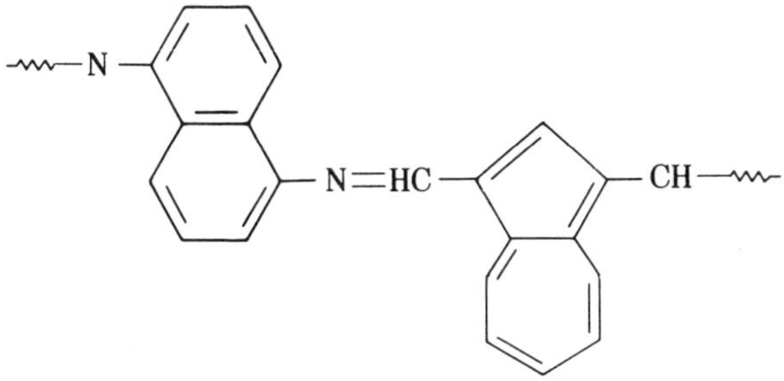

(XII)

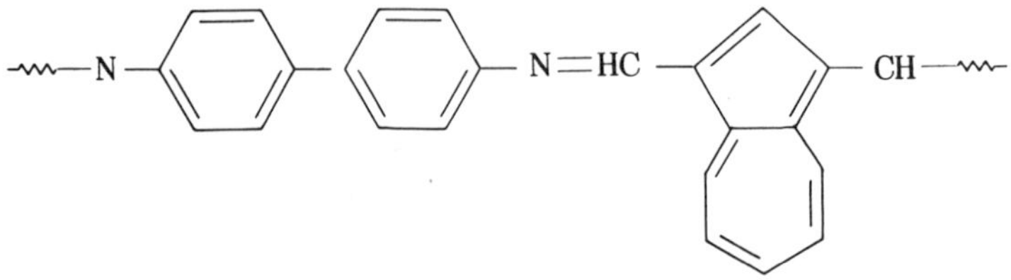

(XIII)

This could be attributed to steric factors.

Volpe and coworkers [29] also investigated isomeric effects for some dibenzoylbenzene/diamine polymers, structure (XIV):

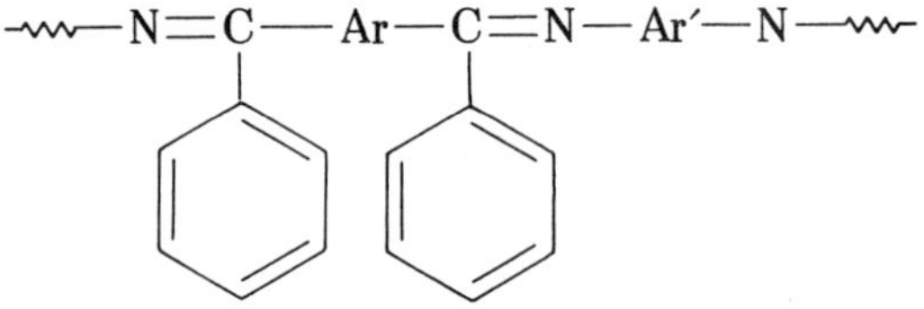

(XIV)

where

$$Ar =$$

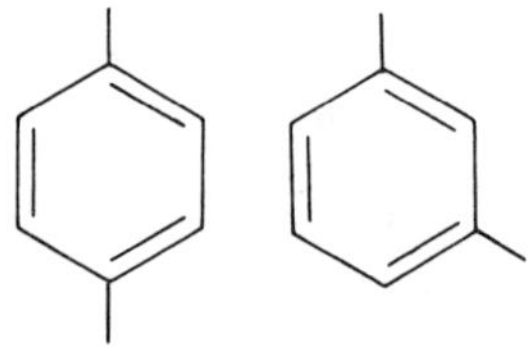

$$Ar' =$$

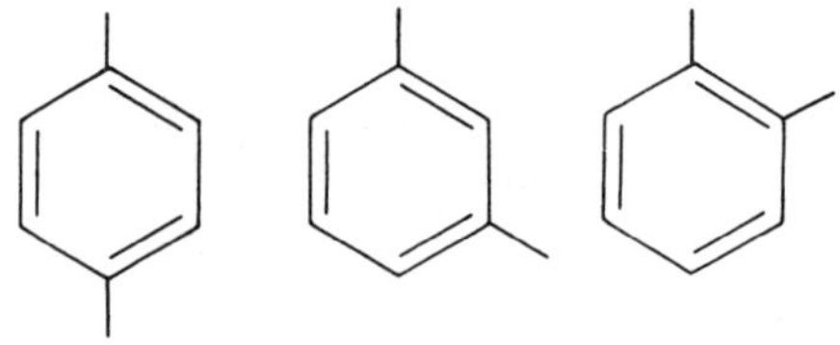

They found that the para-para (Ar—Ar') polymer was more resistant to thermal and oxidative degradation than the para-meta and para-ortho isomers.

Weak Links and End Groups

The behavior of the thermal degradation of some addition polymers, as polystyrene, has been attributed to the presence of weak links by various workers such as Jellinek [30] and Grassie [31]. This concept was advanced to explain the rapid initial decrease in molecular weight during polystyrene decomposition. On the other hand, others [32] have suggested that this behavior was due to intermolecular transfer in which polymer radicals abstracted tertiary hydrogen atoms from polystyrene chains with chain scission subsequently occurring at these points.

The presence of weak links has been ascribed to various causes: the incorporation of oxygen into the polymer molecules, trace contaminants, etc. Strong arguments have been advanced for the weak-link and inter-molecular-transfer concepts. However, it may be apropos here to mention a relatively simple mathematical treatment recently presented by Mac-Callum [33] which tends to discredit the weak-link hypothesis. In the fol-lowing it is assumed that scission of both weak and normal links occurs at random in the polymer molecules. Let P_0 = initial chain length; $N_n{}^o$ and and $N_w{}^o$, the number of normal and weak interunit links, respectively, at the start of degradation. Assuming that scission is a first-order reaction, then

$$-\frac{dN_n}{dt} = k_n N_n \tag{12}$$

and

$$-\frac{dN_w}{dt} = k_w N_w \tag{13}$$

where k_n and k_w are rate constants and $k_w \gg k_n$. From Eqs. (12) and (13),

$$N_n = N_n{}^o e^{-k_n t} \tag{14}$$

and

$$N_w = N_w{}^o e^{-k_w t} \tag{15}$$

Also, we may write (the fraction of weak links initially present is very small)

$$N_n{}^o + N_w{}^o \approx N_n{}^o \tag{16}$$

and

$$N_n{}^o = n_0 P_0 \tag{17}$$

where n_0 = total number of molecules initially present. Then, the total number of breaks B after time t may be written

$$B = N_w{}^o + N_n{}^o \left(1 - e^{-k_n t}\right) \tag{18}$$

or the number of breaks per original molecule becomes

$$\frac{B}{n_0} = P_0 \left(\frac{N_w^o}{N_n^o} + 1 - e^{-k_n t} \right) \tag{19}$$

which may be written approximately as

$$\frac{B}{n_0} \simeq P_0 \left(\frac{N_w^o}{N_n^o} + k_n t \right) \tag{19a}$$

Since $B/n_0 \cong P_0/P_t - 1$, Eq. (19a) becomes

$$\frac{1}{P_t} - \frac{1}{P_0} \simeq \frac{N_w^o}{N_n^o} + k_n t \tag{20}$$

It can be seen from Eq. (20) that if a plot of $1/P_t - 1/P_0$ versus t affords an intercept, the presence of weak links is indicated. MacCallum carried out such plots for various addition polymers, e.g., polystyrene and polyethylene (cf. Fig. 6.4). From this figure and Eq. (20), the presence of weak links is apparent. However, it would also have to be concluded that the concentration of weak links is a function of the degradation temperature—which is an untenable conclusion.

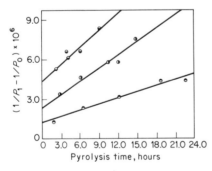

FIG. 6.4 Polystyrene. Change in reciprocal number-average molecular weight as a function of time. Temperature of degradation; ◕ 280°C, ◑ 290°C, ◒ 298°C [33].

Although the concept of weak links during degradation of addition polymers is debatable, this concept appears to be more firmly established in the case of the degradation of condensation polymers. Thus, Gaudiana and Conley [34] studied the oxidation of model compounds of thermally stable aromatic heterocyclic polymers (polybenzimidazolones (PBA), poly-benzimidazoles (PBI), and polybenzimides (PBIM). The oxidative pyrolysis products obtained from PBA and its model compounds were carbon dioxide, carbon monoxide, water, and trace amounts of cyanogen. No condensable degradation products (excluding water) could be detected. The degradation products from model compounds of PBI and PBIM were identical except that aromatic condensable products were detected. However, in every case examined, only the *acid-derived* portions of the molecules were isolated as degradation products. Attempts to trap condensates which originated from the amine-derived portions of the molecules were futile, despite expectations to the contrary. The failure to obtain nitrogen-containing fragments from the amine residues of the PBI and PBIM model compounds, regardless of their position in the molecule, offered convincing evidence that the initial stages of oxidative attack occur preferentially on benzenoid rings bearing the nitrogen function. Thus, the weak link in these systems is presumably the nitrogen-containing heterocyclic and adjacent benzenoid rings, e.g.,

Krasnov and coworkers [27] studied the thermal degradation of isomeric aromatic polyamides and postulated that during the degradation, hydrolysis of amide bonds occurred preferentially with subsequent decarboxylation of carboxyl end groups. Further, Teleshov and coworkers [35] studied the formation of polybenzimidazopyrrolones from polyamido acids. They concluded that the uncyclized amide units (from the polyamido acids) were the weak parts of the polymer chains subjected to thermal degradation. Considerable amounts of carbon dioxide formed, presumably as a result of the hydrolytic decomposition of amide bonds, with subsequent decarboxylation of free carboxyl groups.

Johnston and Gaulin [36] studied the thermal degradation of various model polyimides in vacuo, e.g., *N*-phenylphthalimide. The following mechanism was postulated:

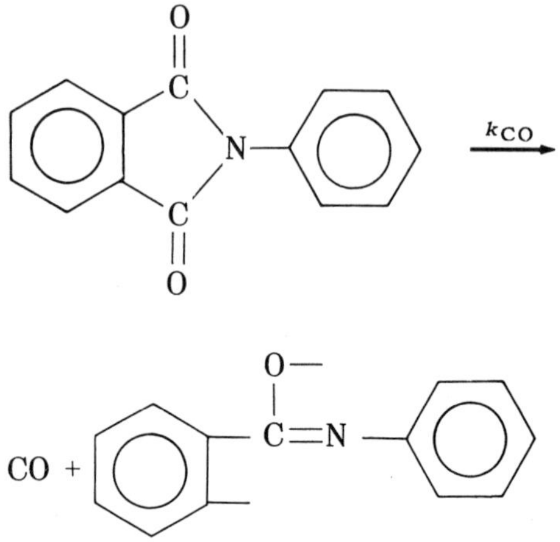

$$CO + (A)$$

(22)

$$A \xrightarrow{k_{1a}} CO + char + HCN + H_2 + minor\ products \qquad (23)$$

$$A + free\ radicals \xrightarrow{k_{1b}} Stabilized\ oxygenated\ product\ (B) \qquad (24)$$

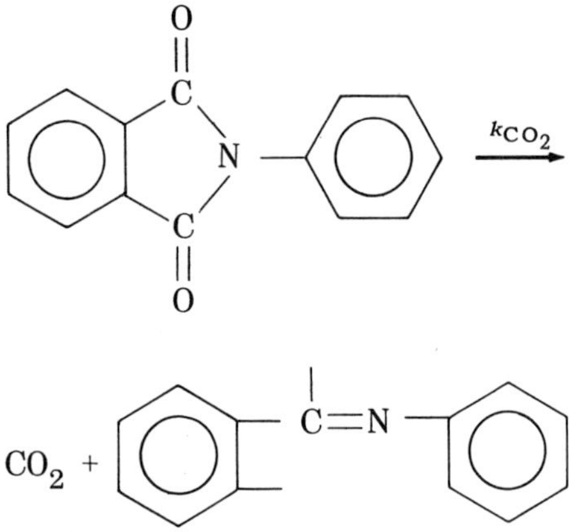

$$CO_2 + (C)$$

(25)

As previously indicated for the degradation of other thermally stable polymers, carbon dioxide and carbon monoxide were important products. From Eqs. (22) to (25),

$$-\frac{d[I]}{dt} = k_{CO} + k_{CO_2}[I] \qquad (26)$$

or

$$k_{CO} + k_{CO_2} = \frac{\ln[I_0]/[I]}{t} \tag{27}$$

where $[I]$ = imide concentrations. Also,

$$\frac{d[CO_2]}{dt} = k_{CO_2}[I] \tag{28}$$

or

$$[CO_2] = k_{CO_2} \int_0^t [I]\,dt \tag{29}$$

Upon substituting Eq. (27) into (29) and integrating,

$$[CO_2] = \frac{k_{CO_2}}{k_{CO} + k_{CO_2}} [I_0] \left\{ 1 - \exp\left[-\left(k_{CO} + k_{CO_2}\right)t \right] \right\} \tag{30}$$

$$[CO_2] = \frac{k_{CO_2}}{k_{CO} + k_{CO_2}} ([I_0] - [I]) \tag{30a}$$

Substituting Eq. (27) into (30a), we obtain

$$\frac{[CO_2]\ln[I_0]/[I]}{[I_0] - [I]} = k_{CO_2}t \tag{31}$$

The ratio k_{CO}/k_{CO_2} can be evaluated from a plot of $[CO_2]$ versus imide decomposed [Eq. (30a)]. From Eq. (31), k_{CO_2} can be evaluated, and consequently the value of k_{CO}. In this manner, values of rate constants for various polyimide model compounds were obtained, as shown in Table 1.

From the scheme, Eqs. (22) to (25), a relationship may also be established between carbon monoxide liberated and the amount of imide decomposed. Thus, we may write

$$\frac{d[CO]}{dt} = k_{CO}[I] + k_{1a}[A] \tag{32}$$

Assuming steady-state conditions for the intermediate A,

$$k_{CO}[I] = (k_{1a} + k_{1b})[A] \tag{33}$$

and

$$\frac{d[CO]}{dt} = k_{CO}[I]\left(1 + \frac{k_{1a}}{k_{1a} + k_{1b}}\right) \tag{34}$$

Dividing Eq. (34) by (28) affords

$$\frac{[CO]}{[CO_2]} = \frac{k_{CO}}{k_{CO_2}}\left(1 + \frac{k_{1a}}{k_{1a} + k_{1b}}\right) \tag{35}$$

Substituting Eq. (30a) into (35) gives

$$[CO] = \frac{k_{CO}}{k_{CO} + k_{CO_2}}\left(1 + \frac{k_{1a}}{k_{1a} + k_{1b}}\right)([I_0] - [I]) \tag{36}$$

It can be seen from Eqs. (30a) and (36) that a plot of the amount of carbon dioxide, or carbon monoxide, liberated versus the amount of imide

TABLE 1   Values of Rate Constants for Various Polyimide Model Compounds [36]

Compound	Temperature, °C	Reaction rate (sec^{-1}) constants	Ratio: k_{CO}/k_{CO_2}
N-Phenylphthalimide	550	$k_{CO} = 5.99 \times 10^{-5}$ $k_{CO_2} = 4.23 \times 10^{-5}$	1.42
	575	$k_{CO} = 3.37 \times 10^{-4}$ $k_{CO_2} = 2.92 \times 10^{-4}$	1.15
(Extrapolated)	600	$k_{CO} = 1.55 \times 10^{-3}$ $k_{CO_2} = 1.8 \times 10^{-3}$	0.86
Phthalimide	600	$k_{CO} = 5.3 \times 10^{-5}$ $k_{CO_2} = 2.12 \times 10^{-4}$	0.25
N-Methylphthalimide	600	$k_{CO} = 2.69 \times 10^{-4}$ $k_{CO_2} = 8.73 \cdot \times 10^{-5}$	3.06
Skybond 700.	550	- - - - - - - - - - - - -	0.67

decomposed should afford a linear relationship, as was observed. Since the intermediate C [Eq. (25)] could not be detected during the formation of carbon dioxide, the decomposition of phthalimide was investigated in order to shed more light on the formation of carbon dioxide. Since the ratio of

$k_{CO}/k_{CO_2} \approx 0.25$, the formation of carbon dioxide should be predominant,

$$\tag{37}$$

Thus, benzonitrile was a major product formed, as anticipated, and the formation of carbon dioxide appears to arise from the imide ring. This tends to discredit conclusions reached by Bruck [37] on the thermal degradation of a polyimide (H-film from Du Pont). Bruck claimed that the relatively large amounts of carbon dioxide liberated arose essentially from impurities (uncyclized polyamide carboxylic acids) present in the polymer. In this connection, it may be mentioned again that Teleshov and coworkers [35] also concluded that large amounts of carbon dioxide, which formed during polybenzimidazopyrrolone degradation, originated from uncyclized amide units used in forming the polymer, namely,

$$\tag{38}$$

However, in this case, carbon dioxide should only arise from the uncyclized polyamido acid precursor of the final polymer product.

In addition to the presence of weak links, end groups may affect polymer stability. Thus, the thermal stability of polyformaldehyde depends upon the nature of the terminal groups. Polyformaldehyde possessing hydroxyl end group commences to degrade at about 170°C, whereas the polymer with acetyl end groups decomposes at about 200°C [38] [cf. Eqs. (39) and (40)]. Presumably, this happens because the degradation occurs by terminal initiation followed by similar depolymerization. The less reactive acetyl group makes terminal initiation more difficult, namely,

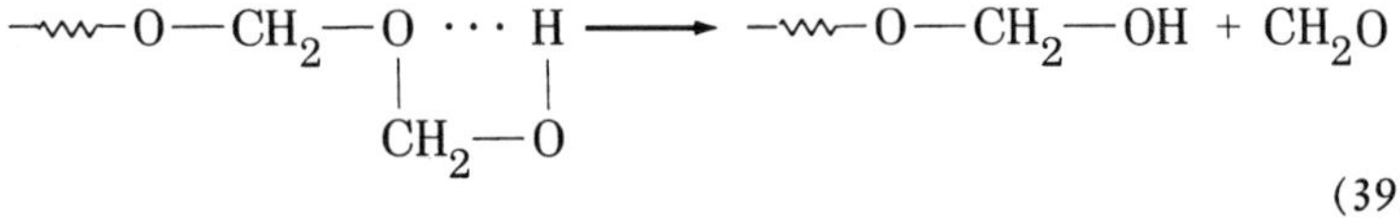

$$\text{—w—O—CH}_2\text{—O} \cdots \text{H} \longrightarrow \text{—w—O—CH}_2\text{—OH} + \text{CH}_2\text{O}$$
$$\qquad\qquad\quad\underset{|}{\text{CH}_2\text{—O}}$$

(39)

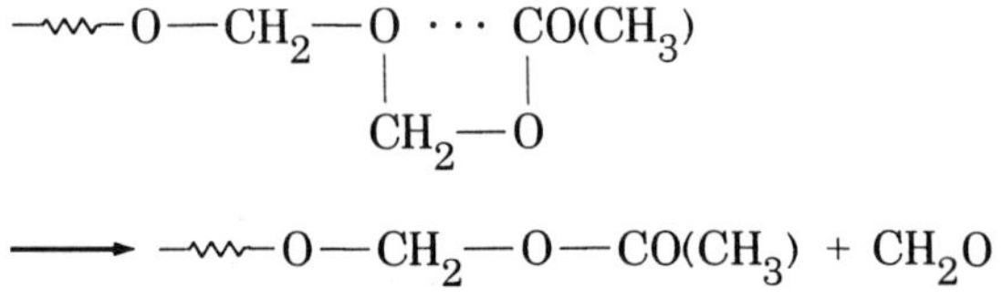

$$\text{—w—O—CH}_2\text{—O} \cdots \text{CO(CH}_3)$$
$$\qquad\qquad\quad\underset{|}{\text{CH}_2\text{—O}}$$

$$\longrightarrow \text{—w—O—CH}_2\text{—O—CO(CH}_3) + \text{CH}_2\text{O} \qquad (40)$$

Presence of Reactive Groups

The low thermal stability of some polymers may be attributed to the presence of reactive moieties in their chain units, e.g., —OH, —Cl, RCO—, which may be eliminated from the molecule in the form of water, hydrogen chloride, alcohol, ammonia, etc. Thus, in the case of poly(vinyl chloride),

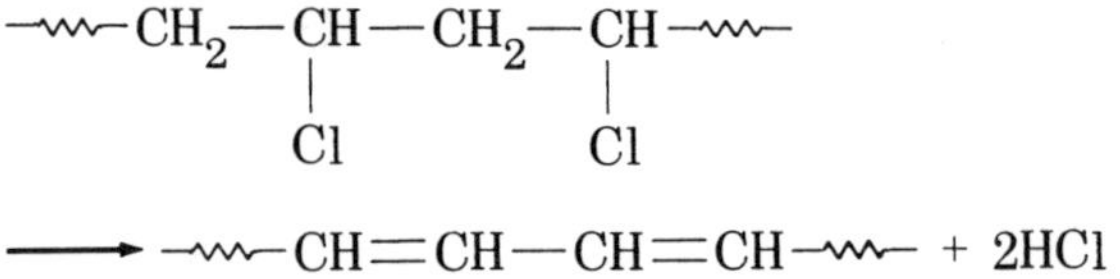

$$\text{—w—CH}_2\text{—CH—CH}_2\text{—CH—w—}$$
$$\qquad\quad\underset{|}{\text{Cl}}\qquad\qquad\underset{|}{\text{Cl}}$$

$$\longrightarrow \text{—w—CH}{=}\text{CH—CH}{=}\text{CH—w—} + 2\text{HCl} \qquad (41)$$

On the other hand, the presence of some reactive groups in the polymer molecule can lead to greater thermal stability. For example, when poly-methylacrylonitrile was heated in the presence of small amounts of acid units (initiators of the reaction), the polymer color changed from yellow to red, the molecular weight did not change much, and no volatile material was produced [39]. Also, infrared measurements indicated the disappearance of C≡N groups. The following reaction mechanism was postulated:

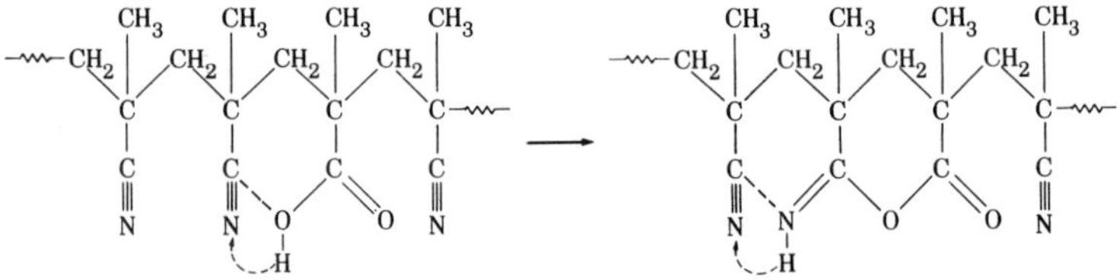

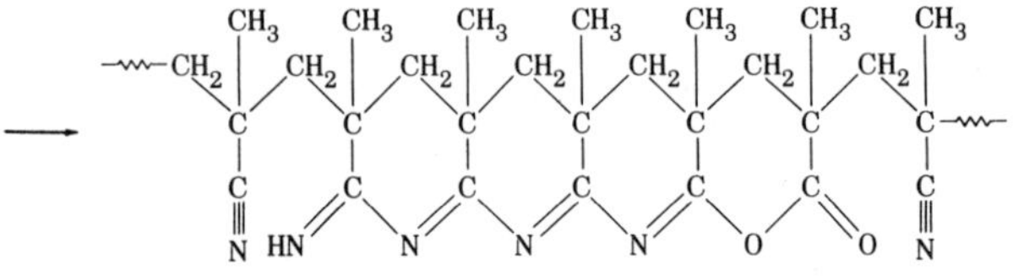

(42)

The changes which occur in the infrared spectrum of polyacrylonitrile as it is heated are essentially similar to those which occur for polymethylacrylonitrile. A major difference is that although the rate of cyclization for polymethylacrylonitrile is closely proportional to the acid concentration, reducing to zero in polymer free of acid, there is a large residual rate for polyacrylonitrile at zero acid content. This has been attributed to the presence of tertiary C—H structures in polyacrylonitrile which can act as internal initiators in much the same way as carboxyl groups [cf. Eq. (42)]. At temperatures higher than about 270°C, the cyclized structure shown in Eq. (42) undergoes dehydrogenation to form aromatic products (ladder polymers consisting of nitrogen heterocycles).

Replacement of Carbon in Main Chain and Metal Chelates

In attempts to increase the thermal stability of polymers, numerous inorganic and semi-inorganic polymers have been prepared. In the former class may be cited the PON polymers which consist of phosphorus, oxygen, and nitrogen, namely,

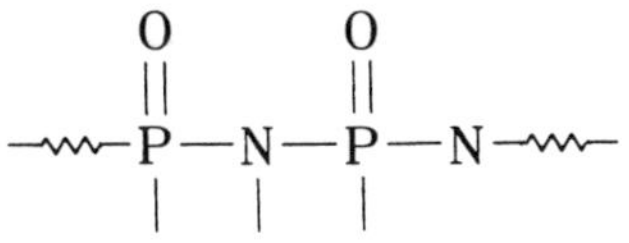

$$\begin{array}{ccc} O & & O \\ \| & & \| \\ \text{---}P\text{---}N\text{---}P\text{---}N\text{---} \\ | & | & | \end{array}$$

(XV)

In the latter class may be cited such polymers as the silicones, namely,

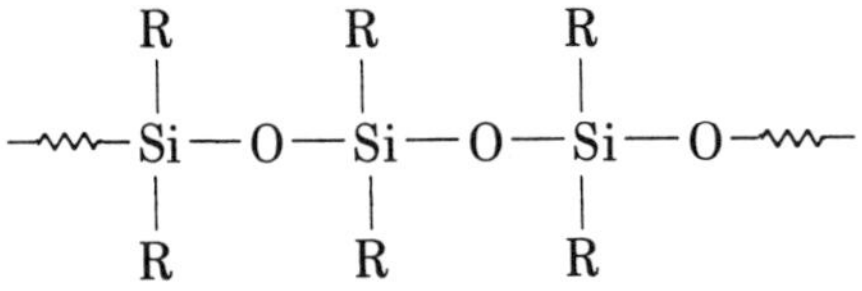

$$\begin{array}{ccc} R & R & R \\ | & | & | \\ \text{---}Si\text{---}O\text{---}Si\text{---}O\text{---}Si\text{---}O\text{---} \\ | & | & | \\ R & R & R \end{array}$$

(XVI)

where R = —CH_3, —C_2H_5, etc. Modified silicones may be formed by replacement of some of the Si by other elements, e.g., Al. In general, it was found [40, 41] that the presence of metal (other than Si) decreased the resistance to oxidation of polymers containing phenyl and vinyl groups. The opposite was found to hold for the polysiloxanes containing methyl and ethyl groups. For polymers containing phenyl and vinyl moieties, the polymer resistance to oxidation increased in the order Ti > Sn > Al; for polymers containing methyl and ethyl groups, the opposite was true. Furthermore, comparison of siloxanes containing Al or Ti atoms in the main chain showed that the Si—O—Ti bond is more stable to hydrolysis with aqueous hydrochloric acid than Si—O—Al bonds. Verkhotin and coworkers [42] found that when boron was incorporated into a siloxane chain, the thermal stability of the resulting polymer [see structure (XVII)] increased as the boron content rose from 0 to 0.20 percent,

$$
-\left(\begin{array}{c} CH_3 \\ | \\ Si-O- \\ | \\ CH_3 \end{array}\right) -B-O-\left(\begin{array}{c} CH_3 \\ | \\ -Si-O- \\ | \\ CH_3 \end{array}\right)-
$$

$$
\begin{array}{c} O \\ | \\ H_3C-Si-CH_3 \\ | \\ O \\ | \end{array}
$$

(XVII)

Thus, whereas the weight loss of polymer with 0 percent boron begins at 220°C, the weight loss of polymer containing 0.20 percent boron commences at about 300°C. Also, the introduction of a carborane residue (carborane is a $C_2B_{10}H_{10}$ polycyclic structure in which a nucleus of 10 boron atoms is bridged by 2 carbon atoms) into a polymer molecule can produce a large increase in thermal stability. When the carborane moiety was introduced into polydimethylsiloxane, much better thermal stability was achieved [43],

$$
\begin{array}{ccc} CH_3 & CH_3 \\ | & | \\ -\!\!\!\wedge\!\!\!\wedge\!\!\!-Si-O-Si-O-\!\!\!\wedge\!\!\!\wedge\!\!\!- \\ | & | \\ CH_3 & (CH_2)_n-C_2B_{10}H_{10} \end{array}
$$

(XVIII)

Some of the problems associated with inorganic-type polymers are: the difficulty in obtaining high-molecular-weight products that can be properly characterized, the reactivity of the inorganic elements in the chains, e.g., the ease of oxidative and hydrolytic degradation, and the tendency of the polymers to decompose into small rings at elevated temperatures [44].

Another class of heat-resistance polymers is metal chelates (or coordination polymers). Inoy [45] determined the thermal stability of a series of coordination polymers of various metals with bis-8-hydroxyquinol-5-ylmethane and found that the zinc chelate exhibited little weight loss (in nitrogen) up to about 400°C.

SOME ENERGY FACTORS AFFECTING POLYMER STABILITY

Resonance Effects

As previously indicated under Polymer Morphology, page 294, and Polymer Chain Structure, page 296, an increase in backbone rigidity usually leads to an increase in polymer thermal stability. This rigidity is enhanced by the presence of aromatic (versus aliphatic) moieties which possess high resonance energy. The higher the degree of conjugation in the main chain, the higher the resonance energy. Thus, in polyimides and polybenzimidazoles, where no atoms are present that are not part of an aromatic or heterocyclic ring, or are otherwise stabilized by resonance considerations, high thermal stability is achieved. Besides increased resonance effects, conjugation is associated with coplanarity, which implies a decrease in the degrees of freedom of the molecules involved, thereby leading to increased chain rigidity. The latter is often associated with increased polymer resistance to thermal, oxidative, and chemical degradation.

Rode and coworkers [46] investigated the thermal stability of the isomeric polymers, poly-1,2,4-oxadiazole (XIX) and poly-1,3,4-oxadiazole (XX):

(XIX)

(XX)

It was found that (XIX) began to lose weight in vacuo at about 220°C whereas (XX) began to lose weight at about 280°C. The workers utilized the Hückel molecular orbital approximation to estimate the π-electron resonance energies of 1,2,4-oxadiazole and of 1,3,4-oxadiazole. It was found that the total π-electron energy of the former compound was 10.29 β while that for the latter was 10.81 β units. This is in accord with the thermal stability of the polyoxadiazoles. The greater stability of (XX) was ascribed not only to the greater stability of the π-electron system but also to the larger conjugation energy for (XX) than for (XIX). Results obtained from the thermal oxidation of the isomeric polyoxadiazoles paralleled those from the thermal degradation in vacuo.

Intramolecular Forces

An increase in the strength of bonds between atoms generally results in increased polymer stability. Thus, since the carbon-fluorine bond strength involves an energy of about 116 kcal mole^{-1} and the carbon-hydrogen bond energy is about 97 kcal mole^{-1}, one might expect polytetrafluorethylene (Teflon) to be more thermally stable than polyethylene, as observed. However, when some of the fluorine is replaced by hydrogen, e.g., poly-(vinyl fluoride), the thermal stability of the polymer declines sharply owing to ready elimination of the hydrogen halide. Similar considerations apply to poly(vinyl chloride), poly(vinylidene chloride), poly(vinyl alcohol), and similar substances. The silicon-oxygen bond strength (106 kcal mole^{-1}) is greater than the straight-chain carbon-carbon bond (83 kcal mole^{-1}), and in this respect it is not surprising that silicon-based polymers are generally more stable than organic polymers with carbon backbones. Furthermore, the carbon-carbon bonds in aromatic ring systems are much more stable than carbon-carbon bonds in a linear chain configuration. Thus C_{arom}—C_{arom} bond strength is 100 kcal mole^{-1} compared with C_{aliph}—C_{aliph} strength of 83 kcal mole^{-1}. Other bond strengths are: C_{arom}—H (100 kcal mole^{-1}), C_{arom}—N (110 kcal mole^{-1}), and C_{arom}—O (107 kcal mole^{-1}), which indicate that ring systems are preferable in respect to good thermal stability.

Cox and coworkers [47] investigated the thermal degradation of various fluorine-containing polymers in vacuum. They observed that the most stable polymers were the fully fluorinated polytetrafluoroethylene (Teflon) and the copolymer of tetrafluoroethylene and hexafluoropropylene. Substitution of one of the fluorine atoms in Teflon by chlorine decreased the stability considerably (under similar conditions the Teflon began to decompose at a temperature which was higher by about 150°C), while substitution with a chlorine and a hydrogen atom caused a further decrease in stability (a drop in temperature of about 190°C). Cox and coworkers [48] also degraded fluorine-containing polymers in an oxygen atmosphere and found that the relative stabilities observed were similar to those observed in the absence of oxygen (in a vacuum). However, it should be mentioned here that the substitution of fluorine for hydrogen atoms in a polymer does not necessarily lead to increased thermal or oxidative stability. Thus, Volpe and coworkers [29] found that structure (XXI) was less stable than structure (XXII) when each was subjected to thermal and oxidative degradation (this illustrates the importance of mechanistic studies during degradation).

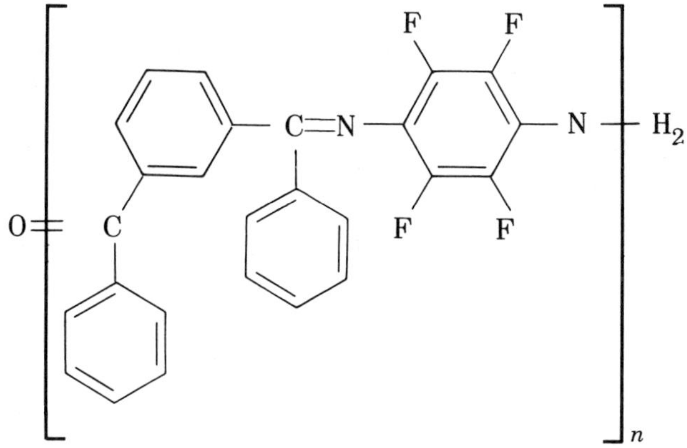

(XXI)

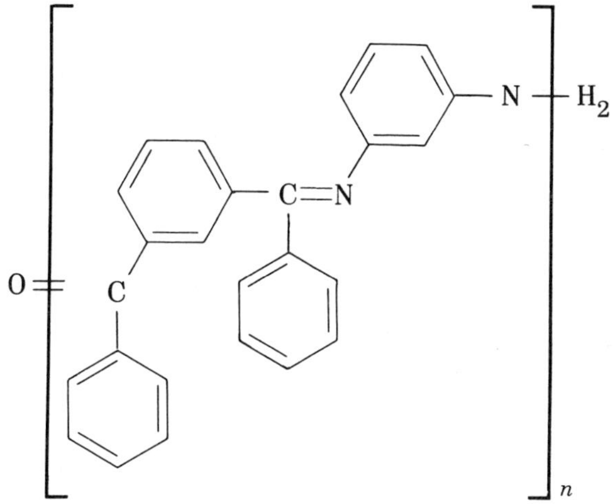

(XXII)

The strength of bonds in polymers can also influence the relative importance of transfer versus depropagation reactions. Thus, the much greater
strength of the carbon-fluorine bond over that of the carbon-hydrogen
bond can account for the greater role that transfer processes play during
polyethylene degradation than during Teflon degradation. When hydrogen
atoms are substituted for some fluorine atoms in Teflon, the predominant
decomposition reaction is no longer one of depropagation. Furthermore, if
a fluorine atom in Teflon is replaced by chlorine, e.g., as in polychlorotri-
fluoroethylene, the weaker carbon-chlorine bond allows transfer reactions
to occur, thereby reducing the monomer yield to about 28 percent. It
should also be mentioned here that the relative importance of depropagation and transfer is also influenced by the reactivity of the degrading

polymer radical as well as by the availability of reactive atoms (generally hydrogen atoms). The behavior of poly(α-methylstyrene) is typical of 1,1-disubstituted vinyl polymers. The degrading radical is relatively unreactive owing to its being trisubstituted and in an α-position relative to an unsaturated group. Abstraction of the primary and secondary hydrogen atoms available cannot compete effectively with depropagation, and monomer is thus a significant volatile product.

The dissociation energy D_{R-R} of polymers into radicals has been correlated with polymer stability [14]. When transfer reactions are absent during thermal degradation, it can be shown (see Chap. 4) that the rate of weight loss for processes involving random or terminal initiation, and depropagation and termination by disproportionation is

$$-\frac{dW}{dt} = f\left[k_d\left(\frac{k_i}{k_t}\right)^{1/2}\right] \tag{43}$$

where $-dW/dt$ = rate of weight loss; and k_d, k_i, k_t denote rate constants for depropagation, initiation, and termination, respectively. Thus, from Eq. (43),

$$E_{exp} = \frac{E_i}{2} + E_d - \frac{E_t}{2} \tag{44}$$

From the energy diagram in Fig. 6.5 and from Eq. (44), it can be shown that

$$D_{R-R} = E_i - E_t = 2(E_{exp} - E_d) \tag{45}$$

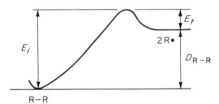

FIG. 6.5 Energy diagram for the dissociation of polymers into radicals.

where E_{exp}, E_i, E_t, and E_d denote activation energy observed for initiation, for termination, and for depropagation, respectively. Various values of E may be obtained by photochemical means. Thus, for photoinduced degradation,

$$E_{photo} = E_d - \frac{E_t}{2} \tag{46}$$

and for postirradiative degradation,

$$E_{post} = E_t - E_d \tag{47}$$

From Eqs. (45) to (47), values of E_i, E_d, and E_t can be estimated, and hence the value of D_{R-R}. In this manner, values of D_{R-R} were obtained for various polymers and presented in Table 2, along with the temperatures at which the polymers decompose (T_d).

TABLE 2   Dissociation Energies of Polymers into Radicals

Polymer	D_{R-R} (kcal mole^{-1})	T_d, °C
Polyethylene	<99	400
Polypropylene	<85	380
Polyisobutylene	<74	340
Polytetrafluorethylene	<70	≈500
Polymethyl methacrylate.	66	330
Polystyrene	64	360
Poly(α-methylstyrene)	60	290
Poly(trifluorchlorethylene).	54	≈300

It can be seen from the table that with the exception of Teflon thermal stabilities of various polymers (as judged by values of T_d) appear to correlate well with values of D_{R-R}.

Ceiling Temperatures

Another parameter which appears to correlate well with thermal stability of polymers (as measured by T_d) is the ceiling temperature T_c. In 1938 Snow and Frey [49] observed that the copolymerization of alkenes with sulfur dioxide would only occur below a fairly well defined temperature, which they termed *ceiling temperature*.

Dainton and Ivin [50] advanced a mathematical treatment for T_c as follows. If in addition to a propagation reaction [Eq. (48)], a depropagation reaction also occurs [Eq. (49)], then we may write [Eq. (50)]:

$$R_n^{\cdot} + M \xrightarrow{k_p} R_{n+1}^{\cdot} \tag{48}$$

$$R_n^{\cdot} \xrightarrow{k_d} R_{n-1}^{\cdot} + M \tag{49}$$

$$-\frac{d[M]}{dt} \equiv R = k_p[R \cdot][M] - k_d[R \cdot] \tag{50}$$

When $R = 0$, then

$$k_d = k_p[M]_e \tag{51}$$

where $[M]_e$ = monomer equilibrium concentration. Using Eq. (51) and the absolute rate theory,

$$\exp\left(\frac{\Delta S_d^*}{R_g} - \frac{E_d}{R_g T_c}\right) = \exp\left(\frac{\Delta S_p^*}{R_g} - \frac{E_p}{R_g T_c}\right)[M]_e \qquad (52)$$

where ΔS^* = entropy of activation and R_g = gas constant. Upon solving Eq. (52) for T_c,

$$T_c = \frac{E_p - E_d}{\Delta S_p^* - \Delta S_d^* + R_g \ln[M]_e} = \frac{\Delta H_p}{\Delta S_0 + R_g \ln[M]_e} \qquad (53)$$

where ΔH_p = heat of polymerization and ΔS_0 = standard entropy change of polymerization. Above T_c the polymer radical is relatively unstable and depropagation will be favored, whereas below T_c the polymer radical is relatively stable and propagation will be favored. Thus, the value of T_c will be some measure of thermal stability, as can be seen from Table 3.

In general, $T_d > T_c$ since the initiation of free radicals by dissociation of primary valence bonds would require more energy than the decomposition of radical bonds. The exceptional behavior of Teflon implies that carbon-carbon bonds in Teflon radicals are more stable than these bonds in the polymer itself. In any event, it should be noted that T_c is a purely

TABLE 3 Decomposition and Ceiling Temperatures [14]

Polymer	$T_d, {}^\circ C$	$T_c, {}^\circ C$
Polytetrafluorethylene	510	680
Polyethylene	400	400
Polypropylene	380	300
Polystyrene	360	230
Polymethyl methacrylate	330	220
Polymethacrylonitrile	>220	177
Polyisobutylene	340	50
Poly(α-methylstyrene)	290	7
Polyformaldehyde	>100	10–58

hypothetical stability index in many cases, e.g., for polyethylene, where monomer formation during decomposition is exceedingly low (<1 percent).

From ceiling-temperature measurements may be determined the heat of polymerization ΔH_p. Thus, employing the Arrhenius relationship and Eq. (50),

$$R = [\text{R·}] \left[A_0 \exp\left(-\frac{E_p}{R_g T}\right) [M] - A_d \exp\left(-\frac{E_d}{R_g T}\right) \right] \tag{54}$$

where A = frequency factor. Equation (54) leads to (cf. Ref. 50)

$$\frac{dR}{dT} = [\text{R·}] \left(k_p[M] \frac{E_p}{R_g T^2} - k_d \frac{E_d}{R_g T^2} \right) + (k_p[M] - k_d) \frac{d[\text{R·}]}{dT} \tag{55}$$

Near the ceiling temperature, $T \cong T_c$, and $k_p[M] = k_d$, and therefore

$$\lim_{T \to T_c} \frac{dR}{dT} = \frac{k_p[M][\text{R·}]}{R_g T_c^{\,2}} (E_p - E_d) = \frac{R_p \Delta H_p}{R_g T_c^{\,2}} \tag{56}$$

where $R_p = k_p[M][\text{R·}]$ and $E_p - E_d \approx \Delta H_p$.
Equation (56) may also be written as

$$\lim_{T \to T_c} \frac{d(k_p[M] - k_d)}{dT} = \frac{k_p[M] \Delta H_p}{R_g T_c^{\,2}} \tag{56a}$$

From Eq. (56), ΔH_p may be theoretically determined from the slope of line A in Fig. 6.6 (the dotted line denotes degree of polymerization as a function of temperature).

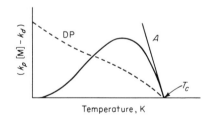

FIG. 6.6 Estimation of ΔH_p from Eq. (56).

The value of ΔH_p may also be obtained as follows. From Eq. (53),

$$\ln[M]_e = \frac{\Delta H_p}{R_g T_c} - \frac{\Delta S_0}{R_g} \tag{57}$$

From Eq. (57), a plot of ln $[M]_e$ versus $1/T_c$ should afford a linear relationship whose slope yields a value of ΔH_p. Such a plot was carried out from data obtained from the rate of pressure drop upon the irradiation of a mixture of gaseous methyl methacrylate with its solid polymer at various temperatures [51] (Fig. 6.7).

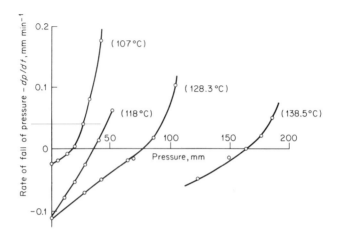

FIG. 6.7 Rate of fall of pressure on irradiation of a mixture of gaseous methyl methacryalate with its solid polymer at various temperatures. The field above the pressure axis corresponds to polymerization and that below to depolymerization.

Values of $[M]_e$ (or pressure P_e) at various temperatures can be obtained from Fig. 6.7 from the intersection of the various curves with the zero axis. A plot of $\ln P_e$ versus $1/T_c$ afforded a value of $\Delta H_p \approx 14$ kcal mole^{-1} (reported 13 kcal mole^{-1} [14]) for methyl methacrylate.

Intermolecular Forces

Owing to van der Waals attractive forces between molecules, every molecule in a liquid or solid possesses potential energy. The cohesive energy density (CED) may be defined as being equal to the potential energy of one cubic centimeter of material. In the following, CED will be briefly discussed along with its relationship to internal pressure P_i, which was employed in polymer morphology (page 294) in connection with intermolecular forces between polymer molecules.

The interaction force of attraction between two isolated molecules can often be represented by a Lennard-Jones expression (LJ) of the form,

$$\text{LJ attraction potential} = -\frac{\epsilon}{r^6} \qquad (58)$$

where r = distance between molecules and ϵ = interaction parameter. The half-sum of attractive pairwise interactions between a given molecule and all other molecules N at a distance r_i from a given molecule may be expressed as

$$\omega = -0.5\epsilon \sum_{r_i=0}^{r_i=\infty} Nr_i^{-6} \tag{59}$$

The potential energy of a mole of material may be expressed as $E = N_0\omega$ where N_0 = Avogrado number. It is necessary here to define a solubility parameter δ as

$$\delta^2 = \text{CED} = -\frac{E}{V} = -\frac{N_0\omega}{V} = \frac{\Delta H_v - RT}{V} \tag{60}$$

where ΔH_v = heat of vaporization per mole and V = molar volume. If we assume that CED is a measure of the strength of the interactions in a material and if hydrogen bonding is neglected, then dispersion D, dipole-dipole DD, and induction I energies must be considered. Expressions for each of these energies may be written for molecules S and P as (cf. Ref. 52)

$$D = -\frac{3I_s I_p \alpha_s \alpha_p}{2(I_s + I_p)r^6} \tag{61}$$

where I = ionization potential and α = polarizability,

$$DD = -\frac{2\mu_s^2 \mu_p^2}{3kTr^6} \tag{62}$$

where μ = dipole moment and k = Boltzmann constant, and

$$I = -\frac{\alpha_s \mu_p^2 + \alpha_p \mu_s^2}{r^6} \tag{63}$$

If the three types of energy are additive, we can write for the total interaction of two identical molecules [cf. Eqs. (58) and (61) to (63)]

$$\epsilon = \frac{2\mu^4}{3kT} + \frac{3I\alpha^2}{4} + 2\alpha\mu^2 \tag{64}$$

Furthermore, it can be seen that CED is proportional to ω [cf. Eqs. (59) and (60)]. The intermolecular forces as expressed by CED can be correlated with polymer properties. Thus, when CED is low (and the intermolecular forces are small) and the molecules possess relatively flexible chains, they have properties generally connected with elastomers. When the CED is relatively high, the corresponding polymers usually exhibit high stability, e.g., high thermal stability, especially where molecular symmetry is favorable for the formation of ordered crystalline regions. Some linear polymers and their corresponding values of CED (cal-cm^3) are: poly(ethylene terephthalate), 114; polyacrylonitrile, 237. (The corresponding values of the glass-transition temperatures T_g of these polymers are 69 and about 104°C.) It may be mentioned here that although the value of CED for polyethylene is only 62 cal-cm^3 ($T_g \cong -120°$C), its simple highly symmetric structure enhances its tendency to crystallize, thereby imparting to it characteristics of a plastic or fiber. Generally, strong intermolecular bonding requires the presence of polar or hydrogen-bonding groups, e.g., >CO, —CN, and ≡CCl. Thus, in poly(vinyl chloride) (CED = 91), the C—Cl dipoles increase the lateral CED of the system and with it the rigidity, softening temperature, and resistance against dissolution and swelling.

Although it is relatively easy to measure values of CED for simple liquids from heats of vaporization ΔH_v (the Clausius-Clapeyron equation may be employed), it is not possible to determine such values for polymers from values of ΔH_v because of their lack of volatility. Instead, a slightly cross-linked polymer is prepared, and samples are placed into a series of liquids of known values of δ. The δ of the solvent in which the polymer swelled the most is taken as the value of δ for the polymer.

It was previously indicated under Polymer Morphology (page 294) that a thermodynamic quantity referred to as internal pressure, $P_i = (\partial U/\partial V)_T$, was a measure of intermolecular forces. Although $\delta^2 = -E/V$ is not equivalent to $\partial U/\partial V$, it has been shown [53] that the numerical values of δ^2 and $\partial U/\partial V$ should be similar if not identical. Since $(\partial U/\partial V)_T \approx T(\partial P/\partial T)_V$

TABLE 4   Internal Pressure and Cohesive Energy Densities of Various Amorphous Polymers at 20°C [54]

Polymer	P_i, cal-cm^3	CED, cal-cm^3	$(P_i/CED) = n$
Polyethylene	80	62	1.3
Polystyrene.	110	76	1.4
Poly(ethyl acrylate)	105	88	1.2
Poly(dimethylsiloxane)	57	58	1.0

and $(\partial P/\partial T)_V = \alpha/\beta$, where α is equal to compressibility and β is the coefficient of thermal expansion, Allen and coworkers [54] determined values of P_i for various polymers, using values of α and β, and compared these values with values of CED (= δ^2) (Table 4). It can be seen from Table 4 that there is a fairly good correspondence between P_i and CED for the polymers listed. Besides the correspondence between P_i and δ^2 values, there also appears to be some correlation between δ and T_g. Thus, polymers of high δ tend to possess high values of T_g, as previously indicated for polyethylene, poly(ethylene terephthalate), and polyacrylonitrile. Hayes [55] obtained a good empirical relationship among molar cohesion (CED multiplied by molar volume), T_g, and polymer structure,

$$H_c = 0.5mRT_g - 25m \tag{65}$$

where H_c = molar cohesion and m = an empirical number obtained from the polymer structure. In this connection, it may also be mentioned that Allen and coworkers [56] have shown that the internal pressure is temperature-sensitive around T_g. Thus, rubbery polymers (above T_g) have values of $\partial U/\partial V \geqslant \delta^2$ whereas glassy polymers (below T_g) have values of $\partial U/\partial V < \delta^2$. Such behavior was attributed to the belief that when glassy polymers are compressed, the applied pressure has little effect on the intermolecular separation between nearest-neighbor segments since these configurations are frozen. However, this does not occur in rubbers where the interacting segments possess much mobility.

Until now, intermolecular forces have been attributed to dispersion, induction, and dipole-dipole interactions. Another important interaction to be considered is hydrogen bonding. The hydrogen bond strength usually lies between the strength of dipole interactions and that of chemical bonds. Hydrogen atoms can approach other suitable interacting units very closely (small size) so that dipole rotation is inhibited and the interaction energy between the aligned dipoles becomes relatively large. A very striking effect of lateral hydrogen bonding between regularly spaced groups is exhibited by linear polyamides whose —CO—NH— groups (amide) are responsible for such bonding. By introducing different paraffinic —CH$_2$— chains of different lengths between the amide groups, polymers may be prepared with various physical properties, e.g., the reduction of lateral hydrogen bonding by large distances between amide groups can lead to low-melting polymers. A method of estimating hydrogen-bonding interaction energies in polyamides has been reported by Slonimskii and coworkers [57] and is discussed in the following.

Substances selected for examination were aromatic polyesters and polyamides. In the former, there can be no hydrogen bonding as there can be in the latter. Further, it was assumed that hydrogen bonds are not static— they are continually formed and broken under equilibrium conditions. Thus, we may write

$$\Delta F = -RT \ln K = \Delta H - T \Delta S \tag{66}$$

where ΔF = change in free energy of the system, K = equilibrium constant, ΔH = enthalpy change, and ΔS = entropy change. Because the volume change of the system is small ($P \Delta V \rightarrow 0$), $-\Delta H = \Delta E$, where ΔE =

hydrogen bonding energy. Thus,

$$K = \frac{N_2}{N_1} = \exp\left(\frac{\Delta S}{R}\right) \exp\left(\frac{\Delta E}{RT}\right) \tag{67}$$

where N_2 = number of units of the molecule linked by hydrogen bonds, and N_1 = number of units not linked by hydrogen bonds. If it is further assumed that the increment in the softening point of a polyamide with hydrogen bonds ΔT compared with the softening point of a polyamide without hydrogen bonds T is proportional to the ratio N_2/N_1, we may write

$$\Delta T = k\frac{N_2}{N_1} = A \exp\left(\frac{\Delta E}{RT}\right) \tag{68}$$

where $A = ke^{\Delta S/R} \approx$ const. In Eq. (68), ΔT was taken as the difference in the softening points of a polyamide-polyester pair, each of which possessed similar repeating structural units (cf. Table 5). From Eq. (68), a plot of ln ΔT versus $1/T$ should afford a linear relationship whose slope yields the value of ΔE. However, from Fig. 6.8, it can be seen that apparently two types of polyamides exist. The first type has relatively low softening points, and in this case ΔT is nearly constant. This implies that at low temperatures, there is nearly complete displacement of the equilibrium in the direction of

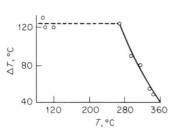

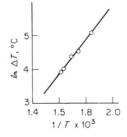

FIG. 6.8 Difference in softening points of aromatic polyamides and polyesters ΔT of similar structure versus softening point of corresponding polyester [57].

FIG. 6.9 Curve of ln ΔT versus $1/T$ for a series of aromatic polyamides and polyesters of the same structure (from thermomechanical data) [57].

hydrogen-bond formation, and the value of ΔT remains approximately constant ($<260°C$). The second type has relatively high softening points, and in this case the value of ΔT decreases exponentially with a rise in the polyester softening temperatures. This implies that at higher temperatures

TABLE 5 Softening Points Determined by a Thermomechanical Method for the Series of Aromatic Polyesters and Polyamides of the Following Structure [57]:

$$[-\underset{\underset{O}{\|}}{C}-R-\underset{\underset{O}{\|}}{C}-O-R'-O-]_n, [-\underset{\underset{O}{\|}}{C}-R-\underset{\underset{O}{\|}}{C}-HN-R'-NH-]_n$$

R	R'	Softening point, °C	
		Polyester	Polyamide
$-(CH_2)_8-$	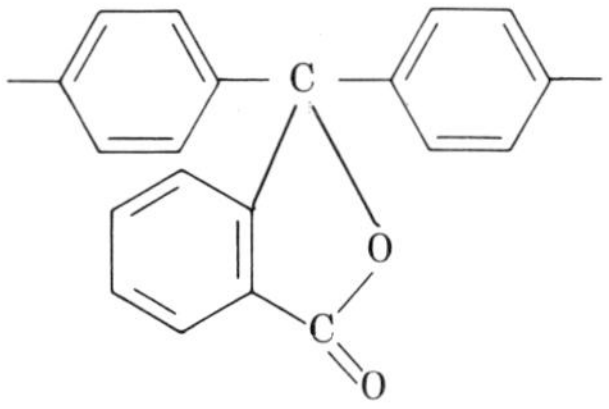	100	220
$-(CH_2)_3-$	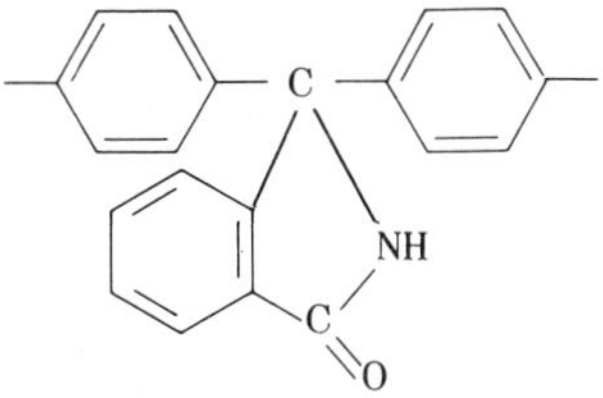	95	225
$-(CH_2)_4-$	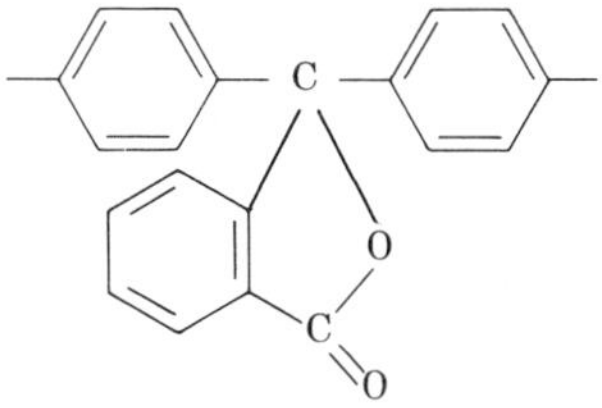	120	240
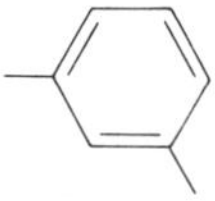	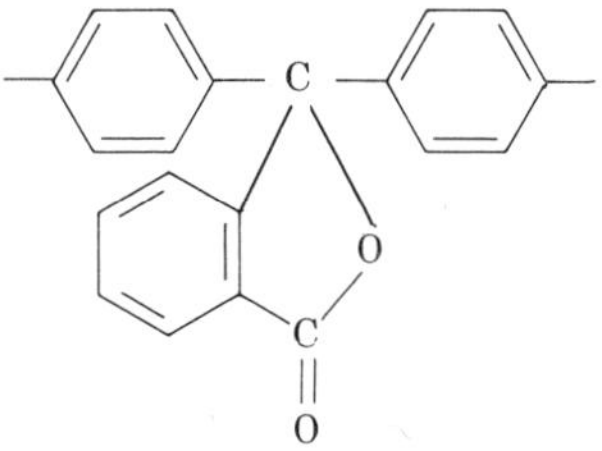	295	385
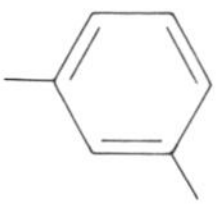	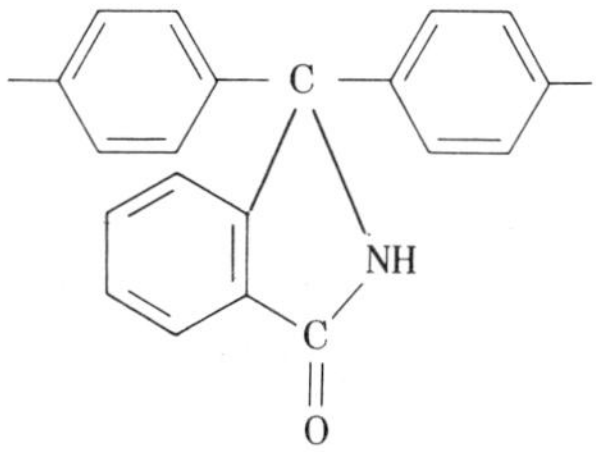	265	390

TABLE 5   Softening Points Determined by a Thermomechanical Method for the Series of Aromatic Polyesters and Polyamides of the Following Structure [57] : Continued

R	R'	Softening point, °C	
		Polyester	Polyamide
	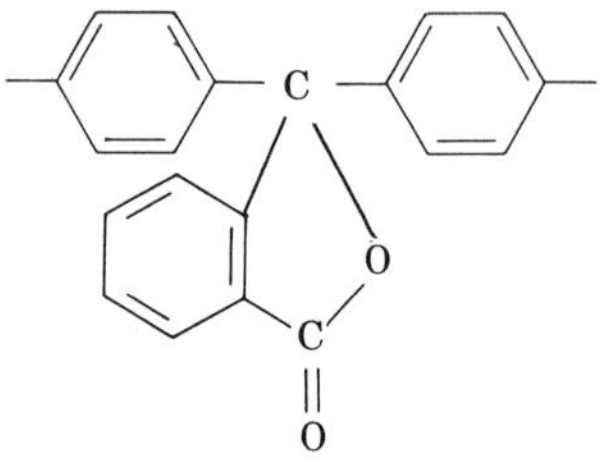	335	390
	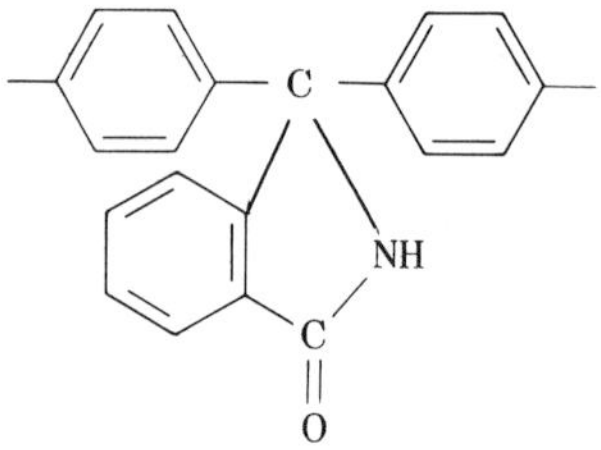	315	395
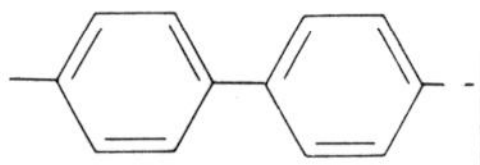	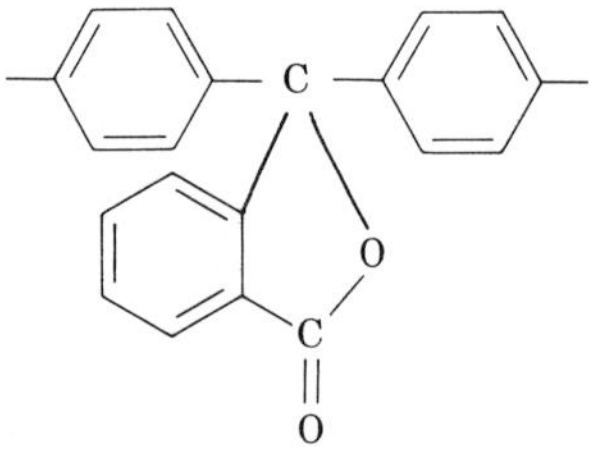	340	395

(>260°C) the equilibrium is displaced in the direction of hydrogen-bond breakage with a consequent decrease in ΔT (with temperature). A plot of Eq. (68) for the second type (Fig. 6.9) afforded a value of $\Delta E \cong 9$ kcal mole^{-1}, which was in good agreement with hydrogen-bond energies in polyamides estimated by other methods, e.g., infrared spectroscopy [58] .

SOME OTHER FACTORS

Sample Size

Sample thickness can affect polymer stability by influencing the rate of oxygen diffusion into the sample, in the case of oxidative degradation, and by causing a thermal barrier (large temperature gradient in sample). The extent of the diffusional control may be strongly influenced by the morphology of the polymer sample. Thus, Smith [59] reported the absence of diffusion control of reaction rate during the thermal decomposition of amorphous polymers. If diffusion control of products from reaction sites existed, rate of degradation would be affected. However, in contrast to results reported for the decomposition of crystalline solids, the overall pyrolysis kinetics of amorphous polymers in the rubbery state, e.g., vinylidene fluoride–hexafluoropropylene copolymers (Viton), did not appear to be seriously affected by sample size (or by heating rate). Recently, Reich and Stivala [60] studied the thermal degradation of relatively thick solid specimens of Teflon (>60 mils). They found that the thermally insulated sample (unidirectional heat flow occurred) afforded decomposition rates which gave low values of the overall activation energy E_a of decomposition. By accounting for heat-transfer effects, an expression could be derived from which values of E_a were obtained which were in reasonably good agreement with reported values for Teflon degradation. This expression indicated that for the relatively thick Teflon samples the degradation involved primarily a surface effect.

When finely powdered samples are employed, diffusional effects would be expected to be minimal (assuming no conglomeration, etc., during sample decomposition). However, as is often the case, polymer films are used for oxidation studies. It is important to know whether oxidation rate is a true indication of chemical reaction or is limited by the relatively slow diffusion of oxygen into the film sample. When the film thickness is below a certain value, the chemical reaction, rather than diffusion, will be the controlling factor. Thus, Stivala, Jadrnicek, and Reich [61] found that during the autoxidation of films of atactic polypropylene, the rate of carbonyl formation per unit thickness of sample was approximately the same for films between 1 to 4 mils thickness. At a film thickness of about 7.5 mils, the rate decreased noticeably. Others have observed similar results [62] for other polyolefins, e.g., polyethylene. When diffusion, without chemical reaction, occurs, the amount of oxygen diffusing into the film sample (DO) may be expressed by [63]

$$DO = \text{const} \frac{t^{1/2}}{L} \tag{69}$$

where t = time and L = sample thickness. Quackenbos [64] studied the effect of thick films on polyethylene oxidation at about $365°C$ (the carbonyl content was observed). Between values of $L = 11$ and 30 mils, the carbonyl content was inversely proportional to L and directly proportional to $t^{1/2}$. These results are in accord with Eq. (69) if it is assumed that the amount of carbonyl formed is directly proportional to the amount of oxygen that diffused into the sample, DO. Recently, Kwei [65] studied the photooxidation of poly(vinyl chloride) films of about 1 mil thickness and concluded that the photochemical oxidation was diffusion-controlled. Thus, when the amount of oxygen absorbed per unit weight of polymer was plotted against $t^{1/2}$, a linear relationship was obtained.

Sample Impurities

Although the presence of sample impurities may be adventitious, they may greatly affect polymer stability. Thus, Schooten and Wijga [66] studied the thermal degradation of polypropylene at 230 to 300°C under rigorous exclusion of oxygen, and attributed the initiation of decomposition to the presence of hydroperoxide groups in the polymer. The presence of such groups was indicated by the apparent low activation energy of decomposition, by the effect of the addition of trace amounts of copper (which catalyze hydroperoxide decomposition), by the influence of polymer preheating at 180°C (which presumably destroys weak links present and thereby increases polymer stability), and by the effect of various antioxidants (which presumably inhibit reactions of oxygen-containing radicals).

Small amounts of metallic impurities which remain in polymers prepared by Ziegler catalysts may have a pronounced effect on oxidation behavior. Thus, Stivala, Yo, and Reich [67] studied the autoxidation kinetics of isotactic polybutene-1 (IPB) which possessed an ash content of 0.17 percent. The ash consisted mainly of Al, Ti, and Si with lesser amounts of Mg, Pb, Fe, Ca, Mn, Sn, and Cu. From the autoxidation results obtained, various autoxidation processes appeared to be occurring more rapidly for IPB than for atactic polybutene-1 (APB) under similar experimental conditions; e.g., induction times for IPB were shorter than for APB, and the value of activation energy for hydroperoxide decomposition E for IPB (25 kcal mole^{-1}) was about 4 kcal mole^{-1} lower than the value for APB (29 kcal mole^{-1}). Such behavior was attributed to the presence of much larger amounts of metallic impurities in IPB (0.17 percent as ash) than in APB (0.04 percent as ash). Others have observed similar effects of metallic impurities. Thus, Lombard and Knopf [68] found that the cobaltic acetyl acetonate–catalyzed decomposition of cumene hydroperoxide in solution gave a value of $E = 25$ kcal mole^{-1}, whereas in the absence of the metal salt a value of $E = 31$ kcal mole^{-1} was obtained. Furthermore, metallic salts can lower autoxidation induction times [61] as well as cause marked changes in polymer viscosity during autoxidation [69].

Oxidation Products

Products that form during oxidation may exert an inhibitory effect. Thus, Yur'ev and coworkers [16a] concluded that a reduction in oxidation rate of cetane at 140°C occurred as a result of the formation of formic acid (and formates) during the oxidation. It was shown that the decomposition of *tert*-butyl hydroperoxide in formic acid (and formates) proceeded at a much higher rate than in hydrocarbons and that the pH of the medium had little effect. Reactions which were postulated were

$$RO\cdot + HCO_2H \longrightarrow ROH + \cdot CO_2H \tag{70}$$

$$\cdot CO_2H + RO_2H \longrightarrow CO_2 + ROH \text{ (or } H_2O)$$
$$+ RO\cdot \text{ (or } HO\cdot) \tag{71}$$

It thus appears that the reduction in oxidation rate is associated with rapid hydroperoxide decomposition due to the accumulation of formic acid (and formates) in the system. This was further verified by the addition of pentadecyl formate to a polymer, prepared from a mixture of diazomethane and diazoethane, before oxidation. A marked reduction in oxidation rate occurred.

Recently Reich and Stivala [70] investigated the kinetic aspects of the autoxidation of poly(ethylene glycol) (PEG) in solution. At relatively large reaction times, there were large discrepancies between calculated and observed values of oxygen absorption. This behavior was attributed to the formation of oxidation products at the high reaction times which inhibit the oxidation.

Such products were also postulated in view of the fact that polyolefin oxidation rates generally increase more rapidly than those observed for PEG. In this connection may be mentioned the report of Mayo and coworkers [71], which commented on results obtained by Kern and coworkers [25] (cf. page 304) for the autoxidation of polyolefin in solution. In the case of atactic and isotactic polybutene-1, Kern and coworkers [72] oxidized these polymers at 70.6°C in brombenzene solution in presence of azobisisobutyronitrile (AIBN) as initiator. They found that the rate of oxygen absorption by the isotactic polymer (and presumably the atactic polybutene) fell off much more rapidly with time than was anticipated from the disappearance of AIBN (low kinetic chain lengths were obtained). Mayo [71] suggested that the drop in rate was due to autoretardation, which could arise as follows. Because of the short chain lengths and the relatively large amounts of alcohol found, there apparently occurs much chain cleavage. Primary and secondary alcohol groups accumulate, as well as aldehydes, ketones, and hydroperoxides. The oxidation of the alcohols can lead to hydrogen peroxide and/or $HO_2 \cdot$ radicals. The latter are good chain terminators and could account for the slowing of the polybutene oxidation. This autoretardation concept was extended by Mayo to the (different) results obtained by Kern and coworkers [25] for the autoxidation of atactic and isotactic polypropylenes in solution at 130 to 170°C, in the absence of initiator, based upon extrapolation of relations obtained during later stages of reaction to zero time. These extrapolated rates corresponded to about one-third of the highest rates recorded, when they should have been zero. Thus, it was suggested that the autocatalysis observed in the later oxidation stages was a superposition on a retardation of a fast and neglected early reaction.

Molecular Weight and Molecular Weight Distribution (cf. pp. 167ff)

During the thermal degradation of polymers, the rate of degradation may depend upon the molecular weight of a homodisperse polymer. This relationship depends upon the type of degradation mechanisms involved. Thus, when the mechanism involves chain-end (or terminal) initiation, depropagation, and bimolecular termination (cf. Chap. 1), the following rate expression may be derived [73]:

$$-\frac{dW}{dt} = k_d \left(\frac{k_i m}{\rho k_t D_P} \right)^{1/2} W \qquad (72)$$

where W = weight of polymer at time t; k_i, k_d, k_t denote rate constants for initiation, depropagation, and termination, respectively; m = monomeric molecular weight; ρ = polymer density; and D_P = number-average degree of polymerization. On the other hand, when initiation occurs at random, and the remaining steps are the same as before, then the degradation rate is no longer a function of polymer molecular weight (or degree of polymerization).

The manner in which a heterodisperse polymer degrades thermally can depend upon its molecular-weight distribution. MacCallum [74] analyzed the kinetics of the depolymerization of addition polymers for various mechanisms. Thus, when initiation occurs at random and all the molecules unzip (depropagate) completely without any termination, then the following expression could be derived:

$$1 - C = M^{n/(n-1)} \qquad (73)$$

where C = fractional conversion; M = fractional molecular weight (M_t/M_0), where M_0 is original number-average molecular weight and M_t is number average molecular weight at time t; and n = const, whose value depends upon the type of molecular-weight distribution. When the polymer is homodisperse, $n = 1$; when the distribution follows an exponential type as

$$N_x^{\,o} = \frac{N^o}{P_0} \exp\left(-\frac{x}{P_0}\right) \qquad (74)$$

where $N_x^{\,o}$ = number of molecules of degree of polymerization x initially in the sample; and N^o and P_0 are the total number of molecules and number-average degree of polymerization, respectively, before depolymerization, then $n = 2$. For a "coupling" type distribution, i.e.,

$$N_x^{\,o} = \frac{4N^o}{P_0^{\,2}} x \exp\left(-\frac{2x}{P_0}\right) \qquad (75)$$

the value of $n = 1.5$. (Samples of poly(methyl methacrylate) and polystyrene prepared in bulk using free-radical initiators possess exponential and coupling types of distribution, respectively.) Equation (73) may be graphically represented for various values of n as shown in Fig. 6.10.

Others have also reported on the effect of polymer dispersity on degradation rate. Wall and Flynn [14] reported on the rate of volatilization of polystyrene versus conversion. They indicated that the rate depended on molecular weight and on whether the polymer sample was fractionated or unfractionated. Bagby and coworkers [75] have also indicated that the manner in which the fractional molecular weight varies with conversion for poly(methyl methacrylate) depends on whether the polymer sample is fractionated or unfractionated and also on the degradation temperature.

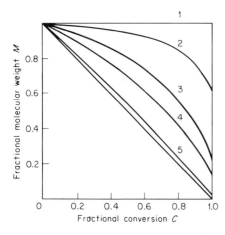

FIG. 6.10 A series of curves drawn according to Eq. (73). The values of n for the curves are: (1) $n = 1$; (2) $n = 1.1$; (3) $n = 1.5$; (4) $n = 2$; (5) $n = 10$ [74].

Although Baum [76] obtained results which indicated that molecular-weight variations did not affect the oxidation rate of a polyethylene resin, Meltzer and coworkers [77] found that molecular-weight distribution may play an important role in polymer oxidation. For example, two polyethylene resins of identical weight-average molecular weight but differing in molecular-weight distribution were subjected to oxidation at temperatures of 139 and 152.5°C. At both temperatures, the resin which possessed a broader distribution oxidized more readily. It was suggested that end groups were responsible for the difference in oxidizability and that the number of these present was governed by molecular-weight distribution.

Effect of Phase Change

In order to minimize diffusional and heat-transfer effects during polymer degradation, polymer samples are often used as fine powders. However, when the degradation is carried out at temperatures which involve a phase change, e.g., sintering, softening, or melting, the polymer sample may conglomerate, thereby causing diffusional and heat-transfer effects to become important.

Grant and Ward [78] reported that the infrared spectra of molten isotactic, molten syndiotactic, and atactic polypropylene were nearly similar. It might be expected that at the melting point, the various types of polypropylene would exhibit similar oxidation characteristics. Similar considerations should apply to other polymers. Hawkins and coworkers [7] found that at the melting point (about 140°C), linear and branched polyethylene oxidized at similar rates (the linear form was more crystalline). Both polymers behave as amorphous above the melting point. However, below the melting point (about 100°C), the branched (less crystalline) polyethylene oxidized more readily than the linear (more crystalline) polymer.

Effect of Temperature and Oxygen Concentration

In general, in the absence of additives such as antioxidants, it may be stated that during the autoxidation of saturated polyolefins, the induction period, if any, decreases, and the oxidation rate increases as temperature and/or oxygen pressure are increased. Similar considerations apply to polymer oxidations in the presence of additives as antioxidants and metal accelerators. (In the preceding, it is assumed that phase changes are absent.)

Temperature may also cause decomposition mechanisms to change. Thus, Barlow and coworkers [79] studied the pyrolysis of poly(methyl methacrylate), and from the dependence of reaction rate on the initial molecular weight at various temperatures they postulated that the pyrolysis mechanism changed with temperature. At the lower temperatures (about 300°C), the depropagating chains are end-initiated, and termination is mainly bimolecular. At the higher reaction temperatures (about 500°C), initiation occurs mainly at random, and the majority of chains are terminated as the depropagation reaches the end of the molecule—the terminal radical distills out of the system. The higher-temperature mechanism may be represented by an expression similar to Eq. (73). (cf. Fig. 6.10.) When the fractional molecular weight was plotted against conversion during the degradation of poly(methyl methacrylate) at 500°C, a curve intermediate to those obtained in Fig. 6.10 for cases 2 or 3 was also obtained for the unfractionated polymer ($n > 1$). For the fractionated poly(methyl methacrylate), no change in molecular weight occurred with conversion [cf. curve 1 of Fig. 6.10 ($n = 1$ for homodisperse polymer)].

Effect of Solvents

Solvents can exert an effect on the degradation of polymers. A case in point is the work carried out by Lal and coworkers [80] on the thermal degradation of polyethylene in α-chloronaphthalene solution. These workers assumed that degradation occurred by random chain scission and that the following corresponding expression was valid [Eq. (76) applies when the degree of degradation, α is much less than unity] :

$$\frac{P_w}{P_0} = \frac{2}{S^2}(e^{-S} + S - 1) \tag{76}$$

where P_w = weight-average degree of polymerization at time t, P_0 = initial degree of polymerization, and S = average number of chain scissions per polymer molecule. For large values of P_0 and for small values of $k_1 t$ the following equation should hold during the initial degradation stages:

$$\alpha = \frac{S}{P_0} = k_1 t \qquad (77)$$

where k_1 = rate constant. Values of S may be obtained from intrinsic viscosity data and Eq. (76). Then from a plot of S versus t, values of k_1 can be obtained at various temperatures from Eq. (77) and subsequently the activation energy for the random degradation process. From Fig. 6.11, it can be seen that Eq. (77) held approximately during the initial stages of the polyethylene degradation at various reaction temperatures. The value of the activation energy E obtained was about 35 kcal mole^{-1}. Values of E for polyethylene degradation in the bulk phase have been reported by Anderson and Freeman [81] to be about 48 kcal mole^{-1} at low conversion (about 3 percent). At this low conversion, zero-order kinetics obtained which implied that weak links were being initially broken (these links may arise from branching). At higher conversions, the value of E rose to about 67 kcal mole^{-1}. A possible explanation advanced by Lal and coworkers for the lower value of E in α-chloronaphthalene as opposed to that in the bulk phase during the initial stages of degradation was that in solution unrestricted vibrational modes for the weak bonds were present. A similar situation may not exist in the bulk phase where the activation of weak bonds for scission may require higher energy.

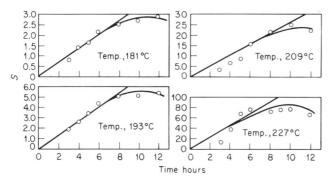

FIG. 6.11 Variation of S with time at various temperatures during polyethylene degradation in α-chloronaphthalene [80].

Carlsson and Wiles [82] subjected powdered, unstabilized polypropylene to air for 1 to 5 min at 225°C, and based upon infrared absorption spectroscopy, they assigned the following structures to two principal ketone products:

$$
-\text{ww}-\underset{\underset{\text{H}}{|}}{\overset{\overset{\text{CH}_3}{|}}{\text{C}}}-\text{CH}_2-\overset{\overset{\text{O}}{||}}{\text{C}}-\text{CH}_2-\underset{\underset{\text{H}}{|}}{\overset{\overset{\text{CH}_3}{|}}{\text{C}}}-\text{ww}- \quad ,
$$

$$(\text{XXIII})$$

$$
-\text{ww}-\underset{\underset{\text{H}}{|}}{\overset{\overset{\text{CH}_3}{|}}{\text{C}}}-\text{CH}_2-\overset{\overset{\text{O}}{||}}{\text{C}}-\text{CH}_3
$$

$$(\text{XXIV})$$

Both products could be formed by cleavage of the following type of radical:

$$
-\text{ww}-\underset{\underset{\text{H}}{|}}{\overset{\overset{\text{CH}_3}{|}}{\text{C}}}-\text{CH}_2-\underset{\underset{\text{O}\cdot}{|}}{\overset{\overset{\text{CH}_3}{|}}{\text{C}}}-\text{CH}_2-\underset{\underset{\text{H}}{|}}{\overset{\overset{\text{CH}_3}{|}}{\text{C}}}-\text{ww}-
$$

$$(\text{XXV})$$

The formation of (XXIII) requires that (XXV) lose a methyl radical instead of the larger alkyl groups. This does not generally occur in solution and implies that the reactivity of polymeric alkoxy radicals (XXV) in the solid phase is much different from the reactivity of alkoxy radicals in solution. Mayo [71] accounted for such a behavior by assuming that the cleavage of an alkyl group from a *tert*-alkoxy radical (XXV) is reversible and that the direction of the cleavage reaction of (XXV) depends upon the mobility and/ or reactivity of the cleaved radical. (Based on the higher bond strength of the C—CH$_3$ bond than the C-alkyl bond, it would be expected a priori that cleavage of alkyl groups would be preferred.) The methyl radical can diffuse and abstract hydrogen from the substrate readily, whereas the polymeric alkyl radical is immobilized in the polymer and is less reactive in hydrogen abstraction. In this connection may be mentioned the work of Reich, Stivala, and coworkers [1]. These workers postulated a mechanism to explain various results obtained from polyolefin autoxidations which assumed that the combination of polymeric peroxy radicals in the bulk phase was negligible owing to the immobility of such radicals. Generally, in solution, such a termination step cannot be neglected. In this respect may be mentioned the work of Bawn and Chaudhri [83, 84] who studied the thermal autoxidation of atactic polypropylene in solution. When the polymer concentration was relatively low ($\ll 1.5$ M), it was necessary to employ a relatively simple scheme which included termination by combination of peroxy radicals. However, when the polymer concentration was increased (> 1.5 M), discordant results were obtained. By neglecting termination by polymeric radicals at the higher concentrations (and higher medium viscosity) and employing a scheme similar to that used by Reich, Stivala, and coworkers for bulk-phase oxidations, Bawn and Chaudhri were able to explain many of the oxidation data obtained. Reich and Stivala [85] utilized their scheme to correlate intrinsic viscosity data obtained by Bawn and Chaudhri [84].

REFERENCES

1. Reich, L., and S. S. Stivala: "Autoxidation of Hydrocarbons and Polyolefins," chap. 6, Marcel Dekker, Inc., New York, 1969.
2. Preston, J., R. W. Smith, and C. J. Stehman: *J. Polymer Sci.*, **C-19**:7 (1967).
3. Millett, M. A., W. E. Moore, and J. F. Seaman: *Ind. Eng. Chem.*, **46**:1493 (1954).
4. Ravens, D. A. S.: *Polymer*, **1**:375 (1960).
5. Farrow, G., D. A. S. Ravens, and I. M. Ward: *Polymer*, **3**:17 (1962).
6. Winslow, F. H., and W. Matreyek: Paper presented at 141st American Chemical Society Meeting, Washington, D.C., Mar. 21-29, 1962.
7. Hawkins, W. L., W. Matreyek, and F. H. Winslow: *J. Polymer Sci.*, **41**:1 (1959).
8. Winslow, F. H., C. J. Aloisio, W. L. Hawkins, W. Matreyek, and S. Matsuoka: *Chem. Ind. (London)*, **1963**:533.
9. Winslow, F. H., and W. Matreyek: "Polymer Preprints," vol. 5, p. 552, American Chemical Society Meeting, Chicago, September, 1964.
10. Bassett, D. C.: *Polymer*, **5**:457 (1964).
10a. Luongo, J. P.: *J. Polymer Sci.*, **B-1**:141 (1963).
11. Lilyquist, M. R., and J. R. Holsten: *J. Polymer Sci.*, **C-19**:77 (1967).
12. Weiss, J. O., H. S. Morgan, and M. R. Lilyquist: *J. Polymer Sci.*, **C-19**:29 (1967).
13. Lancaster, J. M., B. A. Wright, and W. W. Wright: *J. Appl. Polymer Sci.*, **9**:1955 (1965).
14. Wall, L. A., and J. H. Flynn: *Rubber Chem. Technol.*, **35**:1157 (1962).
15. Willbourn, A. H.: *J. Polymer Sci.*, **34**:569 (1959).
16. Yur'ev, V. M., A. N. Pravednikov, and S. S. Medvedev: *Dokl. Akad. Nauk SSSR*, **124**:335 (1959).
16a. Yur'ev, V. M., A. N. Pravednikov, and S. S. Medvedev: *J. Polymer Sci.*, **55**:353 (1961).
17. Madorsky, S. L.: "Thermal Decomposition of Organic Polymers," Interscience Publishers, a division of John Wiley & Sons, Inc., New York, 1964.
18. Van Deusen, R. L.: *J. Polymer Sci.*, **B-4**:211 (1966).
19. Doroshenko, Y. E.: *Russ. Chem. Revs.*, **36**:563 (1967).
20. Lee, H., D. Stoffey, and K. Neville: "New Linear Polymers," McGraw-Hill Book Company, New York, 1967.
21. Berlin, A. A., N. G. Matveeva, and A. I. Sherle: *Izv. Akad. Nauk SSSR, Otd. Khim. Nauk*, **1959**:2261.
22. Winslow, F. H., M. Y. Hellman, W. Matreyek, and R. Salovey: "Polymer Preprints," vol. 5, p. 47, American Chemical Society Meeting, Philadelphia, April, 1964.
23. Mark, H. F.: *J. Polymer Sci.*, **C-9**:1 (1965).
24. Murayana, N., and I. Amagi: *J. Polymer Sci.*, **B-4**:115 (1966).
25. Dulog, L., E. Radlmann, and W. Kern: *Makromol. Chem.*, **60**:1 (1963).
26. Bawn, C. E. H., and S. A. Chaudhri: *Polymer*, **9**:123 (1968).
27. Krasnov, Y. P., V. M. Savinov, L. B. Sokolov, V. I. Logunova, V. K. Belyakov, and T. A. Polyakova: *Vysokomolekul. Soedin.*, **8**:380 (1966).
28. Stivala, S. S., G. R. Sacco, and L. Reich: *J. Polymer Sci.*, **B-2**:943 (1964).
29. Volpe, A. A., L. G. Kaufman, and R. G. Dondero: *J. Macromol. Sci.*, **A3**:1087 (1969).

30.   Jellinek, H. H. G.: *J. Polymer Sci.*, **3**:850 (1948).
31.   Cameron, G. G., and N. Grassie: *Polymer*, **2**:367 (1961).
32.   Simha, R., and L. A. Wall: *J. Phys. Chem.*, **56**:707 (1952).
33.   MacCallum, J. R.: *Makromol. Chem.*, **83**:129 (1965).
34.   Gaudiana, R. A., and R. T. Conley: *J. Polymer Sci.*, **B-7**:793 (1969).
35.   Teleshov, E. N., N. B. Feldblyum, and A. N. Pravednikov: *Vysokomolekul. Soedin.*, **A10**:422 (1968).
36.   Johnston, T. H., and C. A. Gaulin: *J. Macromol. Sci.-Chem.*, **A3**:1161 (1969).
37.   Bruck, S. D.: "Polymer Reprints," vol. 5, p. 148, American Chemical Society Meeting, Philadelphia, April, 1964.
38.   Igarashi, S., I. Mita, and H. Kambe: *Bull. Chem. Soc., Japan,* **37**:1160 (1964).
39.   Grassie, N., and I. C. McNeill: *J. Polymer Sci.*, **27**:207 (1958).
40.   Andrianov, K. A., and I. F. Manucharova: *Izv. Akad. Nauk SSSR, Otd. Khim. Nauk,* **1962**:420.
41.   Petrashko, A. I., and K. A. Andrianov: *Polymer Sci., (USSR),* **6**:1670 (1964).
42.   Verkhotin, M. A., K. A. Andrianov, M. N. Yermakova, S. R. Rafikov, and V. V. Rode: *Vysokomolekul. Soedin.,* **8**:2139 (1966).
43.   Green, J., N. Mayer, A. F. Kotloby, M. M. Fein, E. L. O'Brien, and M. S. Cohen: *J. Polymer Sci.,* **B-2**:109 (1964).
44.   Bawn, C. E. H.: *Proc. Roy. Soc. (London),* **A282**:91 (1964).
45.   Inoy, I.: *J. Inorg. Nucl. Chem.,* **26**:139 (1964).
46.   Rode, V. V., E. M. Bondarenko, V. V. Korshak, A. L. Rusanov, E. S. Krongauz, P. A. Bochvar, and I. V. Stankovich: *J. Polymer Sci.,* **A-1, 6**:1351 (1968).
47.   Cox, J. M., B. A. Wright, and W. W. Wright: *J. Appl. Polymer Sci.,* **8**:2935 (1964).
48.   Cox, J. M., B. A. Wright, and W. W. Wright: *J. Appl. Polymer Sci.,* **8**:2951 (1964).
49.   Snow, R. D., and F. E. Frey: *Ind. Eng. Chem.,* **30**:176 (1938).
50.   Dainton, F. S., and K. Ivin: *Quart. Rev.,* **12**:61 (1958).
51.   Ivin, K.: *Trans. Faraday Soc.,* **51**:1267 (1955).
52.   Small, P. A.: *J. Appl. Chem.,* **3**:71 (1953).
53.   Nanda, V. N., and R. Simha: *J. Phys. Chem.,* **68**:3158 (1964).
54.   Allen, G., G. Gee, D. Mangaraj, D. Sims, and G. J. Wilson: *Polymer,* **1**:467 (1960).
55.   Hayes, R. A.: *J. Appl. Polymer Sci.,* **5**:318 (1961).
56.   Allen, G., and D. Sims: *Polymer,* **4**:105 (1963).
57.   Slonimskii, G. L., A. A. Askadskii, V. V. Korshak, S. V. Vinogradova, Vygodskii, and S. N. Salazkin: *Vysokomolekul. Soedin.,* **A9**:1706 (1967).
58.   Trifan, D. S., and J. E. Terenzi: *J. Polymer Sci.,* **28**:443 (1958).
59.   Smith, D. A.: *Nature,* **208**:1200 (1965).
60.   Reich, L., and S. S. Stivala: *Thermochim. Acta,* **1**:65 (1970).
61.   Jadrnicek, B., S. S. Stivala, and L. Reich: *J. Appl. Polymer Sci.,* **14**:2537 (1970).
62.   Biggs, B. S.: *Natl. Bur. Stds. (U.S.), Arch.,* **525**:137 (1953).
63.   Barrer, R. M.: "Diffusion In and Through Solids," p. 216, Cambridge University Press, New York, 1941.
64.   Quackenbos, H. M.: *Polymer Eng. Sci.,* **6**:117 (1966).
65.   Kwei, K. -P. S.: *J. Polymer Sci.,* **A-1, 7**:1075 (1969).
66.   v. Schooten, J., and P. W. O. Wijga: In Thermal Degradation of Polymers, *Soc. Chem. Ind. (London) Monograph* 13, pp. 432ff., 1961.
67.   Stivala, S. S., G. Yo, and L. Reich: *J. Appl. Polymer Sci.,* **13**:1289 (1969).
68.   Lombard, R., and J. Knopf: *Bull. Chim. Soc. France,* **1966**:3926.
69.   Ryshavy, D. R., and L. B. Balaban: *SPE Trans.,* **2**:25 (1962).
70.   Reich, L., and S. S. Stivala: *J. Appl. Polymer Sci.,* **13**:977 (1969).
71.   Mayo, F. R., T. Mill, E. Niki, and H. Richardson: *Stanford Res. Inst. Proj.* PRC-8012, *Progr. Rept.* 3, Sept. 13–Dec. 12, 1969.
72.   Dulog, L., J. K. Weise, and W. Kern: *Makromol. Chem.,* **118**:66 (1968).
73.   Friedman, H. L.: *J. Polymer Sci.,* **45**:119 (1960).

74. MacCallum, J. R.: *European Polymer J.*, **2**:413 (1966).
75. Bagby, G., R. S. Lehrle, and J. C. Robb: *Makromol. Chem.*, **119**:122 (1968).
76. Baum, B.: *J. Appl. Polymer Sci.*, **2**:281 (1959).
77. Meltzer, T. H., J. J. Kelley, and R. N. Goldey: *J. Appl. Polymer Sci.*, **3**:84 (1960).
78. Grant, I. J., and I. M. Ward: *Polymer*, **6**:223 (1965).
79. Barlow, A., R. S. Lehrle, J. C. Robb, and D. Sunderland: *Polymer*, **8**:537 (1967).
80. Lal, K., M. Singh, and H. L. Bhatnagar: *Indian J. Chem.*, **5**:412 (1967).
81. Anderson, D. A., and E. S. Freeman: *J. Polymer Sci.*, **54**:253 (1961).
82. Carlsson, D. J., and D. M. Wiles: *Macromolecules*, **2**:587 (1969).
83. Bawn, C. E. H., and S. A. Chaudhri: *Polymer*, **9**:113 (1968).
84. Bawn, C. E. H., and S. A. Chaudhri: *Polymer*, **9**:123 (1968).
85. Reich, L., and S. S. Stivala: *J. Appl. Polymer Sci.*, **13**:23 (1969).

NAME INDEX

SUBJECT INDEX